Entire and Subharmonic Functions

Titles in This Series

Volume

11 **B. Ya. Levin,** editor
Entire and subharmonic functions
1992

10 **A. V. Babin and M. I. Vishik,** editors
Properties of global attractors of partial differential equations
1992

9 **A. M. Vershik,** editor
Representation theory and dynamical systems
1992

8 **E. B. Vinberg,** editor
Lie groups, their discrete subgroups, and invariant theory
1992

7 **M. Sh. Birman,** editor
Estimates and asymptotics for discrete spectra of integral and differential equations
1991

6 **A. T. Fomenko,** editor
Topological classification of integrable systems
1991

5 **R. A. Minlos,** editor
Many-particle Hamiltonians: spectra and scattering
1991

4 **A. A. Suslin,** editor
Algebraic K-theory
1991

3 **Ya. G. Sinaĭ,** editor
Dynamical systems and statistical mechanics
1991

2 **A. A. Kirillov,** editor
Topics in representation theory
1991

1 **V. I. Arnold,** editor
Theory of singularities and its applications
1990

Advances in

SOVIET

MATHEMATICS

Volume 11

Entire and Subharmonic Functions

B. Ya. Levin
Editor

American Mathematical Society
Providence, Rhode Island

Translation edited by A. B. SOSSINSKY

1991 *Mathematics Subject Classification.* Primary 14H05, 30B05, 30C80, 30D15, 30D20, 30D35, 30E05, 30E25, 31A02, 31A05, 34A55, 34B24, 42B10, 60E10; Secondary 31A05, 34L10, 34L15, 42A75.

Library of Congress Catalog Card Number: 91-640741

ISBN 0-8218-4110-6; ISSN 1051-8037

The paper used in this book is acid-free and falls within the guidelines
established to ensure permanence and durability. ♾

This publication was typeset using AMS-TEX,
the American Mathematical Society's TEX macro system.

10 9 8 7 6 5 4 3 2 1   97 96 95 94 93 92

Contents

Foreword

The present collection is compiled from papers written by participants of the Research Seminar on the Theory of Functions at Kharkov University. For the most part, the papers are related to the theory of entire and subharmonic functions.

The seminar was founded in 1953 by B. Ya. Levin and is still functioning today. The scope of subjects at the seminar was not restricted. The talks given by its participants were related to various problems in the theory of functions and functional analysis and to adjacent problems in calculus and mathematical physics.

Among the most active participants of the seminar in different years were V. S. Azarin, A. F. Grishin, V. P. Gurariĭ, A. E. Eremenko, M. I. Kadets, V. E. Katsnel′son, Yu. I. Lyubarskiĭ, V. I. Matsaev, V. D. Mil′man, I. V. Ostrovskiĭ, L. I. Ronkin, M. L. Sodin, V. A. Tkachenko, and other well-known mathematicians. Mathematicians from other cities and from abroad often visit the seminar. The present collection contains works reported recently at the seminar.

March 1992 B. Levin

Translated by V. A. TKACHENKO

ADVANCES IN SOVIET MATHEMATICS
Volume 11, 1992

On the Completely Regular Growth of Analytic Functions Having Maximum on a Ray

A. M. ULANOVSKIĬ

Let f be an analytic function in U, U being the complex plane $\mathbb{C}$ or the angle $U(\alpha, \beta) = \{ z \in \mathbb{C} : \alpha < \arg z < \beta \}$. We denote

$$M(r) = M(r, f, U) = \sup_{|z| \leqslant r, z \in U} |f(z)|,$$

$$\rho = \rho(f, U) = \limsup_{r \to \infty} \frac{\log \log M(r)}{\log r},$$

$$\lambda = \lambda(f, U) = \liminf_{r \to \infty} \frac{\log \log M(r)}{\log r}.$$

The values ρ and λ are called the *order* and the *lower order* of f respectively.

The numbers ρ and λ are important characteristics of the growth of f. If $\rho = \lambda$, then f is said to have *regular growth.*

It is well known that, in general, ρ and λ may be arbitrary nonnegative numbers satisfying the inequality $0 \leqslant \lambda \leqslant \rho < \infty$. Some restrictions on ρ and λ may appear if f has no zeros in some angles. If, for instance, f is an entire function and if all zeros of f lie on a ray, then by [1] $\rho \leqslant 1 + [\lambda]$, where $[\lambda]$ is the integer part of λ. If the zeros of f lie on three rays, then by [2] $[\rho] \leqslant 3[\lambda] + 3$, while the difference $\rho - \lambda$ may be arbitrarily large.

Some conditions of another kind involving restrictions on ρ and λ were obtained in [3]. In particular, it was proved that if f is an entire function such that the set of all $z \in \mathbb{C}$ where $|f(z)| M(|z|) \leqslant 1$ has an unbounded component, then $\rho = \lambda$ (possibly, $= \infty$).

In this paper we investigate the growth of an analytic function f (in particular, connections between ρ and λ) satisfying one of the following conditions:

(A) f is analytic in $U(-2\alpha, 2\alpha)$, $0 < \alpha \leqslant \pi/2$; f has no zeros in $U(-\alpha, \alpha)$; $|f|$ attains its maximum at some point of the positive ray $|f(r)| = M(r)$, $r > 0$.

1991 *Mathematics Subject Classification.* Primary 30C80.

1051-8037/92 $1.00 + $.25 per page

(B) f is analytic in $U(-\alpha, \alpha)$, $0 < \alpha \leqslant \pi$, continuous in

$$\overline{U(-\alpha, \alpha)} = \{ z \in \mathbb{C} : -\alpha \leqslant \arg z \leqslant \alpha \}$$

and satisfies $|f(r)|M(r) \leqslant 1$, $r > 0$.

Condition (B) was first stated in connection with [3]. Condition (A) was studied in connection with the following supplement to [4] which was communicated to me by A. M. Vishnyakova: if f satisfies (A), then $\rho - \lambda \leqslant 3\pi/\alpha$.

It appears that if an analytic function f satisfies either condition (A) and (B) and has sufficiently rapid growth

$$\limsup_{r\to\infty} r^{-\pi/\alpha} \log M(r) > 0, \tag{1}$$

then f has not only regular but completely regular growth (in the sense of Levin-Pfluger) in $U(-\pi/\rho, \pi/\rho)$ with respect to r^ρ. Note that condition (1) implies $\rho \geqslant \pi/\alpha$; on the other hand, if $\rho > \pi/\alpha$, then (1) is true.

THEOREM 1. *Let f satisfy* (A) *and* (1). *Then $\rho = \lambda$. If additionally $\rho < \infty$, then there exists a $\sigma \geqslant 0$ such that the asymptotic equality*

$$\log|f(z)| = \sigma \operatorname{Re} z^\rho + o(|z|^\rho), \qquad |z| \to \infty, \tag{2}$$

holds uniformly in $z \in U(-\pi/\rho + \varepsilon, \pi/\rho - \varepsilon)$ for any $0 < \varepsilon < \pi/\rho$.

THEOREM 2. *Let f satisfy* (B) *and* (1). *Then $\rho = \lambda$. If additionally $\rho < \infty$, then there exists a $\sigma \leqslant 0$ such that the asymptotic equality* (2) *holds uniformly in $z \in U(-\pi/\rho, \pi/\rho)$ outside some exceptional set $|z| \notin E$ such that*

$$\lim_{R\to\infty} R^{-1} \int_{(0, R)\cap E} dr = 0.$$

NOTE. Theorems 1 and 2 cease to be true if condition (1) is omitted. For instance, for all $0 \leqslant \lambda \leqslant \rho \leqslant 1$ there exists an entire function f satisfying (A) with $\alpha = \pi/2$. The same is true for a function f satisfying (B) with $\alpha = \pi$. The corresponding examples can be constructed applying methods of [1, Chapter II, §5, Chapter VI, §1]. Note that if an entire function f has only negative zeros, $\rho > 1$ and the maximum of $|f|$ lies on the positive real ray, then $f(z^2)$ satisfies (A) with $\alpha = \pi/2$. Theorem 1 implies $\rho = \lambda$.

In order to prove Theorems 1 and 2 we need two lemmas.

LEMMA 1. *Let f satisfy* (A). *For any δ, $0 < \delta < \alpha$, there exists a constant $C = C(\delta)$ such that*

$$|\log f(z)| \leqslant C \log|f(c|z|)|, \qquad z \in U(-\alpha + \delta, \alpha - \delta).$$

PROOF. It follows from condition (A) that

$$\left.\frac{\partial \log|f(re^{i\varphi})|}{\partial \varphi}\right|_{\varphi=0} = 0.$$

So

$$\partial \arg f(r)/\partial r = 0, \qquad \arg f(r) = \text{const}.$$

Without loss of generality we can assume that $f(r) > 0$, $r > 0$. Consider a function $g_R(\zeta) = \log f(R + \zeta R \sin\alpha)$ which is analytic in the open disk $\{\zeta \in \mathbb{C} : |\zeta| < 1\}$. The Carathéodory inequality [5, p. 28] implies

$$\begin{aligned} |g_R(\zeta)| &\leqslant 2(1-|\zeta|)^{-1}\Big\{\max_{|\zeta|=1} \operatorname{Re} g_R(\zeta) - \operatorname{Re} g_R(0)\Big\} + |g_R(0)| \\ &= 2(1-|\zeta|)^{-1}\{\log f(R(1+\sin\alpha)) - \log f(R)\} + \log f(R) \\ &\leqslant 2(1-|\zeta|)^{-1} \log f(2R) \end{aligned}$$

which proves the lemma.

LEMMA 2. *Let* $u = \log|f|$, $\gamma = \min\{\pi/\lambda, \alpha\} > 0$.

a) *If* f *satisfies* (A), *then* $(\partial u/\partial\varphi)(z) \leqslant 0$, $z \in U(0, \gamma)$.

b) *If* f *satisfies* (B), *then* $u(re^{i(\theta-\varphi)}) + u(re^{i\varphi}) \leqslant 0$, $0 \leqslant |\varphi| \leqslant |\theta| \leqslant \gamma$, $\varphi\theta > 0$.

Besides, $u(re^{i\varphi}) \leqslant 0$, $0 < |\varphi| \leqslant \gamma/2$.

PROOF. a) Denote

$$f_\theta(z) = \overline{f(e^{2i\theta}\overline{z})}/f(z),$$

where $0 < \theta < \gamma$. Condition (A) shows that f_θ is analytic in $U(0, \theta)$,

$$|f_\theta(re^{i\theta})| = 1 \quad \text{and} \quad * \, |f_\theta(r)| \leqslant 1, \qquad r > 0.$$

It follows from Lemma 1 that $\lambda(f_\theta, U(0, \theta)) \leqslant \lambda < \pi/\theta$. Hence, using the Phragmen-Lindelöf principle, we have $|f_\theta(z)| < 1$, $z \in U(0, \theta)$. Hence

$$u(re^{i(2\theta-\varphi)}) \leqslant u(re^{i\varphi}), \qquad 0 < \varphi < \theta,\ 0 < \theta < \gamma.$$

For $\varphi = \theta - \varepsilon$ we have $u(re^{i(\theta+\varepsilon)}) - u(re^{i(\theta-\varepsilon)}) \leqslant 0$, $\varepsilon > 0$, and assertion a) follows.

b) Denote

$$\tilde{f}_\theta(z) = \overline{f(e^{i\theta}\overline{z})} f(z).$$

Condition (B) shows that $|\tilde{f}_\theta(r)| \leqslant 1$ and $|\tilde{f}(re^{i\theta})| \leqslant 1$, $r > 0$. It is easy to prove that $\lambda(\tilde{f}_\theta, U(0, \theta)) \leqslant \lambda < \pi/\theta$. Using the Phragmen-Lindelöf principle, we get $|\tilde{f}_\theta(z)| < 1$, $z \in U(0, \theta)$, and

$$u(re^{i(\theta-\varphi)}) + u(re^{i\varphi}) \leqslant 0, \qquad 0 \leqslant \varphi \leqslant \theta,\ 0 < \theta \leqslant \gamma.$$

In the same way we prove that

$$u(re^{i(\theta-\varphi)}) + u(re^{i\varphi}) \leqslant 0, \qquad 0 \geqslant \varphi \geqslant \theta,\ 0 > \theta \geqslant -\gamma.$$

Putting $\varphi = \theta/2$, we have $u(re^{i\theta/2}) \leqslant 0$, $0 \leqslant |\theta| \leqslant \gamma$, and the proof of assertion b) is completed.

PROOF OF THEOREM 1. We can assume that $\lambda < \infty$. Then it follows from Lemma 2 that the function $v(\zeta) = (\partial u/\partial\varphi)(\zeta^{\gamma/\pi})$ is harmonic and nonpositive in $U(0, \pi)$. Since $|f|$ has a maximum on the real positive axis, $v(r) = 0$, $r > 0$. Hence [6, p. 191], the following representation is true:

$$\exists b \geqslant 0: \quad v(\zeta) = -b\operatorname{Im}\zeta + \int_{-\infty}^{0} v(t)\left\{\operatorname{Im}\frac{1}{t-\zeta}\right\}dt, \qquad \zeta \in U(0, \pi),$$

where

$$\int_{-\infty}^{0} \frac{|v(t)|}{1+t^2}\,dt < \infty.$$

It is easy to see that the asymptotic equality

$$\left|\frac{\partial}{\partial\varphi}\int_{-\infty}^{0} v(t)\left\{\operatorname{Im}\frac{1}{t-z^{\pi/\gamma}}\right\}dt\right| = o(|z|^{\pi/\gamma}), \qquad |z| \to \infty,$$

holds uniformly in $z \in U(0, \gamma - \varepsilon)$ for any $0 < \varepsilon < \gamma$. Since

$$\Delta u = r^{-2}\frac{\partial^2 u}{\partial\varphi^2} + r^{-1}\frac{\partial}{\partial r}r\frac{\partial u}{\partial r} \equiv 0 \quad \text{in } U(0, \gamma),$$

we have

$$\left(\frac{\partial}{\partial r}r\frac{\partial u}{\partial r}\right)(z) = -\frac{1}{r}\frac{\partial^2 u}{\partial\varphi^2}(z) = \frac{b\pi}{\gamma r}\operatorname{Re} z^{(\pi/\gamma)-1} + o(|z|^{(\pi/\gamma)-1}), \qquad |z| \to \infty.$$

Integrating, we obtain the estimate

$$u(z) = (\gamma b/\pi)\operatorname{Re} z^{\pi/\gamma} + o(|z|^{\pi/\gamma}), \qquad |z| \to \infty,$$

which holds uniformly in $z \in U(0, \gamma - \varepsilon)$. We can prove this estimate similarly for $z \in U(-\gamma + \varepsilon, 0)$. Now it remains to show that $\rho = \lambda$. It is obvious that $\rho \leqslant \pi/\gamma = \max\{\lambda, \pi/\alpha\}$. If $\rho > \pi/\alpha$, then $\pi/\gamma = \lambda$, $\rho = \lambda$. If $\rho = \pi/\alpha$, then it follows from (1) that $b > 0$ and so $\lambda = \rho$.

PROOF OF THEOREM 2. We can assume that $\lambda < \infty$. Then it follows from Lemma 2 that the function $\tilde{v}(z) = u(z^{\gamma/\pi})$ is subharmonic and nonpositive in $U(-\pi/2, \pi/2)$. Thus by [7], there exist $b > 0$ and a set $F \subset [1, \infty)$ such that $\int_F d\log r < \infty$ and

$$\tilde{v}(z) = -b\operatorname{Re} z + o(|z|), \qquad |z| \to \infty,\ |z| \notin F,\ z \in U(-\pi/2, \pi/2),$$
$$\tilde{v}(z) \leqslant -b\operatorname{Re} z + o(|z|), \qquad |z| \to \infty,\ z \in U(-\pi/2, \pi/2).$$

Hence it follows that

$$u(z) = -b\operatorname{Re} z^{\pi/\gamma} + o(|z|^{\pi/\gamma}), \quad |z| \to \infty,\ |z|^{\pi/\gamma} \notin F,\ z \in U(-\pi/(2\gamma), \pi/(2\gamma)), \tag{3}$$

$$u(z) \leqslant -b\operatorname{Re} z^{\pi/\gamma} + o(|z|^{\pi/\gamma}), \quad |z| \to \infty,\ z \in U(-\pi/(2\gamma), \pi/(2\gamma)). \tag{4}$$

Using (3), the monotonicity of $M(r)$ and the inequality $\log M(r) \leqslant -u(r)$, it is easy to show that $\rho \leqslant \pi/\gamma = \max\{\lambda, \pi/\gamma\}$. From this inequality and

(1), just as in the proof of Theorem 1, we obtain $\lambda = \rho = \pi/\alpha$. Now it remains to show that (3) holds in $U(-\pi/\rho, \pi/\rho)$.

Denote by

$$h(\varphi) = \limsup_{r\to\infty} r^{-\rho} u(re^{i\varphi})$$

the indicator function of f. It follows from (4) that $h(\varphi) = -b\cos\rho\varphi$, $-\pi/(2\rho) < \varphi < \pi/(2\rho)$, and from Lemma 2 that

$$h((\pi/\rho) - \varphi) \leqslant -h(\varphi) = b\cos\rho\varphi = h(-\varphi), \qquad 0 < \varphi < \pi/(2\rho).$$

From these equations and the main identity for indicators [5, Chapter 1, §16] we obtain $h(\varphi) = -b\cos\rho\varphi$, $0 < |\varphi| < \pi/\rho$. Hence, the function f has a trigonometric indicator in $U(-\pi/\rho, \pi/\rho)$ and by (3) f has completely regular growth in $U(-\pi/(2\rho), \pi/(2\rho))$. Therefore [5, p. 213], f has completely regular growth in $U(-\pi/\rho, \pi/\rho)$, i.e., there exists a set $E \subset (0, \infty)$ such that

$$\lim_{R\to\infty} R^{-1} \int_{E\cap(0,R)} dr = 0,$$

and the asymptotic equality

$$u(z) = -b\operatorname{Re} z^{\rho} + o(|z|^{\rho}), \qquad |z| \to \infty,\ |z| \notin E,$$

holds in $U(-\pi/\rho, \pi/\rho)$.

The following two corollaries of Theorem 1 were proved by I. V. Ostrovskiĭ.

COROLLARY 1. *If a function f satisfies the conditions of Theorem 1, then f has completely regular growth with respect to r^{ρ} in $U(-2\pi/\rho, 2\pi/\rho)$ and its indicator equals $h(\varphi) = \sigma\cos\rho\varphi$, where σ is defined in (2).*

COROLLARY 2. *If a function f satisfies the conditions of Theorem 1 and has no zeros in $U(-\beta, \beta)$, where $\alpha \leqslant \beta \leqslant 2\alpha$, then the asymptotic equality (2) holds uniformly in $z \in U(-\beta+\delta, \beta-\delta)$ for every $0 < \delta < \beta$.*

PROOF OF COROLLARY 1. Consider the function

$$\tilde{f}(z) = \overline{f(e^{2\pi\rho i}\overline{z})}/f(z), \qquad z \in U(0, \pi/\rho).$$

It follows from (A) that $\tilde{f}$ is analytic in $U(0, \pi/\rho)$ and is bounded on the sides of the angle. It is easy to see that the growth of $\tilde{f}$ is not larger than a mean type of order ρ. Hence, $\tilde{f}$ has completely regular growth in $U(0, \pi/\rho)$ and $\tilde{h}(\varphi) = b\sin\rho\varphi$, $b \geqslant 0$. Since f has also completely regular growth in $U(0, \pi/\rho)$ and $h(\varphi) = \sigma\cos\rho\varphi$, we conclude that the function $\overline{f(e^{2\pi i/\rho}\overline{z})}$ has completely regular growth in $U(0, \pi/\rho)$ with indicator $h((2\pi/\rho) - \varphi) = \sigma\cos\rho\varphi - b\sin\rho\varphi$. This means that f has completely regular growth in $U(\pi/\rho, 2\pi/\rho)$. Since the indicator h satisfies the inequality $h(\varphi) + h(\varphi + \pi/\rho) \geqslant 0$, we see that $b = 0$. This proves Corollary 1.

To prove Corollary 2, we consider the function $g(z) = z^{-\rho}\log f(z)$ which is analytic in $U(-\beta, \beta)$. It follows from Lemma 2 (with β substituted for

α) and from Theorem 1 that g is bounded in $U(-\beta+\delta, \beta-\delta)$, $0<\delta<\beta$. It follows from (2) that the $\lim_{r\to\infty} \operatorname{Re} g(r)$ exists. Since $\arg f(r) = \text{const}$, the limit $\lim_{r\to\infty} g(r)$ exists. This implies [8, p. 51] that $g(z)$ tends to the same limit when $z\to\infty$, $z\in U(-\beta+\delta, \beta-\delta)$.

Note that propositions analogous to Theorems 1, 2 and Corollaries 1, 2 are true for a subharmonic function u satisfying one of the following conditions:

(A′) u is subharmonic in $U(0, 2\alpha)$ and in $U(-2\alpha, 0)$, $0<\alpha\leqslant\pi$; u is harmonic in $U(-\alpha, \alpha)$ and

$$u(z)\leqslant u(|z|), \qquad z\in U(-2\alpha, 0)\cup U(0, 2\alpha);$$

(B′) u is subharmonic in $\overline{U(-\alpha, \alpha)}$, $0<\alpha\leqslant\pi$, and

$$u(z)+u(|z|)\leqslant 0, \qquad z\in U(-\alpha, \alpha).$$

In conclusion we formulate the following proposition.

THEOREM 3. *Let*

$$f(z)=\sum_{0}^{\infty} a_k z^k$$

be an entire function such that $a_k\geqslant 0$, $k=0, 1, \ldots$. *If* f *satisfies* (1), *where* $0<\alpha\leqslant\pi/2$, *and has no zeros in* $U(-\alpha, \alpha)$, *then the following limit exists*

$$\lim_{n\to\infty} (n\log n)^{-1} \log\left\{1\Big/\sum_{k=n}^{\infty} a_k\right\} = \frac{1}{\rho}.$$

This means that if f satisfies the assumptions of Theorem 3, then the coefficients a_k of f have regular asymptotic behavior.

The proof of Theorem 3 is based on the asymptotic equality (2).

REFERENCES

1. A. A. Goldberg and I. V. Ostrovskiĭ, *Distribution of values of meromorphic functions*, "Nauka", Moscow, 1970. (Russian)

2. E. V. Gleizer, *On the growth of entire functions having zeros on a system of rays*, Ukrain. Mat. Zh. **38** (1986), no. 3, 297–302; English transl. in Ukrainian Math. J. **38** (1986).

3. W. K. Hayman and B. Kjellberg, *On the minimum of a subharmonic function on a connected set*, Studies in Pure Math. to the memory of Paul Turán (Paul Erdös, eds.), Akadémiai Kiadó, Budapest, 1983, pp. 291–322.

4. A. M. Vishnyakova, I. V. Ostrovskiĭ, and A. M. Ulanovskiĭ, *On a conjecture of Yu. V. Linnik*, Algebra i Analiz **2** (1990), no. 4, 71–78; English transl. in Leningrad Math. J. **2** (1991).

5. B. Ya. Levin, *Distribution of zeros of entire functions*, GITTL, Moscow, 1956; English transl., Amer. Math. Soc., Providence, R.I., 1964.

6. K. Hoffman, *Banach spaces of analytic functions*, Prentice-Hall, Englewood Cliffs, 1962.

7. W. Hayman, *Questions of regularity connected with the Phragmen-Lindelöf principle*, J. Math. Pures Appl. **35** (1956), 115–126.

8. R. Nevanlinna, *Analytic functions*, Springer-Verlag, New York, Heidelberg, and Berlin, 1970.

INSTITUTE FOR LOW TEMPERATURE PHYSICS & ENGINEERING, 47 LENIN PR., KHARKOV 310164 UKRAINE

Translated by THE AUTHOR

ADVANCES IN SOVIET MATHEMATICS
Volume 11, 1992

Value Distribution of Sequences of Rational Functions

M. SODIN

The classical Nevanlinna theory deals with the value distribution of an arbitrary meromorphic function. In this paper we find analogs of some theorems of this theory for arbitrary sequences of rational functions whose degrees are increasing. For this purpose we use the potential theory approach. We would like to mention here that our results were partly inspired by the well-known papers [10, 4].

Now we give the basic definitions and state the principal results.

Let R be a rational function. Define the *proximity value* of the function R with respect to the value $a \in \overline{\mathbb{C}}$ by

$$d(\infty, R) = \int \log^+ |R(z)|\, \sigma(dz), \qquad d(a, R) = d(\infty, (R-a)^{-1}),$$

where

$$\sigma(dz) = \frac{1}{\pi}\frac{\rho\, d\rho\, d\theta}{(1+\rho^2)^2}, \qquad z = \rho e^{i\theta},$$

is the spherical area measure. We omit the limits of integration when an integral is taken over the entire Riemann sphere $\overline{\mathbb{C}}$. The quantity introduced above possesses the basic properties of the Nevanlinna proximity function ([9, 5]), for example

$$d(\infty, R_1 + R_2) \leqslant d(\infty, R_1) + d(\infty, R_2) + \log 2,$$
$$d(\infty, R_1 R_2) \leqslant d(\infty, R_1) + d(\infty, R_2).$$

To characterize the distribution of roots of the equation $R(z) = a$, $a \in \overline{\mathbb{C}}$, we define the measure μ_R^a on $\overline{\mathbb{C}}$ such that $\mu_R^a(E)$ is equal to the number of these roots (with their multiplicities taken into account) on the set $E \subseteq \overline{\mathbb{C}}$. For any finite measure η on $\overline{\mathbb{C}}$ we define its lifting $R^*\eta$ in the following way:

$$R^*\eta(E) = \int \mu_R^a(E)\, \eta(da).$$

The definitions of $d(a, R)$ and μ_R^a are the same for the case when $a = a(z)$ is a rational function.

1991 *Mathematics Subject Classification*. Primary 30D35; Secondary 31A05.

Consider a sequence $\mathfrak{R} = (R_m)$ of rational functions with increasing degrees: $r_m = \deg R_m$, $r_1 < r_2 < \cdots$. Roughly speaking, the sequence of degrees (r_m) plays a role of the Nevanlinna characteristic, and the sets

$$\mathbf{E}(\mathfrak{R}) = \left\{a \in \mathbb{C} : \limsup_{m\to\infty} \frac{d(a, R_m)}{r_m} \geqslant 0\right\},$$

$$\mathbf{D}(\mathfrak{R}) = \left\{a \in \overline{\mathbb{C}} : \liminf_{m\to\infty} \frac{d(a, R_m)}{r_m} \geqslant 0\right\},$$

are analogs of the sets of Valiron and Nevanlinna exceptional values.

THEOREM 0.1. *Let $\varphi(t)$, $t > 0$, be an increasing function, $\varphi(0) = 0$,*

$$\int_0 \frac{\varphi(t)}{t}\, dt < \infty,$$

and let H_φ be the corresponding Hausdorff measure. Then $H_\varphi(\mathbf{E}(\mathfrak{R})) = 0$. If the sequence $\mathfrak{R}$ is "lacunar" in the sense that $\sum_m 1/r_m < \infty$, then the set $\mathbf{E}(\mathfrak{R})$ has zero logarithmic measure $H_{\log}$, and consequently, zero logarithmic capacity.

Using the notion of the Hyllengren measure ([7, 8, 6]) we will prove a more precise theorem, but the question whether the set $\mathbf{E}(\mathfrak{R})$ always has zero capacity remains open. We will also prove that the set $\mathbf{E}(\mathfrak{R})$ is at most countable.

Now we describe our results directly related to the distribution of values. We use the approach developed in [2, 3, 11]. If a rational function $a(z)$ is fixed, then the sequence of measures $(\mu_m^a) = (\mu_{R_m}^a)$ is precompact with respect to the weak topology in the space of measures (further on we shall only use this topology for measures). Consider the set of all limit measures for (μ_m^a) as $m \to \infty$. It is convenient to use this set for the description of the asymptotic distribution of a-points of functions R_m (compare [1]). The following results are devoted to an investigation of the dependence of such limit sets on the rational function $a(z)$.

We say that a sequence of measures μ_m is equidistributed with respect to $R_m^*\sigma$ (we recall that σ is the spherical area measure), if

$$(R_m^*\sigma - \mu_m)/r_m \to 0, \qquad m \to \infty. \tag{0.1}$$

THEOREM 0.2. *The following statements are true:*

(1) *if $a(z)$ is a rational function such that $d(a, R_m) = o(r_m)$, $m \to \infty$, then (0.1) is valid for $\mu_m = \mu_m^a$;*
(2) *if η is an arbitrary probability measure on $\overline{\mathbb{C}}$ which does not load values from $\mathbf{E}(\mathfrak{R})$ (this means that $\eta(a) = 0$ for every $a \in \overline{\mathbb{C}}$), then (1.2) is valid for $\mu_m = R_m^*\eta$.*

The following proposition states the refined precompactness of the sequence of measures μ_m^a and the countability of the set $\mathbf{E}(\mathfrak{R})$.

PROPOSITION 0.3. *Let* $\mathfrak{R} = (R_m)$, $m \in \mathbb{N}$, *be an arbitrary sequence of rational functions with increasing degrees. Then there exists a subsequence* $\mathfrak{R}_1 = (R_m)$, $m \in \mathbb{N}_1 \subseteq \mathbb{N}$, *with the following properties*:

(1) $r_m^{-1} R_m^* \sigma \to \nu$, $m \to \infty$, $m \in \mathbb{N}_1$;

(2) *for any rational function* $a(z)$ *the limit* $r_m^{-1} \mu_m^a \to \infty$, $m \to \infty$, $m \in \mathbb{N}_1$, *exists*;

(3) *for all rational functions* $a(z)$, *except perhaps a countable set, we have* $\nu^a = \nu$.

The last theorem is an analog of the Nevanlinna Second Main Theorem.

THEOREM 0.4. *Let* $\mathfrak{R}_1$ *be the subsequence from the previous theorem. Then*

$$\sum(\nu - \nu^a) \leqslant 2\nu,$$

where the sum is taken over all rational functions such that $\nu^a \neq \nu$, *and the inequality* $\lambda \leqslant \eta$ *between measures* λ *and* η *means that* $\lambda - \eta$ *is a nonnegative measure.*

The paper is organized as follows. In §1 we prove preliminary results on the quantity $d(a, R)$, and prove the first assertion of Theorem 0.2. In §2 we study the size of the set $\mathbf{E}(\mathfrak{R})$ and, in particular, we prove Theorem 0.1. In §3 we prove Proposition 0.3, the second assertion of Theorem 0.2 and Theorem 0.4.

Acknowledgments. The author thanks A. Eremenko for a valuable discussions. This work was finished when the author was guest of the DFG Forschungsschwerpunkt Komplexe Mannigfaltigkeiten. The author would like to thank A. Huckleberry and his colleagues from the Ruhr-Universität Bochum for their generous hospitality.

§1. The spherical potential and its properties

We recall some definitions from [9]. Let $[w, a]$ be the chordal distance between values $w, a \in \overline{\mathbb{C}}$, namely

$$[w, a] = \begin{cases} |w - a|(1 + |w|^2)^{-1/2}(1 + |a|^2)^{-1/2}, & a \neq \infty, \\ (1 + |w|^2)^{-1/2}, & a = \infty. \end{cases}$$

We denote by $U[\eta]$ the spherical potential of the measure η:

$$U[\eta](z) = \int \log \frac{1}{[z, w]} \eta(dw).$$

This function is δ-subharmonic (i.e., it is the difference of two subharmonic functions). The Riesz charge (i.e., the generalized Laplacian) of $U[\eta]$ equals $\eta(\overline{\mathbb{C}})\sigma - \eta$.

It is more convenient to use the spherical form of the proximity value

$$d^*(a, R) = \int \log \frac{1}{[R(z), a]} \sigma(dz).$$

As usual, $0 \leqslant d^*(a, R) - d(a, R) \leqslant \log\sqrt{2}$. That is why we denote the spherical form of the proximity value by $d(a, R)$. Now we give two useful expressions for $d(a, R)$.

Let λ be a measure on $\overline{\mathbb{C}}$. Define the image $R_*\lambda$ of λ under R, namely $R_*\lambda(E) = \lambda(R^{-1}E)$, where $E \subseteq \overline{\mathbb{C}}$ is an arbitrary Borel set, $R^{-1}E$ being its preimage under R. In other words,

$$\int f(z)R_*\lambda(dz) = \int f(R(z))\lambda(dz).$$

Thus

$$d(a, R) = U[R_*\sigma](a), \qquad a \in \overline{\mathbb{C}}. \tag{1.1}$$

Now, let $a = a(z)$ be a rational function. Define the δ-subharmonic function

$$w_{R,a}(z) = \log\frac{1}{[R(z), a(z)]}, \qquad z \in \overline{\mathbb{C}};$$

then

$$d(a, R) = \int w_{R,a}(z)\,\sigma(dz). \tag{1.2}$$

Let $R = P/Q$, $a = a_1/a_2$, where P, Q and a_1, a_2 are pairs of polynomials with no common zeros. To simplify notation, we assume that ∞ is not a root of the equation $R(z) - a(z) = 0$, so

$$\deg(a_2P - a_1Q) = \deg R + \deg a.$$

Then

$$\begin{aligned} w_{R,a} &= (1/2)\log(|P|^2 + |Q|^2) + (1/2)\log(|a_1|^2 + |a_2|^2) - \log|a_2P - a_1Q| \\ &= u_R + u_a - u_{R,a}, \end{aligned} \tag{1.3}$$

where the functions u_R, u_a, $u_{R,a}$ are subharmonic. We denote by $\mu[w]$ the Riesz charge of a δ-subharmonic function w.

LEMMA 1.2. *In the previous notation,*

$$\mu[w_{R,a}] = R^*\sigma + a^*\sigma - \mu_R^a, \tag{1.4}$$

$$w_{R,a} = U[\mu_R^a] - U[R^*\sigma] - U[a^*\sigma] + d(a, R). \tag{1.5}$$

(*If* $a = \mathrm{const}$, *then* $a^*\sigma = 0$.)

PROOF. Equality (1.4) follows from (1.3) since $R^*\sigma$, $a^*\sigma$, and μ_R^a are the Riesz measures of subharmonic functions u_R, u_a and $u_{R,a}$ respectively. Since

$$(R^*\sigma + a^*\sigma - \mu_R^a)(\overline{\mathbb{C}}) = 0, \tag{1.6}$$

we get $w_{R,a}(z) = U[\mu_R^a - R^*\sigma - a^*\sigma] + c_{R,a}$, where $c_{R,a}$ is a constant. Integrating this equality over $\sigma(dz)$ and using (1.1) and (1.2), we obtain

$$\begin{aligned} d(a, R) &= \int U[\mu_R^a - R^*\sigma - a^*\sigma]\sigma(dz) + c_{R,a} \\ &= \int (\mu_R^a(ds) - R^*\sigma(ds) - a^*\sigma(ds)) \int \log\frac{1}{[z, s]}\sigma(dz) + c_{R,a}. \end{aligned}$$

But the integral

$$\int \log \frac{1}{[z\,,\,s]}\,\sigma(dz)$$

does not depend on $s \in \overline{\mathbb{C}}$. Using (1.6) again, we conclude that $d(a\,,\,R) = c_{R\,,\,a}$. Lemma 1.1 is proved.

Let $\|\cdot\| = \|\cdot\|_{L^1_\sigma}$. The following lemmas from potential theory are standard.

LEMMA 1.2. *Let* $\{w_n\}$ *be a sequence of* δ*-subharmonic functions on* $\overline{\mathbb{C}}$,

$$|\mu[w_n](\overline{\mathbb{C}})| \leqslant c < \infty. \tag{1.7}$$

Then there exists a subsequence (which we also denote by $\{w_m\}$*) such that* $\|w_m - w\| \to 0$. *Here* w *is a* δ*-subharmonic function on* $\overline{\mathbb{C}}$, *possibly equal to* $+\infty$ *or* $-\infty$.

LEMMA 1.3. *Let* $\{w_m\}$ *be a sequence of* δ*-subharmonic functions in* $\overline{\mathbb{C}}$ *such that*

$$\|w_m\| \to 0. \tag{1.8}$$

Then $\mu[w_m] \to 0$.

PROOF OF THE FIRST ASSERTION OF THEOREM 0.2. Let us consider the nonnegative δ-subharmonic functions $w_m = r_m^{-1} w_{R_m\,,\,a}$. Condition (1.6) is evidently fulfilled and (1.7) follows from $d(a\,,\,R_m) = o(r_m)$, $m \to \infty$. So the first assertion of Theorem 0.2 follows from Lemma 1.3 and (1.4).

§2. The characterization of the set $\mathbf{E}(\mathfrak{R})$

Let us introduce the following notation:

$$\mathbb{D}(z\,,\,t) = \{s : |z - s| < t\}, \qquad \mathbb{D}_t = \mathbb{D}(0\,,\,t),$$
$$\mathbb{D}^*(z\,,\,t) = \{s : [z\,,\,s] < t\}.$$

In what follows K will denote certain constants, maybe different.

Let R be a rational function, $r = \deg R$. Consider the set

$$\mathbf{E}_\varepsilon(R) = \{a \in \overline{\mathbb{C}} : d(a\,,\,R) > \varepsilon r\}.$$

THEOREM 2.1. *The set* $\mathbf{E}_\varepsilon(R)$ *may be covered by no more than* $K(\varepsilon) = (K/\varepsilon)\log(1/\varepsilon)$ *spherical disks* $\mathbb{D}_j^*$ *whose radii do not exceed* $\exp(-\varepsilon r)$.

For the proof of this theorem we need some simple lemmas from potential theory.

LEMMA 2.2 (see for example [6]). *Let* $U \subset \mathbb{D}_1$. *Then for any* $z \in \mathbb{D}_2$

$$\int_U |\log|z - s||\,\sigma(ds) \leqslant K\sigma(U)\log\frac{e}{\sigma(U)}. \tag{2.1}$$

LEMMA 2.3. *Let* $w(z)$ *be a* δ*-subharmonic function,* $|\mu[w](\overline{\mathbb{C}})| < \infty$, *and*

$$\int_{\overline{\mathbb{C}}} w(z)\,\sigma(dz) = 0. \tag{2.2}$$

Then for any set $U \subseteq \overline{\mathbb{C}}$

$$\int_U |w(z)|\,\sigma(ds) \leqslant K|\mu[w](\overline{\mathbb{C}})|\sigma(U)\log\frac{e}{\sigma(U)}. \tag{2.3}$$

PROOF. We have

$$w(z) = \int_{\mathbb{D}_2} \log|z-s|\,\mu(ds) + \int_{\mathbb{C}\setminus\mathbb{D}_2} \log|1-z/s|\,\mu(ds) + c_w. \tag{2.4}$$

Let $U_1 = U \cap \mathbb{D}_1$, $U_2 = U \setminus \mathbb{D}_1$. We shall prove the inequality (2.3) for the set U_1, the proof of this estimate for the set U_2 being completely similar. We shall estimate each of the three terms on the right-hand side of (2.4).

By (2.2) we get

$$\begin{aligned} |c_w| &= \left|\int \sigma(dz)\left\{\int_{\mathbb{D}_2} \log|z-s|\,\mu(ds) + \int_{\mathbb{C}\setminus\mathbb{D}_2} \log|1-z/s|\,\mu(ds)\right\}\right| \\ &\leqslant |\mu[w](\overline{\mathbb{C}})|\left\{\max_{s\in\mathbb{D}_2}\left|\int \log|z-s|\,\sigma(dz)\right|\right. \\ &\qquad\left. + \max_{s\in\mathbb{C}\setminus\mathbb{D}_2}\left|\int \log|1-z/s|\,\sigma(dz)\right|\right\} \\ &\leqslant |\mu[w](\overline{\mathbb{C}})|\left\{\int \log^+(|z|+2)\,\sigma(dz) + \int \log\left(1+\frac{|z|}{2}\right)\sigma(dz)\right\} \\ &\leqslant K|\mu[w](\overline{\mathbb{C}})|. \end{aligned} \tag{2.5}$$

To estimate the second term, we suppose that $z \in U_1 \subset \mathbb{D}_1$, $|s| \geqslant 2$. In this case,

$$\log(1/2) \leqslant \log(1-|z/s|) \leqslant \log|1-z/s| \leqslant \log(1+|z/s|) \leqslant \log(3/2).$$

Hence

$$\int_{U_1} \sigma(dz) \int_{\mathbb{C}\setminus\mathbb{D}_2} |\log|1-z/s||\,\mu(ds) \leqslant K|\mu[w](\overline{\mathbb{C}})|\sigma(U_1). \tag{2.6}$$

We use Lemma 2.2 to estimate the first term in (2.4):

$$\int_{U_1} \sigma(dz) \int_{\mathbb{D}_2} |\log|z-s||\,\mu(ds) \leqslant K|\mu[w](\overline{\mathbb{C}})|\sigma(U_1)\log\frac{e}{\sigma(U_1)}. \tag{2.7}$$

Combining estimates (2.4)–(2.7), we get (2.3). Lemma 2.3 is proved.

PROOF OF THEOREM 2.1. Let $a \in \mathbf{E}_\varepsilon(R)$, i.e., $d(a, R) \geqslant \varepsilon r$. Let us estimate the spherical area of the set

$$U_a = \left\{z \in \overline{\mathbb{C}} : \log\frac{1}{[R(z), a]} \geqslant \frac{d(a, R)}{4}\right\}.$$

We have

$$\begin{aligned} d(a, R) &= \int_{\overline{\mathbb{C}}\setminus U_a} \log\frac{1}{[R(z), a]}\,\sigma(dz) + \int_{U_a} \log\frac{1}{[R(z), a]}\,\sigma(dz) \\ &\leqslant \frac{d(a, R)}{4} + \int\left\{\log\frac{1}{[R(z), a]} - d(a, R)\right\}\sigma(dz) \\ &\quad + d(a, R)\sigma(U_a). \end{aligned}$$

The integrand $w_{R,a}(z) - d(a, R)$ satisfies the conditions of Lemma 2.3, and by (1.4) the total variation of the charge of this function equals $2r$. Therefore, by Lemma 2.3

$$\int_{U_a} |w_{R,a}(z) - d(a, R)|\, \sigma(dz) \leqslant Kr\sigma(U_a) \log \frac{e}{\sigma(U_a)},$$

and

$$d(a, R) \leqslant \left(\frac{1}{4} + \sigma(U_a)\right) d(a, R) + Kr\sigma(U_a) \log \frac{e}{\sigma(U_a)}. \tag{2.8}$$

Let $\sigma(U_a) \leqslant 1/4$. Since $a \in \mathbf{E}_\varepsilon(R)$, it follows from (2.8) that

$$\begin{aligned} \varepsilon/2 &\leqslant K\sigma(U_a) \log(e/\sigma(U_a)), \\ \sigma(U_a) &\geqslant \varepsilon K^{-1} (\log(e/\varepsilon))^{-1}. \end{aligned} \tag{2.9}$$

Let $a_1, \dots, a_N \in \mathbf{E}_\varepsilon(R)$ be a maximal set of numbers such that $U_{a_j} \cap U_{a_k} \neq 0$ when $j \neq k$. Then by virtue of estimate (2.9) we obtain

$$N \leqslant K\varepsilon^{-1} \log(e/\varepsilon). \tag{2.10}$$

For any $a \in \mathbf{E}_\varepsilon(R)$ there exists j, $1 \leqslant j \leqslant N$, such that $U_{a_j} \cap U_a \neq \varnothing$. Then for any $a \in U_a \cap U_{a_j}$ we have

$$[a, a_j] \leqslant 2\max([a, R(z)], [a_j, R(z)]) \leqslant 2\exp(-\varepsilon r/4).$$

Increasing the constant K in (2.10) (or, which is the same, decreasing ε) we may assume that

$$[a, a_j] \leqslant \exp(-\varepsilon r). \tag{2.11}$$

Now the assertion of Theorem 2.1 follows from (2.10) and (2.11).

Now we collect some necessary facts concerning the notion of μ-measure of Hyllengren [7, 8, 6]. Let us fix an increasing sequence of numbers $(\rho_n) \subset \mathbb{N}$. Following Hyllengren, we say that the sequence $(\exp(-c\rho_n))$ majorizes the set $E \subset \mathbb{C}$ if there exists a sequence of complex numbers $(a_n) \subset \mathbb{C}$ such that

$$E \subset \bigcap_{N \geqslant 1} \bigcup_{n \geqslant N} \mathbb{D}(a_n, \exp(-c\rho_n)). \tag{2.12}$$

This means that every point of E is covered by an infinite number of disks $\mathbb{D}(a_n, \exp(-c\rho_n))$. Then, by definition,

$$\mu_{(\rho_n)}(E) = \inf\{1/c : (\exp(-c\rho_n)) \text{ majorizes } E\}.$$

We write μ instead of $\mu_{(\rho_n)}$ when $(\rho_n) = \mathbb{N}$.

We will use the following dependence relations between the Hyllengren and the Hausdorff measures [8].

Let $\varphi(t)$, $t \geqslant 0$, be an increasing function, $\varphi(0) = 0$.

(a) If
$$\sum \varphi(\exp(-\rho_n)) < \infty, \tag{2.13}$$
then $\mu_{(\rho_n)}(E) < \infty$ implies $H_\varphi(E) = 0$.

(b) If
$$\liminf_{n\to\infty} n\varphi(\exp(-\rho_n)) > 0,$$
then $H_\varphi(E) = 0$ implies $\mu_{(\rho_n)}(E) = 0$.

If $(\rho_n) = \mathbb{N}$, then the condition (2.13) is equivalent to
$$\int_0 \frac{\varphi(t)}{t}\, dt < \infty.$$
If
$$\sum r_n^{-1} < \infty,$$
then the function $\varphi(t) = 1/\log(1/t)$ satisfies the condition (2.13). Thus, by the Lindeberg theorem (see, for example, [9]), E has zero logarithmic capacity if the $\mu_{(\rho_n)}$-measure of E is finite.

THEOREM 2.4. *Let $\mathfrak{R} = (R_n)$ be a sequence of rational functions whose degrees $r_n = \deg R_n$ are increasing. Then the set $\mathbf{E}(\mathfrak{R})$ is contained in a countable union of sets of a finite $\mu_{(r_n)}$-measure. Further, if*
$$\mathbf{E}_\varepsilon(\mathfrak{R}) = \left\{a \in \overline{\mathbb{C}} : \limsup_{n\to\infty} \frac{d(a, R_n)}{r_n} \geqslant \varepsilon\right\},$$
then
$$\mu(\mathbf{E}_\varepsilon(\mathfrak{R})) \leqslant K\varepsilon^{-2}\log(1/\varepsilon). \tag{2.14}$$

By virtue of the above-mentioned relationship between Hyllengren and Hausdorff measures this theorem implies Theorem 0.1.

PROOF OF THEOREM 2.4. By the definitions of the sets $\mathbf{E}_\varepsilon(\mathfrak{R})$ and $\mathbf{E}_\varepsilon(R_n)$, and by Theorem 2.1, we have
$$\begin{aligned}\mathbf{E}(\mathfrak{R}) &= \bigcup_{m=1}^{\infty} \mathbf{E}_{1/m}(\mathfrak{R}) = \bigcup_{m\geqslant 1}\bigcap_{N\geqslant 1}\bigcup_{n\geqslant N} \mathbf{E}_{1/m}(R_n)\\ &\subseteq \bigcup_{m\geqslant 1}\bigcap_{N\geqslant 1}\bigcup_{n\geqslant N}\bigcup_{j=1}^{K(\varepsilon)} \mathbb{D}(a_{m,j,n}, \exp(-r_n/m))\\ &= \bigcup_{m\geqslant 1}\bigcup_{j=1}^{K(\varepsilon)}\bigcap_{N\geqslant 1}\bigcup_{n\geqslant N} \mathbb{D}(a_{m,j,n}, \exp(-r_n/m)).\end{aligned}$$

Thus the first assertion of Theorem 2.4 is proved.

Let $s = s(n, j) = K(\varepsilon)n + j$. Then
$$\mathbf{E}_\varepsilon(\mathfrak{R}) \subseteq \bigcap_{N\geqslant 1}\bigcup_{n\geqslant N}\bigcup_{j=1}^{K(\varepsilon)} \mathbb{D}(a_{n,j}, \exp(-\varepsilon n)) \subseteq \bigcap_{N\geqslant 1}\bigcup_{s\geqslant N} \mathbb{D}(a_s, \exp(-\varepsilon s/K(\varepsilon))).$$

This implies (2.14). Theorem 2.4 is proved.

Now we shall show that the estimate (2.14) is rather precise.

THEOREM 2.5. *Let* $E \subset \overline{\mathbb{C}}$ *be a set such that for some increasing sequence* $(r_n) \subseteq \mathbb{N}$ *the inequality* $\mu_{(r_n)}(E) < \infty$ *holds. Then there exists a sequence of polynomials* $\mathfrak{P} = (P_n)$, $\deg P_n = r_n$, *such that* $E \subseteq \mathbf{E}_\varepsilon(\mathfrak{P})$ *with* $\varepsilon = \pi[4\max(2, \mu_{(r_n)}(E))]^{-1}$.

PROOF. For technical reasons, we set $r(n) = r_n$. Fix $c < [\mu_{(r(n))}(E)]^{-1}$. Then a sequence $(\exp(-cr(n)))$ majorizes E. Hence there exists a sequence $(a_n) \subset \mathbb{C}$ such that (2.12) is fulfilled. Let us consider a sequence of polynomials $\mathfrak{P} = (P_n)$, where $P_n(z) = z^{r(n)} + a_n$. If $a \in E$, then for some subsequence $\mathbb{N}_1 \subseteq \mathbb{N}$ the inequality $|a - a_n| \leqslant \exp(-cr(n))$, $n \in \mathbb{N}_1$, holds. Then

$$\begin{aligned} d(a, P_n) &= \int \log^+ \frac{1}{|z^{r(n)} + a_n - a|}\,\sigma(dz) \\ &\geqslant \frac{1}{4}\int_{\mathbb{D}_1} \log^+ \frac{1}{|\rho^{r(n)} e^{ir(n)\omega} + a_n - a|}\,\rho\,d\rho\,d\omega \\ &\geqslant \frac{\pi}{2}\int_0^1 \log^+ \frac{1}{\max(\rho^{r(n)}, |a_n - a|)}\,\rho\,d\rho \\ &\geqslant \frac{\pi r(n)}{2}\max\left(\int_0^1 \left(\log\frac{1}{\rho}\right)\rho\,d\rho, \frac{c}{2}\right) \\ &= \frac{\pi r(n)}{4}\max\left(\frac{1}{2}, c\right), \end{aligned}$$

if $n \in \mathbb{N}_1$. Theorem 2.5 is proved.

REMARK 2.1. Let $F : \overline{\mathbb{C}} \to \overline{\mathbb{C}}$ be some Borel map,

$$\deg F := \max\{\operatorname{card}(F^{-1}a) : a \in \overline{\mathbb{C}}\}.$$

Let us define, as above,

$$d(a, F) = \int \log \frac{1}{[F(z), a]}\,\sigma(dz) = U[F\sigma](a).$$

According to the well-known Cartan-Ahlfors estimate (see, for example, [9]), given an increasing function $h(t)$, $h(0) = 0$, $h(\tau) = 1$, the set

$$\left\{a \in \overline{\mathbb{C}} : U[F\sigma](a) > \int_0^\tau \log\frac{1}{\tau}\,h(d\tau)\right\}$$

may be covered by spherical disks $\mathbb{D}^*(z_j, r_j)$ such that $r_j \leqslant \tau$ and

$$\sum_j h(r_j) \leqslant K.$$

Using this estimate one can prove that if $\mathfrak{F} = (F_n)$ is a sequence of such maps with increasing degrees, then for any $\alpha > 1$ $H_{(\log(1/t))^{-\alpha}}(\mathbf{E}(\mathfrak{F})) = 0$, where

$$\mathbf{E}(\mathfrak{F}) = \left\{ a \in \overline{\mathbb{C}} : \limsup_{n\to\infty} \frac{d(a, F_n)}{\deg F_n} > 0 \right\}.$$

In other words, the logarithmic dimension of the set $\mathbf{E}(\mathfrak{F})$ does not exceed one. Of course, in this situation the values from $\overline{\mathbb{C}} \setminus \mathbf{E}(\mathfrak{F})$ are not equidistributed.

REMARK 2.2. It is not difficult to prove that for any sequence of rational functions $\mathfrak{R} = (R_n)$ the set $\{a : d(a, R_m) \to \infty \text{ as } m \to \infty\}$ has zero logarithmic capacity.

§3. Refined precompactness

Let $f \vee g$ and $f \wedge g$ be the upper and the lower envelopes of the functions f and g. Further we will use two lemmas which are similar to lemmas from [11].

LEMMA 3.1. *Let a, b be distinct rational functions, and let $u_{R,a}$, $u_{R,b}$, u_R be the functions defined by* (1.3). *Then for any rational function R the following estimate is valid $\|u_{R,a} \vee u_{R,b} - u_R\| < K(a, b)$.*

PROOF. Using definitions from §1 we have

$$\begin{aligned} u_{R,a} = \log|a_2P - a_1Q| &\leqslant \log(|a_1| \vee |a_2|) + \log(|P| \vee |Q|) \\ &\leqslant \log(|a_1| \vee |a_2|) + u_R + \log 2. \end{aligned}$$

Hence

$$u_{R,a} \vee u_{R,b} - u_R \leqslant \log(|a_1| \vee |a_2| \vee |b_1| \vee |b_2|) + \log 2.$$

Denoting $a_2P - a_1Q$ and $b_2P - b_1Q$ by P^* and Q^*, respectively, and expressing P, Q via P^*, Q^*, we get the lower estimate for $u_{R,a} \vee u_{R,b} - u_R$. Lemma 3.1 is proved.

Let $\mathfrak{R} = (R_m)$, $r_m = \deg R_m$. We assume that

$$r_m^{-1} R_m^* \sigma \to \nu, \tag{3.1}$$

$$r_m^{-1} \mu_m^a \to \nu^a, \tag{3.2}$$

for some $a = a(z)$. Passing to a subsequence, we assume also that

$$r_n^{-1} w_{R_n,a} \to w^a, \tag{3.3}$$

where w^a is a nonnegative δ-subharmonic function (it is possible that $w^a \equiv +\infty$).

LEMMA 3.2. *Let the conditions of the previous lemma be fulfilled. If the limits* (3.1)–(3.3) *exist, then*

$$w^a \wedge w^b = 0. \tag{3.4}$$

PROOF. Using (1.2) and Lemma 3.1, we have

$$\begin{aligned}\|w_{R_n,a}\wedge w_{R_n,b}\| &= \|(u_{R_n}+u_a-u_{R_n,a})\wedge(u_{R_n}+u_b-u_{R_n,b})\|\\ &< \|u_{R_n,a}\vee u_{R_n,b}-u_{R_n}\|+\|u_a\vee u_b\|<K(a,b).\end{aligned}$$

Let us observe now that the operation $\vee$ in the space of δ-subharmonic functions is continuous in the L^1_σ-topology. Lemma 3.2 is proved.

PROOF OF PROPOSITION 0.3. We assume that the condition (3.1) is satisfied. By the first assertion of Theorem 0.2 which was proved in §1, for all rational functions $a(z)$ such that $d(a, R_m) = o(r_m)$, $m\to\infty$, the limit (3.2) exists and $\nu^a = \nu$. Using a "normal family argument" we construct inductively a certain subsequence $\mathfrak{R}^1\subset\mathfrak{R}$ that satisfies the following conditions:

(i) the set of rational functions

$$\mathbf{E}^*(\mathfrak{R}^1)=\left\{a(z):\ \limsup_{m\to\infty,\ R_m\in\mathfrak{R}^1}\frac{d(a,R_m)}{r_m}>0\right\}$$

is at most countable;

(ii) for every function $\mathbf{E}^*(\mathfrak{R}^1)$ the limit (3.2) exists with $\nu^a\neq\nu$.

For some subsequence $\mathfrak{R}^1\subset\mathfrak{R}$ and for a rational function $a(z)$, suppose that the limit (3.2) exists with $\nu^a\neq\nu$. (If such a subsequence does not exist, then $\mathfrak{R}^1=\mathfrak{R}$, $\mathbf{E}^*(\mathfrak{R}^1)=0$ and everything is proved.) By Lemma 1.2 we may assume that the limit (3.3) also exists. We set

$$\begin{aligned}&S(a,\mathfrak{R}^1)=\sigma(\{z: w^a(z)>0\}),\\ &\mathbf{E}^*(\varepsilon,\mathfrak{R}^1)=\{\,b\in\mathbf{E}^*(\mathfrak{R}^1):\ \exists\mathfrak{R}(b)\subset\mathfrak{R}\text{ such that for the function}\\ &\qquad b(z)\text{ limits (3.2), (3.3) do exist and }S(b,\mathfrak{R}(b))\geqslant\varepsilon\,\}.\end{aligned}$$

STEP 1. Choose the least number $k_1\in\mathbb{N}$ such that $\mathbf{E}^*(1/k_1,\mathfrak{R})\neq\varnothing$. Then there exist a function $a^1\in\mathbf{E}^*(1/k_1)$ and a subsequence $\mathfrak{R}_1=\mathfrak{R}(a^1)\subset\mathfrak{R}$ such that $S(a^1,\mathfrak{R}(a^1))\geqslant 1/k_1$. Set $G_1=\{z: w^{a^1}(z)>0\}$, then

$$\sigma(G_1)=S(a^1,\mathfrak{R}_1)\geqslant 1/k_1. \tag{3.5}$$

. . .

STEP m. Suppose that the functions $a^1, a^2, \dots, a^{m-1}$, the subsequences $\mathfrak{R}_{m-1}\subset\mathfrak{R}_{m-2}\subset\cdots\subset\mathfrak{R}_1\subset\mathfrak{R}$ and the sets $G_j=\{z: w^{a^j}(z)>0\}$, $\sigma(G_j)=S(a_j,\mathfrak{R}_j)\geqslant 1/k_j$, are chosen. Then there are two possibilities: either for all rational functions $a(z)$ different from $a^1,\dots,a^{m-1}$ the limit (3.2) exists with $\nu^a=\nu$ and the proof is over, or there exists a function $a(z)$ for which (3.2) is not valid with $\nu^a=\nu$. In the second case, choose the least number $k_m\in\mathbb{N}$ such that the set

$$\mathbf{E}_m=\mathbf{E}^*(1/k_m,\mathfrak{R}_{m-1})\setminus\{a^1,\dots,a^{m-1}\}$$

is not empty, then choose $a_m \in \mathbf{E}_m$ and a subsequence $\mathfrak{N}_m \subset \mathfrak{N}_{m-1}$ such that (3.2) is valid with $a = a^m$, $\nu^{a^m} \neq \nu$. Set $G_m = \{z : w^{a^m}(z) > 0\}$; then $\sigma(G_m) \geqslant 1/k_m$.

According to Lemma 3.2, $G_l \cap G_j = \varnothing$ when $l \neq j$, hence

$$1 = \sigma(\overline{\mathbb{C}}) \geqslant \sum_{j=1}^{m} \sigma(G_j) \geqslant \frac{m}{k_m},$$

and

$$k_m \geqslant m, \qquad m \in \mathbb{N}. \tag{3.6}$$

Now we consider a sequence $\mathfrak{N}^1$ obtained by the diagonal process from $\mathfrak{N}_1, \dots, \mathfrak{N}_m, \dots$ and show that the conditions (i) and (ii) are satisfied.

Let $a \in \mathbf{E}^*(\mathfrak{N}^1)$. Choose a subsequence $\mathfrak{N}^2$ to satisfy (3.2). We choose a number $m \in \mathbb{N}$ so large that

$$k_m \geqslant m > [1/S(a, \mathfrak{N}^2)] + 2. \tag{3.7}$$

Then, by construction, $a \in \mathbf{E}^*(\mathfrak{N}^1) \subset \mathbf{E}^*(\mathfrak{N}_m)$. If the function $a(z)$ differs from $a^1, \dots, a^{m-1}$, then $S(a, \mathfrak{N}^2) \leqslant 1/(k_m - 1)$ by the definition of the number k_m. This contradicts (3.7), hence the function $a(z)$ coincides with one of the functions $a^1, \dots, a^{m-1}$. So the set $\mathbf{E}^*(\mathfrak{N}^1)$ is countable at most, and condition (i) is fulfilled. Condition (ii) is fulfilled by construction. Proposition 0.3 is proved.

PROOF OF THE SECOND ASSERTION OF THEOREM 0.2. Assume that there exists a measure μ such that $\mu(\{a\}) = 0$ for any $a \in \mathbf{E}(\mathfrak{N})$, and

$$r_n^{-1} R_n^* \sigma \to \nu, \qquad r_n^{-1} R_n^* \mu \to \lambda.$$

Choose a subsequence $\mathfrak{N}^1 \subset \mathfrak{N}$ to satisfy (i) and (ii). Let $F \subset \overline{\mathbb{C}}$ be some ν-normal Borel set, i.e., $\nu(\partial F) = 0$. If we show that the limit

$$\lim_{n\to\infty,\, R_n \subset \mathfrak{N}^1} r_n^{-1} R_n^* \mu(F)$$

exists and equals $\nu(F)$, then $r_n^{-1} R_n^* \mu \nu$, and Theorem 0.2 will be proved.

By the definition of the lifting of a measure we have

$$\begin{aligned} \frac{1}{r_n} R_n^* \mu(F) &= \frac{1}{r_n} \int R_n^* \delta^a(F)\, \mu(da) \\ &= \frac{1}{r_n} \left(\int_{\overline{\mathbb{C}} \setminus \mathbf{E}(\mathfrak{N}^1)} + \int_{\mathbf{E}(\mathfrak{N}^1)} \right) R_n^* \delta^a(F)\, \mu(da), \end{aligned}$$

where δ^a is the Dirac measure concentrated at the point a. The set $\mathbf{E}(\mathfrak{N}^1)$ is countable, and $\mu(\{a\}) = 0$ for every $a \in \mathbf{E}(\mathfrak{N}) \supset \mathbf{E}(\mathfrak{N}^1)$. Thus the second integral equals zero. In view of the inequality $0 \leqslant R_n^*\delta^a(F) \leqslant r_n$, it is possible to pass to the limit in the first integrand. By virtue of the first assertion of Theorem 0.2 and the condition $\nu(\partial F) = 0$, we have

$$r_n^{-1} R_n^* \delta^a(F) \to \nu(F),$$

and

$$\frac{1}{r_n} R_n^* \mu(F) \to \int_{\overline{\mathbb{C}}\setminus \mathbf{E}(\mathfrak{N}^1)} \nu(F)\,\mu(da) = \nu(F).$$

The second assertion of Theorem 0.2 is proved.

PROOF OF THEOREM 0.4. In this proof we assume that the subsequence $\mathfrak{N}^1 \subset \mathfrak{N}$ with properties (i) and (ii) from the proof of previous proposition has already been chosen. Let $G_a = \{z : w^a(z) > 0\}$. It will suffice to prove inequality (0.2) for the restrictions of measures on the unit disk $\mathbb{D}$. Let $\mathbb{G}[\lambda]$ be the Green potential of the measure λ in $\mathbb{D}$. For any rational function $a(z)$, we can write $w^a = \mathbb{G}[\nu] - (\mathbb{G}[\nu] - w^a)$. By Lemma 1.1 the function $\mathbb{G}[\nu] - w^a$ is subharmonic, its Riesz measure equals ν^a. If a and b are distinct rational functions, then by Lemma 3.2,

$$(\mathbb{G}[\nu] - w^a) \vee (\mathbb{G}[\nu] - w^b) = \mathbb{G}[\nu] - w^a \wedge w^b = \mathbb{G}[\nu]. \tag{3.8}$$

We denote by $\underline{\nu}^a$ the restriction of the measure ν^a to the set $\overline{\mathbb{C}} \setminus G_a$. In view of (3.8) the family of subharmonic functions $\{\mathbb{G}[\nu] - w^a\}$ satisfies the condition of a theorem from [3], namely: the upper envelope of any pair of these functions is the same. So, by this theorem, the following equality

$$\sum_a (\nu - \underline{\nu}^a) = 2\nu - \bigwedge_{\substack{a,b \\ a\neq b}} (\underline{\nu}^a + \underline{\nu}^b).$$

is valid. (The largest lower bound $\wedge$ in the space of measures is introduced by the ordering $\lambda_1 \geqslant \lambda_2$ if $\lambda_1 - \lambda_2$ is a nonnegative measure.) Theorem 0.4 is proved.

REFERENCES

1. V. S. Azarin, *On the asymptotic behavior of subharmonic functions of finite order*, Mat. Sb. **108** (1979), no. 2, 147–167; English transl. in Math. USSR-Sb. **36** (1980), 135–154.

2. A. Eremenko and M. Sodin, *The value distribution of meromorphic functions and meromorphic curves from the point of view of potential theory*, Algebra i Analiz **3** (1991), no. 1, 131–164; English transl. in Leningrad Math. J. **3** (1992).

3. ———, *On the distribution of values of meromorphic function of finite order*, Dokl. Akad. Nauk SSSR **316** (1991), 538–541; English transl. in Soviet Math. Dokl. **43** (1991).

4. T. Ganelius, *Sequences of analytic functions and their zeros*, Ark. Mat. **3** (1950), 1–50.

5. A. A. Gol′dberg and I. V. Ostrovskiĭ, *Distribution of values of meromorphic functions*, "Nauka", Moscow, 1970. (Russian)

6. W. K. Hayman, *Value distribution and exceptional sets*, Sém. Math. Sup. (1982), 79–147.

7. A. Hyllengren, *On the lower order of $f(z)\exp(\varphi(z))$*, Ark. Math. **6** (1966), 443–458.

8. ———, *Connections between two sets of functions*, Ark. Math. **13** (1975), 273–285.

9. R. Nevanlinna, *Analytic functions*, Springer-Verlag, New York, Heidelberg, and Berlin, 1970.

10. P. C. Rosenbloom, *Distribution of zeros of polynomials*, Lectures on Functions of Complex Variable (W. Kaplan, ed.), Univ. of Michigan Press, Ann Arbor, 1955, pp. 265–285.

11. M. Sodin, *On value distribution of meromorphic functions of finite order with respect to their arguments*, Sibirsk. Mat. Zh. **31** (1990), no. 2, 169–179; English transl. in Siberian Math. J. **31** (1990).

INSTITUTE FOR LOW TEMPERATURE PHYSICS & ENGINEERING, 47 LENIN PR., KHARKOV 310164 UKRAINE

Translated by THE AUTHOR

ADVANCES IN SOVIET MATHEMATICS
Volume 11, 1992

Inverse Problem for the Sturm-Liouville Operator with Almost-Periodic Potential Having Only Positive Fourier Exponents

M. SIMBIRSKIĬ

§1. Introduction

1.1. Let M be a countable set of positive real numbers closed with respect to addition:

$$M = \{\Lambda_1, \Lambda_2, \dots, \Lambda_n, \dots\}, \qquad \Lambda_n > 0, \; n \in \mathbb{N}. \tag{1.1}$$

The main object of this article is to study the spectral properties and the inverse problem for the Sturm-Liouville operator

$$L = -\frac{d^2}{dx^2} + q(x), \qquad -\infty < x < \infty, \tag{1.2}$$

in the space $L^2(\mathbb{R})$ with a potential $q(x)$ of the form

$$q(x) \sim \sum_{n\in\mathbb{N}} q_n \exp(i\Lambda_n x), \qquad \Lambda_n \in M. \tag{1.3}$$

We do not assume the numbers q_n to be real but in any case the operator L is nonselfadjoint excluding the trivial case $q_n \equiv 0$.

For periodic potentials (1.3) (i.e., $\Lambda_n = n$) this problem was studied by M. G. Gasymov [1] and by V. Guillemin and A. Uribe [2] under the condition $\sum |q_n| < \infty$. In both cited papers special solutions of the equation

$$-y'' + q(x)y = \lambda^2 y \tag{1.4}$$

were constructed and spectral data $\{s_n\}_{n\in\mathbb{N}}$ were determined. It was shown that if a sequence $\{s_n\}$ is sufficiently small (for example,

$$\sum |s_n|(n+1)^{-1} < 1$$

in [1]), then the inverse problem of reconstructing the operator L from this sequence has a solution. Afterwards, the results of [1] were extended

---

1991 *Mathematics Subject Classification*. Primary 34A55, 34B24; Secondary 34L15, 42A75.

by A. D. Orudzhev [3] to almost-periodic potentials of the form (1.3) under strong restrictions (the main being that the set $\{\Lambda_n\}$ has no finite limit points). An operator L generated by a finite sum in (1.3) was studied by P. Sarnak [4] who, in particular, showed that the spectrum of L always coincides with $[0, +\infty)$. Finally, L. A. Pastur and V. A. Tkachenko [5] found necessary and sufficient conditions for a set $\{s_n\}_{n\in\mathbb{N}}$ to be the spectral data of the periodic potential (1.3).

In the present paper we give a criterion of solvability of the inverse problem with spectral data $\{s_n\}_{n\in\mathbb{N}}$ under rather mild assumptions on the almost-periodic potential (1.3). We also establish a detailed connection between analytic properties of the potential and of spectral data in terms of special Banach algebras.

Now let the submultiplicative weight $\rho(\cdot)$ on $(0, +\infty)$ be fixed:

$$\rho(\alpha+\beta) \leqslant \rho(\alpha)\cdot\rho(\beta), \qquad \alpha, \beta > 0.$$

We require this weight not to be "very small" either at zero or at infinity. To be exact, we suppose that there exists a positive constant C such that

$$\rho(y) > C(1+y), \qquad C > 0. \tag{1.5}$$

We define a Banach algebra $A^\rho(M)$ of almost-periodic functions

$$A^\rho(M) = \left\{f : f(t) = \sum_{g\in M} f_g \exp(igt)\right\}, \qquad \|f\|_\rho = \sum_{g\in M} |f_g|\rho(g).$$

This algebra will be of prime importance in our study. Inequality (1.5) implies that every element of this algebra and its derivative are representable by absolutely convergent Fourier series of the form (1.3).

We also define the class $Q^\rho(M)$ of Besicovich almost-periodic functions such that a second primitive of (1.3) exists and belongs to $A^\rho(M)$. This means that

$$Q^\rho(M) = \left\{q : q(x) \sim \sum_{n\in\mathbb{N}} q_n \exp(i\Lambda_n x), \ \sum_{n\in\mathbb{N}} |q_n|\Lambda_n^{-2}\rho(\Lambda_n) < \infty\right\}. \tag{1.6}$$

The latter inequality in (1.6) will be called the *basic condition.*

For further purposes we rearrange the set of integers. We write $\alpha \gg \beta$ (or $\alpha > \beta$) if $\Lambda_\alpha \geqslant \Lambda_\beta$ ($\Lambda_\alpha > \Lambda_\beta$ respectively). Of course, the symbol

$$\sum_{\alpha\,:\,\alpha>\beta}$$

means summing over all α such that $\Lambda_\alpha > \Lambda_\beta$, etc. We also write $\alpha\oplus\beta = \gamma$ if $\Lambda_\alpha + \Lambda_\beta = \Lambda_\gamma$ and $\alpha\ominus\beta = \delta$ if $\Lambda_\alpha - \Lambda_\beta = \Lambda_\delta$.

The fact that all the numbers $\{\Lambda_n\}_{n\in\mathbb{N}}$ are positive allows us to construct directly a special solution of (1.4) (cf. [1, 5]):

THEOREM 1. *Equation* (1.4) *with the potential* $q(x) \in Q^\rho(M)$ *has a solution of the form*

$$e(x, \lambda) = \exp(i\lambda x)\left(1 + \sum_{n\in\mathbb{N}}(\lambda + \Lambda_n/2)^{-1} \sum_{\alpha : \alpha\ll\beta} \varphi_{n,\alpha} \exp(i\Lambda_\alpha x)\right), \qquad (1.7)$$

where

$$\sum_{n\in\mathbb{N}} \Lambda_n^{-1} \sum_{\alpha : \alpha\gg n} |\varphi_{n,\alpha}|\rho(\Lambda_\alpha) < \infty. \qquad (1.8)$$

In the case of a periodic potential the function $e(x, \lambda)$ is a Floquet solution.

It evidently follows from (1.7)–(1.8) that for nonreal λ's both functions $e(x, \lambda)$ and $e(x, -\lambda)$ are defined, and the Wronskian of these functions is equal to $-2i\lambda$. The function $e(x, \lambda)$ is defined and is bounded for positive λ's. It follows that the spectrum of the operator L coincides with $[0, +\infty)$.

1.2. For negative λ's the function $e(x, \lambda)$, generally speaking, is not defined. Nevertheless for every positive λ_0 the following limit

$$\angle\text{-}\lim_{\lambda\to\lambda_0/2} e(x, -\lambda)(\lambda - \lambda_0/2) \equiv \begin{cases} 0, & \lambda \notin M, \\ e_n(x), & \lambda_0 = \Lambda_n, \ n \in \mathbb{N}, \end{cases} \qquad (1.9)$$

exists and is uniform in x. In what follows the symbol $\angle$-lim means the limit in a nontangent direction when λ tends to $\lambda_0/2$ in such a way that $\delta < \arg(\lambda - \lambda_0/2) < \pi - \delta$ for an arbitrary but fixed $\delta > 0$.

The function $e_n(x)$ is a solution of (1.4) with $\lambda = \Lambda_n/2$ linearly dependent on $e(x, \Lambda_n/2)$. Consequently, there exists a complex number s_n, $n \in \mathbb{N}$, such that

$$e_n(x) \equiv s_n e(x, \Lambda_n/2), \qquad n \in \mathbb{N}. \qquad (1.10)$$

As in [5], we can give the following definition

DEFINITION 1. The set $\{s_n\}_{n\in\mathbb{N}}$ is called the *spectral data set* of the operator L (see (1.2)).

The following theorem gives necessary and sufficient conditions for a set $\{s_n\}_{n\in\mathbb{N}}$ to be the spectral data set of the operator L with almost-periodic potential (1.3).

THEOREM 2. *The set* $\{s_n\}_{n\in\mathbb{N}}$ *is the spectral data set of the operator* L *with potential* $q(x) \in Q^\rho(M)$ (*see* (1.6)) *if and only if the following conditions are satisfied simultaneously*:

i) *the function*

$$f(x) \sim \sum_{n\in\mathbb{N}} s_n \Lambda_n \exp(ix\Lambda_n)$$

belongs to the class $Q^\rho(M)$, *i.e.*,

$$\sum_{n\in\mathbb{N}} |s_n|\Lambda_n^{-1}\rho(\Lambda_n) < \infty; \qquad (1.11)$$

ii) *the infinite determinant*

$$D(z) = \det \left\| \delta_{n,k} + \frac{2s_n}{\Lambda_n + \Lambda_k} \exp\left(\frac{\Lambda_n + \Lambda_k}{2} iz\right) \right\|_{n,k=1}^{\infty} \tag{1.12}$$

which, by virtue of (1.11), *lies in the Banach algebra* $A^\rho(M)$, *is its invertible element.*

NOTE. Varying the weight $\rho(\cdot)$ in the definition of the class $Q^\rho(M)$ we obtain a rather detailed scale of correspondence between analytic properties (smoothness, decaying at $+i\infty$, etc.) of the potential and of its spectral data.

Acknowledgment. I would like to express my gratitude to Professor V. A. Tkachenko for helpful discussions and his attention to this work.

§2. Proof of Theorem 1

To construct the special solution (1.7) of equation (1.4) we use, as in [5], the fact that every function from the class $Q^\rho(M)$ may be continued analytically to the upper half-plane. Substituting $x = it$, $\lambda = -i\mu$, $y(x) = Y(t)$ in (1.4) we arrive at the equation

$$-Y(t)'' + p(t)Y(t) = \mu^2 Y(t), \qquad t > 0, \tag{2.1}$$

where

$$p(t) \equiv -q(it) \sim -\sum_{n\in\mathbb{N}} q_n \exp(-\Lambda_n t). \tag{2.2}$$

By virtue of the basic condition (1.6), the function $p(t)$ is summable with respect to the weight t on the half-axis $\mathbb{R}^+ = \{t : t > 0\}$. Hence scattering theory techniques may be applied to (2.1).

LEMMA 1. *The kernel* $K(t, u)$ *of the* $+\infty$*-related transformation operator for* (2.1) *with potential* (2.2) *may be represented in the form*

$$K(t, u) = \sum_{n\in\mathbb{N}} \exp(-\Lambda_n(u-t)/2) \sum_{\alpha : \alpha \gg n} \varphi_{n,\alpha} \exp(-\Lambda_\alpha t), \tag{2.3}$$

where the sequence $\{\varphi_{n,\alpha}\}_{n\in\mathbb{N},\alpha\gg n}$ *satisfies inequality* (1.8).

PROOF. We use the well-known equations for $K(t, u)$ (see [6], p. 126)

$$K_{tt}(t, u) = K_{uu}(t, u) + p(t)K(t, u), \qquad u \geqslant t > 0, \tag{2.4}$$

$$K(t, t) = \frac{1}{2}\int_t^\infty p(s)ds, \qquad t > 0. \tag{2.5}$$

Let us substitute $K(t, u)$ given by (2.3) into (2.4) and (2.5). Comparing coefficients of the same exponentials, we obtain the infinite system of equations:

$$\varphi_{n,\alpha} = -\frac{1}{\Lambda_\alpha(\Lambda_\alpha - \Lambda_n)} \sum_{\{\beta,\gamma : \beta\oplus\gamma=\alpha\}} \varphi_{n,\beta} q_\gamma, \qquad n \in \mathbb{N},\ \alpha \gg n, \tag{2.6}$$

$$\varphi_{\alpha,\alpha} = -\frac{q_\alpha}{2\Lambda_\alpha} - \sum_{n : n<\alpha} \varphi_{n,\alpha}, \qquad \alpha \in \mathbb{N}. \tag{2.7}$$

Let us prove that system (2.6)–(2.7) has a unique solution $\{\varphi_{n,\alpha}\}$ and this solution satisfies inequality (1.8).

Let us fix a number θ, $0<\theta<1$. By virtue of (1.4), we can find a $\sigma>0$ such that

$$2\sum_{n\in\mathbb{N}}\exp(-\sigma\Lambda_n)|q_n|\Lambda_n^{-2}<\theta<1. \tag{2.8}$$

Consider the Banach space of two-index sequences $\mathfrak{B}_\sigma=\{\{\varphi_{n,\alpha}\}_{n\in\mathbb{N},\alpha\gg n}\}$ with the norm

$$\|\{\varphi_{n,\alpha}\}\|=\sup_\alpha\left\{\exp(-\sigma\Lambda_n)\sum_{n:\,n\ll\alpha}\Lambda_n^{-1}|\varphi_{n,\alpha}|\right\} \tag{2.9}$$

and let us define the operator A in the space $\mathfrak{B}_\sigma$ by the formulas:

$$(A\Phi)_{n,\alpha}=-\frac{1}{\Lambda_\alpha(\Lambda_\alpha-\Lambda_n)}\sum_{\{\beta,\gamma:\,\beta\oplus\gamma=\alpha\}}\varphi_{n,\beta}q_\gamma,\quad n\in\mathbb{N},\ \alpha\gg n, \tag{2.10}$$

$$(A\Phi)_{\alpha,\alpha}=-\sum_{n:\,n<\alpha}(A\Phi)_{n,\alpha},\qquad \alpha\in\mathbb{N}. \tag{2.11}$$

Now relations (2.6)–(2.7) may be written in the form $\Phi=R+A\Phi$ in the space $\mathfrak{B}_\sigma$, where $R=\{-(1/2)q_\alpha\Lambda_\alpha^{-1}\delta_{n,\alpha}\}_{n\in\mathbb{N},\alpha\gg n}$, δ being the Kronecker symbol. Let us verify that the operator A is contracting in the space $\mathfrak{B}_\sigma$. Indeed, using (2.9)–(2.11), we have:

$$\begin{aligned}
\|A\Psi\|_\sigma&=\sup_\alpha\left\{\exp(-\sigma\Lambda_\alpha)\sum_{n:\,n\ll\alpha}\Lambda_n^{-1}|(A\Psi)_{n,\alpha}|\right\}\\
&\leqslant\sup_\alpha\left\{\exp(-\sigma\Lambda_\alpha)\left(\Lambda_\alpha^{-1}|(A\Psi)_{\alpha,\alpha}|+\sum_{n:\,n\ll\alpha}\Lambda_n^{-1}|(A\Psi)_{n,\alpha}|\right)\right\}\\
&\leqslant\sup_\alpha\left\{2\exp(-\sigma\Lambda_\alpha)\sum_{n:\,n<\alpha}\Lambda_n^{-1}|(A\Psi)_{n,\alpha}|\right\}\\
&\leqslant\sup_\alpha\left\{2\exp(-\sigma\Lambda_\alpha)\Lambda_\alpha^{-1}\sum_{n:\,n<\alpha}\frac{1}{\Lambda_n(\Lambda_\alpha-\Lambda_n)}\sum_{\gamma:\,\gamma\ll\alpha}|\varphi_{n,\gamma}||q_{\alpha\ominus\gamma}|\right\}.
\end{aligned}$$

First of all, we interchange the summation order over α and γ, then use the inequalities $\Lambda_\gamma\geqslant\Lambda_n$, $\Lambda_\alpha-\Lambda_n\geqslant\Lambda_\alpha-\Lambda_\gamma$ for $\gamma\gg n$, and, afterwards, replace the sum over some β's by the sum over all β's. We obtain

$$\begin{aligned}
\|A\Psi\|&\leqslant\sup_\alpha\left\{2\Lambda_\alpha^{-1}\sum_{\gamma:\,\gamma<\alpha}|q_{\alpha\ominus\gamma}|(\Lambda_\alpha-\lambda_\gamma)^{-1}\exp(-\sigma(L_\alpha-\Lambda_\gamma))\right.\\
&\qquad\qquad\left.\times\sum_{n:\,n\ll\gamma}\Lambda_n^{-1}|\varphi_{n,\gamma}|\exp(-\sigma\Lambda_\gamma)\right\}\\
&\leqslant 2\left(\sum_{\beta\in\mathbb{N}}|q_\beta|\Lambda_\beta^{-1}\exp(-\sigma\Lambda_\beta)\right)\sup_\gamma\left\{\sum_{n\ll\gamma}\Lambda_n^{-1}|v_{n,\gamma}|\exp(-\sigma\Lambda_\gamma)\right\}\\
&\leqslant\theta\|\Psi\|.
\end{aligned}\tag{2.12}$$

Thus, the operator A is contracting and, consequently, the system (2.6)–(2.7) has the unique solution $\{\varphi_{n,\alpha}\}_{n\in\mathbb{N},\alpha\gg n}$ in the space $\mathfrak{B}_\sigma$.

To prove (1.8), we consider the values

$$W_\alpha = \sum_{n\,:\,n\ll\alpha} \Lambda_n^{-1}|\varphi_{n,\alpha}|, \qquad a\in\mathbb{N}. \tag{2.13}$$

By virtue of (2.9), the values W_α, $\alpha\in\mathbb{N}$, are finite and the sequence $W = \{W_\alpha\}_{\alpha\in\mathbb{N}}$ belongs to the Banach space L_σ of sequences

$$L_\sigma = \{\{W_\alpha\}\}, \qquad \|\{W_\alpha\}\|_\sigma = \sup_\alpha \left\{\exp(-\sigma\Lambda_\alpha)\sum_{n\,:\,n\ll\alpha}\Lambda_n^{-1}|W_\alpha|\right\}.$$

Using (2.6)–(2.7) again and the argument similar to that used to estimate the norm of the operator A, we obtain for all $\alpha\in\mathbb{N}$

$$0\leqslant W_\alpha \leqslant \frac{1}{2}|q_\alpha|\Lambda_\alpha^{-2} + 2\Lambda_\alpha^{-1}\sum_{\{\alpha,\gamma\,:\,\beta\oplus\gamma=\alpha\}}|W_\gamma||q_\beta|\Lambda_\beta^{-1}. \tag{2.14}$$

First of all, let us introduce the partial ordering " $\ll$ " in the space L_σ so that $X\ll Y$ means $X_\alpha < Y_\alpha$ for all $\alpha\in\mathbb{N}$, and, secondly, consider the operator J acting in this space according to the formula

$$(JX)_\alpha = 2\Lambda_\alpha^{-1}\sum_{\{\beta,\gamma\,:\,\beta\oplus\gamma=\alpha\}}|X_\gamma||q_\beta|\Lambda_\beta^{-1}, \qquad \alpha\in\mathbb{N}. \tag{2.15}$$

Inequality (2.12) shows that the operator J is contracting, $\|J\|\leqslant\theta<1$, whereas, due to (2.15), J evidently preserves the partial ordering: $X\ll Y$ implies $JX\ll JY$.

Defining the sequence $Q = \{Q_\alpha\}_{\alpha\in\mathbb{N}} = \{\frac{1}{2}|q_\alpha|\Lambda_\alpha^{-2}\}_{\alpha\in\mathbb{N}}$ in L_σ, we may rewrite (2.14) in the form

$$0\ll W\ll Q+JW. \tag{2.16}$$

Iterating (2.16), we get

$$0\ll W\ll V, \tag{2.17}$$

where the sequence $V=\{V_\alpha\}_{\alpha\in\mathbb{N}}$ is the solution of the equation

$$V = Q+JV \tag{2.18}$$

in L_σ. To find V explicitly, we use generating functions of sequences. Let

$$V(x) = \sum_{\alpha\in\mathbb{N}} V_\alpha\exp(\Lambda_\alpha x), \tag{2.19}$$

$$Q(x) = \frac{1}{2}\sum_{\alpha\in\mathbb{N}}|q_\alpha|\Lambda_\alpha^{-2}\exp(\Lambda_\alpha x). \tag{2.20}$$

Now, taking into account (2.15), (2.19) and (2.20), we obtain from (2.18)

$$V(x) = Q(x) + 4\int_{-\infty}^{x} Q'(\tau)V(\tau)\,d\tau,$$

whence one easily finds

$$V(x) = (1/4)[\exp(4Q(x)) - 1]. \tag{2.21}$$

The basic condition (1.6) guarantees that the function $Q(ix)$ belongs to the algebra $A^\rho(M)$. By virtue of (2.21), the function $V(ix)$ also lies in $A^\rho(M)$ which means

$$\sum_{\alpha\in\mathbb{N}} V_\alpha \rho(\Lambda_\alpha) < \infty. \tag{2.22}$$

Going back along the chain (2.22) $\to$ (2.17) $\to$ (2.13) $\to$ (1.8), we find that (2.22) implies (1.8).

Now we conclude that the series (2.3) converges absolutely and defines the twice differentiable function $K(t, u)$ in the domain $u \geqslant t > 0$. This function satisfies (2.5) and all regularity conditions at $+\infty$ required in [6], p. 126. Therefore (see [6], p. 126), the function $K(t, u)$ is a kernel of the $+\infty$-related transformation operator for (2.1) with potential (2.2), and the lemma is proved.

PROOF OF THEOREM 1. It is known ([6, p. 126]) that if a potential of (2.1) is summable with respect to the weight t, then the Jost solution $E(t, \mu)$, which is close to $e^{i\mu t}$ for $\operatorname{Im}\mu \geqslant 0$, $t \to +\infty$, is represented in the form

$$E(t, \mu) = e^{i\mu t} + \int_t^\infty K(t, u) e^{i\mu u} du, \quad t \geqslant 0.$$

Lemma 1 gives

$$E(t, \mu) = \exp(i\mu t)\Big(1 + \sum_{n\in\mathbb{N}} (-i\mu + \Lambda_n/2)^{-1} \sum_{\alpha : \alpha \ll \beta} \varphi_{n,\alpha} \exp(-\Lambda_\alpha t)\Big).$$

Since the coefficient $q(x)$ in equation (1.4) is analytic in the half-plane $\mathbb{C}^+ = \{z : \operatorname{Im} z > 0\}$, the function

$$\begin{aligned} e(x, \lambda) &= E(-ix, i\lambda) \\ &= \exp(i\lambda x)\Big(1 + \sum_{n\in\mathbb{N}} (\lambda + \Lambda_n/2)^{-1} \sum_{\alpha : \alpha \ll \beta} \varphi_{n,\alpha} \exp(i\Lambda_\alpha x)\Big). \end{aligned}$$

satisfies (1.4) and (1.8). Theorem 1 is proved.

§3. Spectral data set $\{s_n\}$ and its properties

Throughout §3 we assume that the potential $q(x)$ belongs to the class $Q^\rho(M)$. For a fixed $n \in \mathbb{N}$ let us consider the expression

$$\begin{aligned} &e(x, -\lambda)(\lambda - \Lambda_n/2) \\ &\quad = \exp(-i\lambda x)\Big(\sum_{\alpha : \alpha \gg n} \varphi_{n,\alpha} \exp(i\Lambda_\alpha x) \\ &\qquad\qquad + \Big[\lambda - \frac{\Lambda_n}{2} + \sum_{k\in\mathbb{N},\, k\neq n} \frac{\lambda - \Lambda_n/2}{-\lambda + \Lambda_k/2} \sum_{\alpha : \alpha \gg k} \varphi_{k,\alpha} \exp(i\Lambda_\alpha x)\Big]\Big). \end{aligned}$$

Our aim is to prove that the expression in square brackets tends to zero uniformly in x as λ tends to $\Lambda_n/2$ in a nontangent direction, which means that $\delta < \arg(\lambda - \Lambda_n/2) < \pi - \delta$ for an arbitrary but fixed $\delta > 0$. Indeed, if $k \neq n$, then

$$\left| \frac{\lambda - \Lambda_n/2}{-\lambda + \Lambda_k/2} \right| < \sin^{-1} \delta$$

and the summands in square brackets are majorized uniformly in x and λ by terms of an absolutely convergent series. Consequently one can pass to the limit termwise and by the Weierstrass criterion the convergence to zero will be uniform in x. Similarly, if $\lambda_0 \notin M$, then

$$\angle\text{-}\lim_{\lambda \to \lambda_0/2} e(x, -\lambda)(\lambda - \lambda_0/2) = 0.$$

Hence the following limit exists:

$$\angle\text{-}\lim_{\lambda \to \Lambda_n/2} e(x, -\lambda)(\lambda - \lambda_0/2) \stackrel{\text{def}}{\equiv} e_n(x) = -\exp\left(-\frac{i}{2}\Lambda_n x\right) \sum_{\alpha : \alpha \gg n} \varphi_{n,\alpha} \exp(i\Lambda_\alpha x).$$

Obviously, the function $e_n(x)$ is analytic, at least, for $\operatorname{Im} x > 0$ and satisfies (1.6) for $\lambda = \Lambda_n/2$. Besides, the Wronskian of solutions $e(x, \Lambda_n/2)$ and $e_n(x)$ is a constant;

$$W\{e(x, \Lambda_n/2), e_n(x)\} \equiv \text{const} \equiv \lim_{\operatorname{Im} x \to +\infty} W\{e(x, \Lambda_n/2), e_n(x)\} = 0.$$

Therefore, there exist complex numbers s_n, $n \in \mathbb{N}$, such that $e_n(x) \equiv s_n e(x, \Lambda_n/2)$ and Definition 1 is justified.

LEMMA 2. *The sequence* $\{s_n\}_{n \in \mathbb{N}}$ *satisfies the following condition*:

$$\sum_{n \in \mathbb{N}} |s_n| \Lambda_n^{-1} \rho(\Lambda_n) < \infty. \qquad (1.11)=(3.1)$$

In particular, this proposition contains assertion i) *of Theorem* 2.

PROOF OF LEMMA 2. It follows from (1.9)–(1.10) that

$$\exp(-i\Lambda_n x/2) \sum_{\alpha : \alpha \gg n} \varphi_{n,\alpha} \exp(i\Lambda_\alpha x)$$
$$\equiv s_n \exp(i\Lambda_n x/2) \left(1 + \sum_{\nu \in \mathbb{N}} \frac{2}{\Lambda_n + \Lambda_\nu} \sum_{\alpha : \alpha \gg \nu} \varphi_{\nu,\alpha} \exp(i\Lambda_\alpha x)\right)$$

and by virtue of the uniqueness theorem for almost-periodic functions, we have

$$\varphi_{n,n} = -s_n, \qquad n \in \mathbb{N}, \qquad (3.2)$$

$$\varphi_{n,\alpha} = -s_n \sum_{k : k \ll \alpha \ominus n} \frac{2}{\Lambda_n + \Lambda_k} \varphi_{k, \alpha \ominus n}, \qquad \alpha > n. \qquad (3.3)$$

Thus (3.1) follows from (3.2) and inequality (1.8) proved above. The lemma is proved.

LEMMA 3. i) *The kernel* $K(t, u)$ *of the* $+\infty$*-related transformation operator for* (2.1) *constructed in Lemma* 1 *satisfies the fundamental integral equation* (*the Marchenko equation*)

$$K(t, s) + F(t+s) + \int_t^\infty K(t, u)F(u+s)du = 0, \tag{3.4}$$

or, in an equivalent form,

$$K(t, t+r) + F(2t+r) + \int_0^\infty K(t, t+v)F(v+r+2t)\,dv = 0, \tag{3.4$'$}$$

where

$$F(y) = \sum_{n\in\mathbb{N}} s_n \exp(-\Lambda_n y/2), \tag{3.5}$$

and $\{s_n\}_{n\in\mathbb{N}}$ *is the spectral data set defined above.*

ii) *The homogeneous equation*

$$g(r) + \int_0^\infty g(v)F(v+r+2t)\,dv = 0, \qquad r > 0, \tag{3.6}$$

corresponding to (3.4$'$) *has only the trivial solution in* $L^2[0, +\infty)$. *Hence the Fredholm determinant* $\Delta(t)$ *of* (3.4), (3.4$'$) *and* (3.6) *does not vanish for* $t \geqslant 0$.

PROOF OF LEMMA 3. i) The fundamental integral equation (3.4) is an easy consequence of relations (3.2)–(3.3). Indeed, let us define the function $F(t)$ for $t \geqslant 0$ by the series (3.5), which is convergent by virtue of Lemma 2. Then

$$\begin{aligned}
N(t, s) &\overset{\text{def}}{\equiv} K(t, s) + F(t+s) + \int_t^\infty K(t, u)F(u+s)\,du \\
&= \sum_{n\in\mathbb{N}} \exp\left(-\frac{\Lambda_n}{2}(s-t)\right) \sum_{\alpha\,:\,\alpha\gg n} \varphi_{n,\alpha}\exp(-\Lambda_\alpha t) \\
&\quad + \sum_{n\in\mathbb{N}} s_n \exp\left(-\frac{\Lambda_n}{2}(s+t)\right) \\
&\quad + \int_t^\infty \sum_{k\in\mathbb{N}} \exp\left(-\frac{\Lambda_k}{2}(u-t)\right) \\
&\qquad \times \sum_{\beta\,:\,\beta\gg k} \varphi_{k,\beta}\exp(-\Lambda_\beta t) \sum_{n\in\mathbb{N}} s_n \exp\left(-\frac{\Lambda_n}{2}(u+t)\right) du.
\end{aligned}$$

Separating the diagonal $\alpha = n$ in the first sum, we collect similar terms and obtain

$$\begin{aligned}
N(t, s) &\equiv \sum_{n\in\mathbb{N}} \{\varphi_{n,n} + s_n\}\exp(-(\Lambda_n/2)(s+t)) \\
&\quad + \sum_{n\in\mathbb{N}} \sum_{\alpha\,:\,\alpha>n} \left\{\varphi_{n,\alpha} + s_n \sum_{k\,:\,k\ll n\ominus\alpha} \frac{2}{\Lambda_n+\Lambda_k}\varphi_{k,n\ominus\alpha}\right\} \\
&\qquad \times \exp(-(\Lambda_n/2)(s-t) - \Lambda_\alpha t) \equiv 0
\end{aligned}$$

because by virtue of (3.2) and (3.3), the expressions in braces are identically equal to zero. The derivation of equation (3.4′) is now trivial.

ii) The proof of this part is based on an idea of V. Ljance [7].

Let $g \in L^2[0, +\infty)$ be a solution of equation (3.6) and let the function h be a solution of the Volterra equation

$$h(r) + \int_0^r h(\zeta)K(\zeta + t, t + r)\,d\zeta = g(r), \qquad r \geqslant 0. \tag{3.7}$$

Substituting (3.7) in (3.6) and using (3.4), we get

$$\begin{aligned} h(r) &+ \int_0^r h(\zeta)K(\zeta + t, t + r)\,d\zeta \\ &+ \int_0^\infty \left(h(v) + \int_0^r h(\zeta)K(\zeta + t, v + t)\,d\zeta\right)F(v + r)\,dv \\ &= h(r) + \int_r^\infty h(\zeta)\Big(F(\zeta + r + 2t) \\ &\qquad + \int_\zeta^\infty K(z + t, v + t)F(v + r + 2t)\,dv\Big)d\zeta \\ &+ \int_0^r h(\zeta)\Big(K(\zeta + t, t + r) + F(\zeta + r + 2t) \\ &\qquad + \int_\zeta^\infty K(\zeta + t, v + t)F(v + r + 2t)\,dv\Big)d\zeta \\ &= h(t) + \int_r^\infty h(\zeta)H_t(\zeta, r)\,d\zeta = 0, \end{aligned} \tag{3.8}$$

where

$$\begin{aligned} H_t(\zeta, r) &\stackrel{\text{def}}{=} F(\zeta + r + 2t) + \int_\zeta^\infty K(\zeta + t, v + t)F(v + r + 2t)\,dv \\ &= \sum_{n\in\mathbb{N}} s_n \exp\left(-\frac{\Lambda_n}{2}(r + \zeta + 2t)\right) + \sum_{n\in\mathbb{N}}\sum_{k\in\mathbb{N}}\sum_{\alpha\,:\,\alpha\gg k} \frac{2s_n}{\Lambda_n + \Lambda_k}\varphi_{k,\alpha} \\ &\qquad \times \exp(-(\Lambda_n/2)(r + \zeta + 2t) - \Lambda_\alpha(\zeta + t)). \end{aligned} \tag{3.9}$$

By virtue of estimates (1.8), (3.1), the series for $H_t(\zeta, r)$ converges, at least, for $t, r, \zeta > 0$, and hence $H_t(\zeta, r)$ is well defined. In addition, using (3.8) and estimates (1.8), (3.1) again, we obtain the estimate

$$\int_r^\infty |H_t(\zeta, r)|\,d\zeta \leqslant \sum_{n\in\mathbb{N}} \frac{2|s_n|}{\Lambda_n}\left(1 + \sum_{k\in\mathbb{N}}\sum_{\alpha\,:\,\alpha\gg k} \frac{1}{\Lambda_k}|\varphi_{k,\alpha}|\right)\exp(-\Lambda_n r),$$

and the right-hand side tends to zero as r tends to infinity. It follows that there exists a number $r_0 > 0$ such that

$$\int_r^\infty |H_t(\zeta, r)|\,d\zeta \leqslant \theta < 1 \quad \text{for all } r \geqslant r_0. \tag{3.10}$$

Since $h(r)$ satisfies (3.8) and $h \in L^2(\mathbb{R}_+)$, we have

$$|h(r)| \leqslant \int_r^\infty |h(\zeta)||H_t(\zeta, r)|\,d\zeta. \tag{3.11}$$

Applying the Cauchy-Bunyakovskiĭ inequality, we find that the function $h(r)$ is bounded, $|h(r)| \leqslant C < \infty$. Substituting this estimate into integral (3.11) and applying (3.10), we get $|h(r)| \leqslant C\theta$ for all $r_0 \geqslant r$. By iterating, we obtain $|h(r)| \leqslant C\theta^k$ for all $r \geqslant r_0$ and all $k \in \mathbb{N}$. But this means that $h(r)$ equals zero identically for $r \geqslant r_0$. Consequently, the upper limit of integration in (3.8) is not an infinity but a finite number r_0. Hence (3.8) is a Volterra equation. It has only the trivial solution $h(r) \equiv 0$. The lemma is proved.

The Fredholm determinant of equation (3.6) is of great importance in our investigation. Now we will link this determinant with the function $D(z)$ defined in Theorem 2 (see (1.12)). To this end we need an abstract lemma.

LEMMA 4. *Let* $\{\psi_k\}_{k\in\mathbb{N}}$, $\{\vartheta_k\}_{k\in\mathbb{N}}$ *be arbitrary systems of vectors in a Hilbert space* $\mathbb{H}$ *satisfying the condition*

$$\sum_{k\in\mathbb{N}} \|\psi_k\| \cdot \|\vartheta_k\| < \infty.$$

Then the operator A *defined by the formula*

$$A = \sum_{k\in\mathbb{N}} (\cdot\,, \psi_k)\vartheta_k$$

is trace class (*for definition see* [8, §VI.6]), *and, besides,*

$$\det(E + A) = \det \|\delta_{j,k} + (\psi_j\,, \vartheta_k)\|_{j,k=1}^{\infty}.$$

PROOF OF LEMMA 4. For a one-dimensional operator P defined by the formula $P = (\cdot\,, \varphi)\psi$, its S_1-norm and the usual operator norm are equal: $\|P\|_{S_1} = \|P\|_{\mathbb{H}} = \|\varphi\| \cdot \|\psi\|$. The first assertion of the lemma now follows from the triangle axiom for S_1-norm. The second assertion is proved in [9, p. 204].

Now let us denote by $\Phi(z)$ an operator in $L^2[0, +\infty)$, acting according to the formula

$$(\Phi(z)g)(s) = \int_0^\infty g(u)F(u + s - 2iz)\,du. \tag{3.12}$$

We shall also need the number

$$\eta = \overline{\lim}\, \frac{\ln\rho(y)}{y} = \sup_y \frac{\ln\rho(y)}{y}.$$

Using (1.5), it is easy to prove that $0 \leqslant \eta < +\infty$.

LEMMA 5. i) *For* $z \in \{z : \operatorname{Im} z > -\eta\}$ *the operator* $\Phi(z)$ *is trace class and the determinant* $\det(E + \Phi(z))$ *is identically equal to* $D(z)$.

ii) *For* $z = it$, $t > 0$, *the Fredholm determinant* $\Delta(t)$ *of equations* (3.4) *and* (3.4′) *coincides with the function* $D(z)$. *Moreover, we have the identity*

$$(d/dt)\ln\Delta(t) \equiv K(t, t), \qquad t > 0. \tag{3.13}$$

PROOF OF LEMMA 5. Let us consider two systems of vectors

$$\vartheta_k(x) \stackrel{\text{def}}{=} \exp(-\Lambda_k x/2), \qquad \psi_k(x) \stackrel{\text{def}}{=} s_k \exp(-\Lambda_k x/2 + i\Lambda_k z), \qquad k \in \mathbb{N},$$

in $L^2[0, +\infty)$. It is easy to see that

$$\Phi(z) = \sum_{k \in \mathbb{N}} (\,\cdot\,, \psi_k)\vartheta_k$$

and that all assumptions of Lemma 4 are satisfied. Assertion i) now follows from Lemma 4.

The coincidence of $D(z)$ and $\Delta(t)$ for $z = it$, $t > 0$, may be proved by comparing the definitions of the operator $\Phi(z)$ according to (3.12) and of $\Delta(t)$ according to (3.6).

Identity (3.13) appears in different aspects of spectral theory of Sturm-Liouville operators. For periodic potentials of the form (1.3) it was established by L. Pastur and V. Tkachenko in [5]. We repeat their reasoning for the general case.

Let $\Gamma_t(u, v)$ be the kernel of the Fredholm resolvent of fundamental integral equation (3.4). By the well-known formula of Fredholm theory [10],

$$\Gamma_t(t, s) - F(t+s) + \int_t^\infty \Gamma_t(s, u)F(u+s)\,du = 0.$$

Comparing this relation with (3.4), using its unique solvability and the uniqueness theorem for almost-periodic functions, we find that $\Gamma_t(t, t) = -K(t, t)$. On the other hand, by the Fredholm formula [10], we have

$$\Gamma_t(t, t) = -(d/dt)\ln\Delta(t).$$

Combining two last relations we complete the proof of Lemma 5.

Now let us denote

$$Q(t) \stackrel{\text{def}}{=} -\int_t^\infty K(\tau, \tau)\,d\tau.$$

By virtue of Lemma 2, the function $Q(-it)$ belongs to the algebra $A^\rho(M)$. Integrating (3.13), which is possible since $\Delta(t) \neq 0$ for $t > 0$, we obtain for $z = it$, $t > 0$

$$D(z) = \exp Q(-iz). \tag{3.14}$$

But the functions on both sides of the latter equation are analytic in the half-plane

$$z \in \{z : \operatorname{Im} z > -\eta\}, \qquad \eta = \sup_y y^{-1}\ln\rho(y),$$

and we can continue this equation to the entire half-plane $\{z : \operatorname{Im} z > -\eta\}$. Since the right-hand side of (3.14) lies in the Banach algebra $A^\rho(M)$ and is invertible in it, we conclude that the function $D(z)$ satisfies condition ii) of Theorem 2.

§4. Proof of Theorem 2

The main subject of this section is the fundamental integral equation (3.4). Let us consider this equation in the form (3.4′):

$$K(t, t+r) + F(2t+r) + \int_0^\infty K(t, t+y)F(2t+r+y)\,dy = 0, \tag{4.1}$$

where the function $F(y)$ is defined by (3.5):

$$F(y) = \sum_{n\in\mathbb{N}} s_n \exp(-\Lambda_n y/2). \tag{4.2}$$

LEMMA 6. *If condition* i) *of Theorem* 2 *is satisfied, then the function* $D(z)$ *defined by* (1.12) *belongs to the algebra* $A^\rho(M)$.

To prove Lemma 6 we need an auxiliary proposition.

LEMMA 7. *The determinant*

$$d = \det \left\| \frac{2}{a_j + b_l} \right\|_{j,l=1}^k,$$

where all numbers a_γ *and* b_γ *are positive, satisfies the inequality*

$$|d| \leqslant \prod_{\gamma=1}^k \sqrt{a_\gamma^{-1} b_\gamma^{-1}}.$$

PROOF OF LEMMA 7. The determinant d has a clear geometric meaning. Namely, let us consider two systems of functions $G = \{g_\gamma\}$ and $H = \{h_\gamma\}$ in $L^2[0, +\infty)$, where $g_\gamma(t) = \exp(-a_\gamma t/2)$, $h_\gamma(t) = \exp(-b_\gamma t/2)$, $\gamma = 1, \dots, k$. We denote the scalar product in $L^2[0, +\infty)$ by $\langle\cdot, \cdot\rangle$. Using an elementary property of real determinants

$$|\det(AB)| = |\det A| \cdot |\det B| = \sqrt{\det(A^* A)\det(B^* B)},$$

we obtain

$$\left|\det \left\| \frac{2}{a_\gamma + b_\delta} \right\|_{\gamma,\delta=1}^k \right| = |\det \|\langle g_\gamma, h_\delta\rangle\|_{\gamma,\delta=1}^k|$$
$$= \sqrt{\det \|\langle g_\gamma, g_\delta\rangle\|_{\gamma,\delta=1}^k \det \|\langle h_\gamma, h_\delta\rangle\|_{\gamma,\delta=1}^k} \leqslant \prod_{\gamma=1}^k \sqrt{a_\gamma^{-1} b_\gamma^{-1}}.$$

Here in the last inequality we have used the fact that the Gramian of a system of functions does not exceed the product of squares of these functions' norms. Lemma 7 is proved.

PROOF OF LEMMA 6. Note that by virtue of part i) of Theorem 2 the function

$$V(t) \stackrel{\text{def}}{=} \sum_{n\in\mathbb{N}} |s_n|\Lambda_n^{-1} \exp(i\Lambda_n t) \tag{4.3}$$

lies in the algebra $A^\rho(M)$. As was shown in Lemma 5, the Fredholm determinant of (4.1) coincides with the function $D(it)$ for $t > 0$. Hence we can use the Fredholm formulas [10]

$$D(it) = \sum_{k=0}^{\infty} \frac{1}{k!} D_k(t),$$

where

$$D_0(t) \stackrel{\text{def}}{=} 1, \qquad D_k(t) \stackrel{\text{def}}{=} \int_0^\infty \cdots \int_0^\infty dt_1 \cdots dt_k \det \|F(2t + t_j + t_l)\|_{j,l=1}^k,$$

and the function $F(t)$ is defined by (4.2).

We intend to prove that the functions $D_k(-it)$, $k \in \mathbb{N}$, belong to the algebra $A^\rho(M)$ and estimate the value of $\|D_k(-it)\|_\rho$. We have:

$$\begin{aligned} D_k(t) &= \int_0^\infty \cdots \int_0^\infty dt_1 \cdots dt_k \det \left\| \sum_{n_j\in\mathbb{N}} s_{n_j} \exp(-\Lambda_{n_j}(2t + t_j + t_l)/2) \right\|_{j,l=1}^k \\ &= \sum_{n_1,\dots,n_k\in\mathbb{N}} s_{n_1} s_{n_2} \cdots s_{n_k} \exp\left(-\sum_{j=1}^k \Lambda_{n_j} t \right) \Delta_{n_1,n_2,\dots,n_k}. \end{aligned} \tag{4.4}$$

Here we have denoted

$$\begin{aligned} \Delta_{n_1,n_2,\dots,n_k} &= \int_0^\infty \cdots \int_0^\infty dt_1 \cdots dt_k \det \| \exp(-\Lambda_{n_j}(t_j + t_l)/2) \|_{j,l=1}^k \\ &= \int_0^\infty \cdots \int_0^\infty dt_1 \cdots dt_k \sum_\sigma \operatorname{sgn} \sigma \prod_{j=1}^k \exp(-\Lambda_{n_j}[t_{\sigma(j)} + t_j]/2), \end{aligned}$$

summation being performed over all permutations σ of k elements, signs being equal to $+1$ for even and -1 for odd permutations σ. Now we collect the same t_j's and then integrate. If π denotes the inverse permutation to σ, we obtain

$$\begin{aligned} \Delta_{n_1,n_2,\dots,n_k} &= \int_0^\infty \cdots \int_0^\infty dt_1 \cdots dt_k \sum_\sigma \operatorname{sgn} \sigma \exp\left\{ -\frac{1}{2} \sum_{j=1}^k (\Lambda_{n_j} + \Lambda_{n_{\pi(j)}}) t_j \right\} \\ &= \sum_\pi \operatorname{sgn} \pi \frac{2}{\Lambda_{n_1} + \Lambda_{n_{\pi(1)}}} \cdots \frac{2}{\Lambda_{n_k} + \Lambda_{n_{\pi(k)}}}. \end{aligned}$$

But the last sum coincides with the determinant

$$\Delta_{n_1,\dots,n_k} = \det \left\| \frac{2}{\Lambda_{n_j} + \Lambda_{n_l}} \right\|_{j,l=1}^k.$$

Lemma 7 implies

$$\left|\det\left\|\frac{2}{\Lambda_{n_j}+\Lambda_{n_l}}\right\|_{j,l=1}^{k}\right| \leqslant \frac{1}{\Lambda_{n_1}\Lambda_{n_2}\cdots\Lambda_{n_k}}. \tag{4.5}$$

Applying (4.5) to (4.4), we find that the Fourier coefficients of the almost-periodic function $D_k(-it)$ are majorized by coefficients of the kth power of the function $V(t)$ defined above. Thus

$$\|D_k(-it)\|_p \leqslant \|V^k(t)\|_p \leqslant \|V(t)\|_p^k. \tag{4.6}$$

Consequently,

$$\|D(z)\|_p \leqslant \sum_{k=0}^{\infty}\frac{1}{k!}\|D_k(-it)\|_p \leqslant \exp(\|V(t)\|_p) < \infty, \tag{4.7}$$

and the function $D(z)$ belongs to the algebra $A^\rho(M)$. Lemma 6 is proved.

Now we are ready to prove Theorem 2.

PROOF OF NECESSITY IN THEOREM 2. If a potential $q(x)$ belongs to the class $Q^\rho(M)$, then Lemma 2 yields i). Lemma 6 shows that i) implies that the function $D(z)$ belongs to the algebra $A^\rho(M)$. The invertibility of $D(z)$ in the algebra $A^\rho(M)$ was proved at the end of the previous section.

SUFFICIENCY. Let a set of numbers $\{s_n\}_{n\in\mathbb{N}}$ satisfying conditions i)–ii) of Theorem 2 be given. We shall construct a potential $q(x)\in Q^\rho(M)$ such that the spectral data set of the operator (1.2) with this potential is identical with the set $\{s_n\}_{n\in\mathbb{N}}$.[1]

To this end we must solve the fundamental integral equation (4.1). By assertion ii), the function $D(z)$ is invertible in the Banach algebra $A^\rho(M)$ and by Lemma 5 it coincides with the Fredholm determinant $\Delta(t)$ of (4.1). Hence, this determinant does not vanish and (4.1) is uniquely solvable. By the well-known formula for the Fredholm resolvent, we have

$$K(t, t+r) = -F(2t+r) + D(it)^{-1}\int_0^\infty D(r, y; t)F(2t+r+y)\,dy. \tag{4.8}$$

Here $D(it)$ is used instead of $\Delta(t)$, and $D(r, y; t)$ denotes the first Fredholm minor, or denominator of the Fredholm resolvent:

$$D(r, y; t) = \sum_{k=1}^{\infty}\frac{1}{k!}D_k(r, y; t), \tag{4.9}$$

[1] Comparing (3.13), (2.5) and (2.2), we find that the potential $q(x)$ may be reconstructed by the formula $q(z) = -2(d^2/dz^2)(\ln D(z))$ but to prove it we would need analysis similar to that given below.

where

$$D_k(r, y; t) \overset{\text{def}}{=} \int_0^\infty \cdots \int_0^\infty dt_1 \cdots dt_k \det \begin{Vmatrix} F(2t+r+y) & \cdots & F(2t+t_k+y) \\ F(2t+r+t_1) & \cdots & F(2t+t_k+t_1) \\ \vdots & \ddots & \vdots \\ F(2t+r+t_k) & \cdots & F(2t+t_k+t_k) \end{Vmatrix}.$$

The following considerations are similar to those used in Lemma 6. Using the definition of $F(y)$ given by (4.2), we have

$$D_k(r, y; t) = \sum_{n_1, \dots, n_{k+1} \in \mathbb{N}} s_{n_1} \cdots s_{n_{k+1}} \exp(-\Lambda_{n_1} t - \Lambda_{n_2} t - \cdots - \Lambda_{n_{k+1}} t) \times \Delta_{n_1, \dots, n_k, n_{k+1}}(r, y). \quad (4.10)$$

As in Lemma 6, the function $\Delta_{n_1 \cdots n_{k+1}}(r, y)$ is equal to

$$\Delta_{n_1, \dots, n_k, n_{k+1}}(r, y) = \int_0^\infty \cdots \int_0^\infty dt_1 \cdots dt_k \times \det \begin{Vmatrix} \exp(\Lambda_{n_1}(r+y)/2) & \cdots & \exp(\Lambda_{n_{k+1}}(t_k+y)/2) \\ \vdots & \ddots & \vdots \\ \exp(\Lambda_{n_1}(r+t_k)/2) & \cdots & \exp(\Lambda_{n_{k+1}}(t_k+t_k)/2) \end{Vmatrix}$$

$$= \det \begin{Vmatrix} \exp\left(-\frac{\Lambda_{n_1} r + \Lambda_{n_1} y}{2}\right) & \frac{2}{(\Lambda_{n_2}+\Lambda_{n_2})} & \cdots & \frac{2}{(\Lambda_{n_2}+\Lambda_{n_{k+1}})} \\ \vdots & \vdots & \ddots & \vdots \\ \exp\left(-\frac{\Lambda_{n_1} r + \Lambda_{n_{k+1}} y}{2}\right) & \frac{2}{(\Lambda_{n_{k+1}}+\Lambda_{n_2})} & \cdots & \frac{2}{(\Lambda_{n_{k+1}}+\Lambda_{n_{k+1}})} \end{Vmatrix}.$$

If we expand the last determinant in terms of its first column, then we will find that the coefficients of the exponentials containing r and y have the form (up to a sign) $\det \|2/(\Lambda_{n_j} + \Lambda_{n_l})\|_{l \neq l_0}$, with j varying between 2 and $k+1$ and l varying between 1 and $k+1$ but missing l_0. Using Lemma 7, we see that $\Delta_{n_1, \dots, n_k, n_{k+1}}(r, y)$ is majorized termwise by the function

$$\sum_{j=1}^{k+1} \exp(-\Lambda_{n_1} r/2 - \Lambda_{n_j} y/2) \sqrt{\Lambda_{n_1} \Lambda_{n_j}}.$$

Substituting this estimate into (4.10), it is easy to conclude that

$$\frac{1}{k!} \int_0^\infty D_k(r, y; t) F(2t+r+y)\, dy$$

is a sum of exponentials and is majorized termwise by the following function

$$\frac{k+1}{k!} [V(it)]^k \sum_{\beta, \gamma \in \mathbb{N}} \frac{2|s_\beta s_\gamma|}{\sqrt{\Lambda_\beta \Lambda_\gamma}} \exp(-(\Lambda_\beta + \Lambda_\gamma) r/2 - \Lambda_\beta t - \Lambda_\gamma t), \quad (4.11)$$

$V(t) \in A^\rho(M)$ being defined by (4.3).

We sum the functions (4.11) over the index k and apply inequality (1.11) to the numbers s_n. It follows that the second term in (4.8) can be represented in the form

$$D(it)^{-1}\sum_{n\in\mathbb{N}}\exp(-\Lambda_n r/2)\sum_{\alpha:\alpha\gg n}\psi_{n,\alpha}\exp(-\Lambda_\alpha t),$$

(cf. (2.5)) and that the numbers $\{\psi_{n,\alpha}\}_{n\in\mathbb{N},\alpha\gg n}$ satisfy inequality (1.8). Using the invertibility of $D(z)$ in the Banach algebra $A^p(M)$, we arrive at the following conclusion: the kernel $K(t, u)$, the solution of the fundamental integral equation, may be written in the form

$$K(t, t+r)=\sum_{n\in\mathbb{N}}\exp(-\Lambda_n r/2)\sum_{\alpha:\alpha\gg n}\varphi_{n,\alpha}\exp(-\Lambda_\alpha t) \tag{4.12}$$

with the sequence $\{\varphi_{n,\alpha}\}_{n\in\mathbb{N},\alpha\gg n}$ satisfying inequality (1.8). We again have proved Lemma 1 but in a different situation.

As a result, we can state some corollaries.

First of all, we substitute (4.12) into (4.1) and compare coefficients of the same exponentials using the uniqueness theorem for almost-periodic functions. We obtain

$$\varphi_{n,n}=-s_n, \qquad n\in\mathbb{N}, \tag{4.13}$$

$$\varphi_{n,\alpha}=s_n\sum_{k:k\ll\alpha\ominus n}\frac{2}{\Lambda_n+\Lambda_k}\varphi_{k,\alpha\ominus n}, \qquad \alpha\gg n, \tag{4.14}$$

which exactly coincides with (3.2) and (3.3).

Secondly, the fundamental integral equation (4.1) implies (see, for example, [6, pp. 170–175]) that the function

$$E(t, \mu)=e^{i\mu t}+\int_t^\infty K(t, u)e^{i\mu u}\,du$$

is a solution of the equation

$$-Y(t)''+p(t)Y(t)=\mu^2 Y(t), \tag{4.15}$$

with the potential $p(t)$ equal to

$$p(t)=-2(d/dt)K(t, t). \tag{4.16}$$

But now it follows from (4.16) and (4.12) that

$$p(t)=-\sum_{n\in\mathbb{N}}p_n\exp(-\Lambda_n t), \qquad t>0, \tag{4.17}$$

where

$$p_n=2\sum_{k:k\ll n}\Lambda_k\varphi_{kn}. \tag{4.18}$$

We continue equation (4.15) analytically into the half-plane $\{t:\operatorname{Re}t>0\}$ which is possible by (4.17), (4.18) and condition (1.8) proved above. Coming back to the notation $t=-ix$, $\mu=i\lambda$, $e(x, \lambda)=E(t, \mu)$, we conclude that the function $e(x, \lambda)$ satisfies the equation $-y'' + q(x)y = \lambda^2 y$, where the almost-periodic potential $q(x)$ has the Fourier series

$$q(x) \sim \sum_{n\in\mathbb{N}} p_n \exp(i\Lambda_n x). \tag{4.19}$$

Let us show that the potential $q(x)$ is the required one. Indeed, first of all, by virtue of (4.18), (4.19) and (1.8), we have

$$\sum_{n\in\mathbb{N}} |q_n| \Lambda_n^{-2} \rho(\Lambda_n) < \infty,$$

i.e., the function $q(x)$ belongs to the class $Q^\rho(M)$.

Secondly, (4.13) and (4.14) show that for every $n \in \mathbb{N}$ the function

$$e_n(x) \overset{\text{def}}{=} -\exp\left(-\frac{i}{2}\Lambda_n x\right) \sum_{\alpha\,:\,\alpha \gg n} \varphi_{n,\alpha} \exp(i\Lambda_\alpha x) = \angle\text{-}\lim_{\lambda\to\Lambda_n/2} e(x, -\lambda)(\lambda - \Lambda_n/2)$$

does exist and is equal to s_n-times the function $e(x, \Lambda_n/2)$. According to Definition 1 (see §1), this means that the set $\{s_n\}_{n\in\mathbb{N}}$ is the spectral data set of the operator (1.2) with potential (4.18). Theorem 2 is proved.

REFERENCES

1. M. G. Gasymov, *Spectral analysis for a class of nonselfadjoint differential operators of second order*, Funktsional. Anal. i Prilozhen. **14** (1980), no. 1, 14–19; English transl. in Functional Anal. Appl. **14** (1980).

2. V. Guillemin and A. Uribe, *Hardy functions and the inverse spectral method*, Comm. Partial Differential Equations **8** (1983), no. 13, 1455–1474.

3. A. D. Orudzhev, *Spectral analysis of a class of nonselfadjoint high-order differential operators*, Akad. Nauk Azerbaĭdzhan. SSR Dokl. **37** (1981), no. 2, 8–11. (Russian)

4. P. Sarnak, *Spectral behavior of quasi-periodic potentials*, Comm. Math. Phys. **84** (1982), 377–401.

5. L. A. Pastur and V. A. Tkachenko, *Spectral theory of the Schrödinger operators with periodic complex-valued potentials*, Funktsional. Anal. i Prilozhen. **22** (1988), no. 2, 85–86; English transl. in Functional Anal. Appl. **22** (1980).

6. V. A. Marchenko, *Spectral theory of Sturm-Liouville operators*, "Naukova Dumka", Kiev, 1972. (Russian)

7. V. E. Ljance, *An analogue of the inverse scattering problem for nonselfadjoint operator*, Mat. Sb. **72(114)** (1967), no. 4, 537–557; English transl. in Math. USSR-Sb. **1** (1967).

8. M. Reed and B. Simon, *Methods of modern mathematical physics*, I. *Functional analysis*, Academic Press, New York, 1972.

9. I. C. Gokhberg and M. G. Kreĭn, *Introduction to the theory of linear nonselfadjoint operators in Hilbert space*, "Nauka", Moscow, 1965; English transl., Amer. Math. Soc., Providence, R.I., 1969.

10. W. V. Lovitt, *Linear integral equations*, McGraw Hill, New York, 1924.

INSTITUTE FOR LOW TEMPERATURE ENGINEERING & PHYSICS, 47 LENIN PR., KHARKOV 310164 UKRAINE

Translated by THE AUTHOR

ADVANCES IN SOVIET MATHEMATICS
Volume 11, 1992

Subharmonic Functions of Completely Regular Growth in a Closed Cone

L. I. RONKIN

In [1–4] the main facts of the theory of subharmonic functions of completely regular growth (f.c.r.g.) in an open cone were determined.[1] In this paper, consisting of two parts, the conditions of c.r.g. in a closed cone are studied. The corresponding problem for holomorphic functions in an angle was previously considered by N. Govorov [5].

§1. Preliminaries

Let $x = (x_1, \dots, x_n) \in \mathbb{R}^n$ and $|x| = (x_1^2 + \dots + x_n^2)^{1/2}$. We set:

$$B_R(x^0) = \{x \in \mathbb{R}^n : |x - x^0| < R\}, \qquad B_R = B_R(0),$$
$$S_R(x^0) = \{x \in \mathbb{R}^n : |x - x^0| = R\}, \qquad S_R = S_R(0).$$

We denote by dS_R the element of $(n-1)$-dimensional volume of the sphere S_R. Let us also set

$$|S_t| = \int_{S_t} dS_t, \quad B_{r,R} = B_R \setminus \overline{B}_r, \quad (E)_{r,R} = E \cap B_{r,R},$$
$$(E)_r = E \cap B_r, \quad \|f\|_\infty = \sup |f(x)|.$$

Let Δ be the Laplace operator in the space $\mathbb{R}^n$ and Δ^* be its spherical part (Laplace-Beltrami operator). Thus

$$\Delta = \frac{\partial^2}{\partial r^2} + \frac{n-1}{r}\frac{\partial}{\partial r} + \frac{1}{r^2}\Delta^*,$$

where $r = |x|$, $x \in \mathbb{R}^n$. Then let Γ be a domain in the sphere S_1 with twice differentiable boundary $\partial\Gamma$. Let us consider the following boundary value problem on Γ:

$$\Delta^*\varphi + \lambda\varphi = 0, \qquad \varphi|_{\partial\Gamma} = 0.$$

1991 *Mathematics Subject Classification.* Primary 31B05.

[1] Some facts concerning rays of c.r.g. are contained in [6].

We denote by $\lambda_j = \lambda_j(\Gamma)$, $0 < \lambda_1 < \lambda_2 \leqslant \lambda_3 \leqslant \cdots$, the eigenvalues of this problem and by $\varphi_j = \varphi_{j,\Gamma}(x)$ the corresponding eigenfunctions normalized by the condition $\int \varphi_j^2 dS_1 = 1$. We assume the domain Γ to be such that $\varphi_j \in C^2(\overline{\Gamma})$, $\forall j$. To be definite let us assume also that $\partial\varphi/\partial n > 0$ on $\partial\Gamma$ (here $\partial/\partial n$ is the interior normal derivative). We set

$$K = K^\Gamma = \{x \in \mathbb{R}^n : x = ty,\ t > 0,\ y \in \Gamma\}.$$

Let $d\sigma$ be an element of $(n-1)$-dimensional volume on ∂K induced by the metric of the space $\mathbb{R}^n$ and correspondingly

$$\sigma(E) = \int_E d\sigma, \qquad E \subset \partial K,\ \Gamma_R = K \cap S_R,$$

$$\varphi_j(x) = \varphi_j(x/|x|), \quad \forall x \in \overline{K} \setminus \{0\},$$

$$\varkappa_j^\pm = \varkappa_j^\pm(\Gamma) = (-n + 2 \pm ((n-2)^2 + 4\lambda_j)^{1/2})/2, \qquad j = 1, 2, \dots,$$

$$\varkappa_0^+ = 0, \qquad \varkappa_0^- = -(n-2).$$

Note that the functions $|x|^{\varkappa_j^\pm}\varphi_j(x)$ are harmonic in K, belong to the class $C^2(\overline{K} \setminus \{0\})$ and vanish on $\partial K \setminus \{0\}$. Further for brevity we shall write φ and $\varkappa^\pm$ instead of φ_1 and $\varkappa_1^\pm$.

We denote by $SH(K)$ the set of all functions subharmonic in K and by $SH^*(K)$ the set of functions $u \in SH(K)$ bounded from above on bounded subsets of the cone K. Now let us denote by $SH(K, \rho]$, $\rho > 0$, the set of functions $u \in SH^*(K)$ satisfying the condition

$$\overline{\lim_{R\to\infty}} \widehat{M}_u(R) R^{-\rho} < \infty,$$

where

$$\widehat{M}_u(R) = \max\{M_u(R), \Phi_u(R)\}, \qquad M_u(R) = \sup_{\Gamma_R} u(x),$$

$$\Phi_u(R) = \int_\Gamma \varphi(x)|u(Rx)|\, dS_1.$$

Let $u \in SH(K, \rho]$. The function $\mathfrak{L}_u^*(x) = \overline{\lim}_{y\to x} \mathfrak{L}_u(y)$, $x \in K$, where $\mathfrak{L}_u(x) = \overline{\lim}_{y\to x} t^{-\rho} u(ty)$, is called the *regularized indicator* (with respect to the order ρ) or, for brevity, simply the indicator of the function u. It is well known that the function $\mathfrak{L}_u^*(x)$ is subharmonic in K and positive homogeneous of degree ρ (i.e., $\mathfrak{L}_u^*(tx) = t^\rho \mathfrak{L}_u^*(x)$).

In [1–4] it was shown that each function $u \in SH^*(K, \rho]$ on $\partial K \setminus [0]$ meets a real measure (charge) which is locally weak (in the $\mathscr{D}'$-topology[2]) boundary value of u. Furthermore, such a function, just as any subharmonic one, meets a Riesz measure μ_u associated with itself, i.e., $\mu_u = (1/\theta_n)\Delta u$,

[2] As usual, we denote by $\mathscr{D}'(G) = \mathscr{D}'$ the space of distributions in the domain G. Correspondingly, $\mathscr{D}(G)$ is the space of test functions in G.

where $\theta_n = (n-2)|S_1|$ for $n > 2$, $\theta_2 = 2\pi$ and the Laplacian Δ must be regarded as an operator in $\mathscr{D}'$.

Just as in [1], we construct the measure τ_u from the measures μ_u and $\mu_{u,\partial}$ and define τ_u on Borel sets $E \Subset \overline{K} \setminus \{0\}$ by the equalities

$$\tau_u(E) = \int_{E\cap K} \varphi \, d\mu_u - \frac{1}{\theta_n} \int_{E\cap\partial K} \frac{\partial\varphi}{\partial n} \, d\mu_{u,\partial}.$$

As usual, for any real measure τ on $E \subset \mathbb{R}^n$, we denote by τ^+, τ^- and $|\tau|$ the positive, negative components of its Jordan decomposition and the variation. Thus $\tau = \tau^+ - \tau^-$, $|\tau| = \tau^+ + \tau^-$. Let us also set

$$\|\tau\|(R) = \int_{(E)_{1,R}} d|\tau|, \qquad R > 1.$$

The following estimate of τ_u is valid.

THEOREM 1 (see [1–3]). *If* $u \in SH(K, \rho]$ *then*

$$\|\tau_u\|(R) \leqslant c_1 R^{\rho+n-2}, \qquad \forall R > 1, \tag{1}$$

for some $c_1 = c_1(u)$.

Let us denote by $G(x, y)$ Green's function of the cone K. If $|x| < |y|$, then the following representation is valid (see [10])

$$G(x, y) = |S_1| \sum_{k=1}^{\infty} \frac{|x|^{\varkappa_k^+} |y|^{\varkappa_k^-}}{\varkappa_k^+ - \varkappa_k^-} \varphi_k(x)\varphi_k(y). \tag{2}$$

We set

$$G_q(x, y) = -G(x, y) + |S_1| \sum_{k=1}^{q} \frac{|x|^{\varkappa_k^+} |y|^{\varkappa_k^-}}{\varkappa_k^+ - \varkappa_k^-} \varphi_k(x)\varphi_k(y), \tag{3}$$

and

$$W_q(x, y) = \begin{cases} G_q(x, y)/\varphi(y), & \forall x, y \in K, \\ \dfrac{\partial G_q(x, y)/\partial n_y}{\partial\varphi(y)/\partial n}, & \forall x \in K, \ y \in \partial K \setminus \{0\}. \end{cases} \tag{4}$$

It was shown in [4] that the estimates

$$\begin{gathered} |W_q(x, y)| \leqslant c_1 |x|^{\varkappa_{q+1}^+} |y|^{\varkappa_{q+1}^-} \varphi(x), \\ x \in K, \ y \in \partial K \setminus \{0\}, \ |x| < t|y|, \ 0 < t < 1, \ c_1 = c_1(t, q), \end{gathered} \tag{5}$$

$$\begin{gathered} W_q(x, y) \leqslant c_2 \frac{|x|^{\varkappa_{q+1}^+} |y|^{\varkappa_{q+1}^-} \varphi(x)}{|x|^{\varkappa_{q+1}^+ - \varkappa_q^+} + |y|^{\varkappa_{q+1}^+ - \varkappa_q^+} \varphi(x)}, \\ x \in K, \ y \in \partial K \setminus \{0\}, \ c_2 = c_2(t, q), \end{gathered} \tag{6}$$

are true.

In [4, 8] the following representation of a function $u \in SH(K, \rho]$ was obtained.

THEOREM 2. *Let* $u \in SH(K, \rho]$ *and let* q *be the minimum integer satisfying the condition*

$$\int^{\infty} t^{\varkappa^-_{q+1}} d\|\tau_u\|(t) < \infty.$$

Then the function $u(x)$ *can be represented in the form*

$$u(x) = \int_{(\overline{K})_{1,\infty}} W_q(x, y) d\tau_u(y) + \sum_{j=1}^{m} a_j \varphi_j(x)|x|^{\varkappa^+_j} + H_u(x) - \int_{(K)_{0,1}} G(x, y)\, d\mu_u(y), \quad (7)$$

where $\varkappa^+_j \leqslant \rho$, $\forall j \leqslant m$ *and the function* $H_u(x)$ *is harmonic in* K *and satisfies* $\lim_{|x|\to\infty} |x|^{-\varkappa^-} H_u(x) = 0$.

Now let us introduce the main subject of this study, namely, functions of c.r.g. in K and $\overline{K}$ (in an open cone and in a closed one).

A set $E \subset \mathbb{R}^n$ is referred to as C^α_0-set if balls $B^{(j)} = B_{r_j}(x^{(j)})$ exist such that $E \subset \bigcup_j B^{(j)}$ and

$$\lim_{R\to\infty} R^{-\alpha} \sum_{|x^{(j)}|<R} r^\alpha_j = 0.$$

A function $u \in SH(K, \rho]$ is referred to as a *function of c.r.g.* in $\overline{K}$ (in a closed cone) if a C^{n-1}_0-set $E \subset K$ exists such that

$$\lim_{\substack{|x|\to\infty \\ x \notin E}} \left| u(x)|x|^{-\rho} - \mathfrak{L}^*_u(x/|x|) \right| = 0.$$

It is easy to see that this definition is equivalent to the following one. A function $u \in SH(K, \rho]$ is referred to as a *function of c.r.g.* in $\overline{K}$ if for every $\varepsilon > 0$ the set $E_\varepsilon(u) = \{x \in K : |u(x) - \mathfrak{L}^*_u(x)| > \varepsilon|x|^\rho\}$ is a C^{n-1}_0-set. A function $u \in SH(K, \rho]$ is referred to as *of c.r.g.* in K (in an open cone) if it is of c.r.g. in any closed cone $\overline{K}' \subset K$. Sometimes, for brevity we shall omit the word "completely" in the word combination "completely regular".

A number of criteria of c.r.g. in K was established in [1–3]. They can be united as in the following theorem.

THEOREM 3. *Let* $u \in SH(K, \rho]$. *Then the following statements are equivalent*:

a) *the function* u *is of c.r.g. in* K;

b) *when* $t \to \infty$, *the functions*

$$u^{[t]} \overset{\text{def}}{=} t^{-\rho} u(tx), \qquad t > 0,$$

tend to the indicator $\mathfrak{L}^*_u$ *as elements of* $\mathscr{D}'(K)$;

c) $\displaystyle\lim_{t\to\infty} \int_{(K)_{1/t,1}} |x|^{-2} u^{[t]}(x)\Psi(x)\, dx = \int_{(K)_1} |x|^{-2} \mathfrak{L}^*_u(x)\Psi(x) dx, \forall \Psi \in C(\overline{K}_1)$;

d) *the measures* $\tau_u^{[t]}$ *defined on Borel sets* $E \Subset \overline{K} \setminus [0]$ *by the equalities*

$$t_u^{[t]}(E) = t^{-\rho-n+2}\tau_u((tE) \setminus B_1)$$

tend to the measure $\widehat{\tau}_u = \tau_{\mathfrak{L}_u^*}$ *as functionals on* $C(\overline{K}_1)$ *when* $t \to \infty$. *In addition to this, if* $\rho \in \{\varkappa_j^+\}_{j=1}^{\infty}$, *then the following limit exists*

$$\lim_{R\to\infty} \int_{(\overline{K})_{1,R}} \varphi_q(x)\varphi^{-1}(x)|x|^{\rho+n+2} d\tau_u(x), \qquad \forall q : \varkappa_q^+ = \rho. \tag{8}$$

Note that the first condition in d) is called the *condition of cone-boundary density*, and the second one is the *condition of cone-boundary balancing*.

§2. Necessary condition of c.r.g. in $\overline{K}$

Let us consider the case $n = 2$ and correspondingly $K = Y_{\alpha,\beta}$, where $Y_{\alpha,\beta} = \{z \in \mathbb{C} : \alpha < \arg z < \beta\}$. In this case a function $f(z)$ holomorphic in $Y_{\alpha,\beta}$ is said to be of c.r.g. in $Y_{\alpha,\beta}$ or in $\overline{Y}_{\alpha,\beta}$ if the subharmonic function $\ln|f(z)|$ belongs to $SH(Y_{\alpha,\beta}, \rho])$ and is of c.r.g. in $Y_{\alpha,\beta}$ or in $\overline{Y}_{\alpha,\beta}$ in the sense of the definition given in §1.

It was mentioned above that the theory of such functions was constructed by N. Govorov [5]. In particular, he proved that for a holomorphic function $f(z)$ to be of c.r.g. in $\overline{Y}_{\alpha,\beta}$ it is necessary and sufficient that it be of c.r.g. in $Y_{\alpha,\beta}$ and its indicator

$$h_f(\theta) \stackrel{\text{def}}{=} \overline{\lim_{r\to\infty}}\, r^{-\rho} \ln|f(re^{i\theta})|$$

satisfy the conditions

$$h_f(\alpha + 0) = \overline{\lim_{r\to\infty}}\, r^{-\rho}\left\{ \overline{\lim_{\substack{z\to re^{i\alpha}\\ z\in Y_{\alpha,\beta}}}} \ln|f(z)| \right\},$$

$$h_f(\beta - 0) = \overline{\lim_{r\to\infty}}\, r^{-\rho}\left\{ \overline{\lim_{\substack{z\to re^{i\beta}\\ z\in Y_{\alpha,\beta}}}} \ln|f(z)| \right\}.$$

In [9] this result was extended to subharmonic functions in $Y_{\alpha,\beta}$. The method, used in [5] and [9] for obtaining the above-mentioned results as well as some others, cannot be extended to the case $n > 2$. The use of conformal mappings is one of the reasons for this. To study the conditions of c.r.g. in $\overline{K}$ for $n > 2$, we shall use the relationship of c.r.g. with weak convergence of the corresponding function families and measures, and their estimates. This relationship was established in [1, 2].

An important condition appears in the statements on necessary and sufficient conditions of c.r.g. in $\overline{K}$. We prefix the statement of the corresponding theorems with the formulation of this condition.

A function $u \in SH(K, \rho]$ is said to satisfy *condition* A if we have the following equality

$$\lim_{R\to\infty} R^{-\rho-n+1}\|\mu_{u,\partial} - \mu_{\mathfrak{L}_u^*,\partial}\|(R) = 0. \tag{9}$$

It is easy to see that this condition is equivalent to the following one: the measures $\mu_{u,\partial}^{[R]}$ defined on ∂K by the equalities

$$\mu_{u,\partial}^{[R]}(E) = R^{-\rho-n+1}\mu_{u,\partial}((K)_{1,\infty} \cap RE), \qquad E \subset \partial K, \tag{10}$$

converge in variation to $\mu_{\mathfrak{L}_u^*,\partial}$ as $R \to \infty$.

Here we shall obtain two theorems which show that condition A is necessary under some additional assumptions. It may be that these assumptions are needless, but we cannot do without them.

THEOREM 4. *If* $u \in SH(K, \rho]$ *and u is of c.r.g. in* $\overline{\widetilde{K}}$, *where* $\widetilde{K} = K^{\widetilde{\Gamma}}$, $\widetilde{\Gamma}$ *is a domain on* S_1 *with smooth boundary and* $\widetilde{\Gamma} \Subset \Gamma$, *then* $u(x)$ *satisfies condition A as a function belonging to* $SH(\widetilde{K}, \rho]$.

PROOF. As mentioned in [2], the equality $d\mu_{u,\partial} = u\,d\sigma$ is valid for any function $u \in SH(K, \rho]$ subharmonic in some neighborhood of the cone K. Therefore in this situation condition A has the form

$$\lim_{R\to\infty} R^{-\rho-n+1}\int_{(\partial\widetilde{K})_{1,R}} |u(x) - \mathfrak{L}_u^*(x)|\,d\tilde{\sigma}(x) = 0, \tag{11}$$

where $d\tilde{\sigma}$ is an element of $(n-1)$-dimensional volume on $\partial\widetilde{K}$, compatible with the Euclidean metric of $\mathbb{R}^n$. To prove this equality we shall use a number of lemmas.

LEMMA 1.[3] *Let* $\lambda(t)$ *be a monotone decreasing nonnegative function on* $\mathbb{R}_+$. *Then for any open set* E *in* $\mathbb{R}^m$, $m \geqslant 1$, *the inequality*

$$\int_E \lambda(|x|)\,dx \leqslant \int_{B_R} \lambda(|x|)dx$$

is valid for $R = R_E$, *where* R *is defined by the inequality*

$$\int_E dx \leqslant \int_{B_R} dx.$$

PROOF. We set

$$S^E(t) = \int_{E\cap S_t} dS_t \quad \text{and} \quad S(t) = |S_t|.$$

Then

$$\int_0^\infty S^E(t)\,dt = \int_0^R S(t)\,dt, \quad \text{so that} \quad \int_0^R (S^E(t) - S(t))\,dt + \int_R^\infty S^E(t)\,dt = 0.$$

[3] In the case $n = 1$ and $K = \partial K = \mathbb{R}_+$, the corresponding version of Lemma 1 (with another proof) appears in [10].

Taking this into account together with the fact that $S^E(t) \leqslant S(t)$ for all t, we get

$$\begin{aligned}\int_E \lambda(|x|)\,dx &= \int_0^\infty \lambda(t)S^E(t)\,dt \\ &= \int_0^r \lambda(t)S(t)\,dt + \int_0^R \lambda(t)(S^E(t) - S(t))\,dt + \int_R^\infty S^E(t)\,dt \\ &\leqslant \int_{B_R} \lambda(|x|)\,dx + \lambda(R)\int_0^R (S^E(t) - S(t))\,dt + \lambda(R)\int_R^\infty S^E(t)\,dt \\ &= \int_{B_R} \lambda(|x|)\,dx.\end{aligned}$$

This completes the proof of the lemma.

LEMMA 2. *The following estimates are valid for the kernel* $W_q(x, y)$, $q = 0, 1, \dots$, $x \in K$, $y \in \overline{K} \setminus \{0\}$, *parallel with the above-mentioned estimates* (5) *and* (6)[4]:

a) *if* $|x| > t|y|$, $t > 1$, *then*

$$|W_q(x, y)| \leqslant c_1\varphi(x)|x|^{\varkappa_q^+}|y|^{\varkappa_q^-}, \qquad c_1 = c_1(t); \tag{12}$$

b) *if* $t_1|x| < |y| < t_2|x|$, *then*

$$-c_2\varphi(x)|x|^{\varkappa_q^+} - c_3\varphi(x)|x||y||x - y|^{-n} \leqslant W_q(x, y) \leqslant c_2\varphi(x)|x|^{\varkappa_q^+}|y|^{\varkappa_q^-},$$
$$c_2 = c_2(t_1, t_2),\ c_3 = c_3(t_1, t_2), \tag{13}$$

$$-c_2\varphi(x)|x|^{\varkappa_q^+}|y|^{\varkappa_q^-} - \varphi^{-1}(y)|x - y|^{-n+2} \leqslant W_q(x, y) \leqslant c_2\varphi(x)|x|^{\varkappa_q^+}|y|^{\varkappa_q^-},$$
$$x \in K,\ y \in K, \tag{14}$$

$$-c_2\varphi(x)|x|^{\varkappa_q^+}|y|^{\varkappa_q^-} - |y||x - y|^{-n+1} \leqslant W_q(x, y) \leqslant c_2\varphi(x)|x|^{\varkappa_q^+}|y|^{\varkappa_q^-},$$
$$x \in K,\ y \in \partial K \setminus \{0\}; \tag{15}$$

c) *if* $0 < t_1 < |x|/|y| < t_2 < \infty$, $\big|x/|x| - y/|y|\big| > \varepsilon > 0$, *then*

$$|W_q(x, y)| \leqslant c_4\varphi(x)|y|^{-n+2} \leqslant c_5\varphi(y)|y|^{-n+2}, \tag{16}$$
$$c_4 = c_4(t_1, t_2, \varepsilon),\ c_5 = c_5(t_1, t_2, \varepsilon).$$

PROOF. The estimate a) follows from the form of the kernel $W_q(x, y)$, representations (2), (3), and the symmetry property of Green's function. Estimates (13) and (14) in b) are consequences of the estimates of Green's function (see [12], Lemma 4 and the remark to it) and the obvious estimate of the value $|W_q(x, y) + G(x, y)|$. Estimate (14) directly implies estimate (15). Estimate c) immediately follows from estimate (13) and the inequality $\big|x/|x| - y/|y|\big| > \varepsilon$.

[4]The numeration of constants differs in different lemmas (theorems).

LEMMA 3. *Let* $u \in SH(K, \rho]$ *and suppose the set* $E \subset \partial \widetilde{K}$ *can be covered by a system of balls* $B^{(j)} = B_{r_j}(x^{(j)})$ *satisfying the condition*

$$R^{-n+1} \sum_{|x^{(j)}|<R} r_j^{n-1} \stackrel{\text{def}}{=} \gamma(R) \to 0 \quad as \ R \to \infty$$

(i.e., E is a C_0^{n-1}-set). Then there exists a constant $c = c(\widetilde{K}, u)$ *such that*

$$\int_{(E)_R} |u(x)|\, d\tilde{\sigma} \leqslant c(\gamma(2R))^{1/(n-1)} R^{\rho+n-1}. \tag{17}$$

PROOF. It follows from Theorem 2 that it is sufficient to consider separately the case $u(x) = \varphi_j(x)|x|^{\varkappa_j^+}$, where $\varkappa_j^+ \leqslant \rho$, and the case

$$u(x) = \int_{(\overline{K})_{1,\infty}} W_q(x, y)\, d\tau_u(y),$$

where the number q is defined by the measure τ_u just as in Theorem 2.

Let us consider the first case. Note that since $\partial\widetilde{\Gamma}$ is smooth, a constant c_1 exists such that

$$\tilde{\sigma}(\partial \widetilde{K} \cap B^{(j)}) \leqslant c_1 r_j^{n-1}. \tag{18$'$}$$

Besides, since $r_j/|x^{(j)}| \to 0$ as $j \to \infty$, we have

$$(E)_R \subset \bigcup_{|x^{(j)}|<2R} B^{(j)}.$$

Therefore

$$\begin{aligned}\int_{(E)_R} |x|^{\varkappa_j^+}\, d\tilde{\sigma}(x) &\leqslant R^{\varkappa_j^+} \int_{(E)_R} d\tilde{\sigma} \leqslant R^{\varkappa_j^+} \sum_{|x^{(j)}|<2R} \tilde{\sigma}(E \cap B^{(j)}) \\ &\leqslant c_1 R^{\varkappa_j^+} \sum_{|x^{(j)}|<2R} r_j^{n-1} \leqslant c_1 2^{n-1} R^{\varkappa_j^+ + n-1} \gamma(2R).\end{aligned} \tag{18}$$

Therefore

$$\int_{(E)_R} |u(x)|\, d\tilde{\sigma}(x) \leqslant \|\varphi_j\|_\infty c_1 2^{n-1} \gamma(2R) R^{\rho+n-1}$$

when $u(x) = \varphi(x)|x|^{\varkappa_j^+}$, $\varkappa_j^+ \leqslant \rho$. Hence inequality (17) holds in the case considered.

The proof of the lemma in the case

$$u(x) = \int_{(\overline{K})_{1,\infty}} W_q(x, y)\, d\tau_\mu(y)$$

will be carried out under the assumption that $\varkappa_q^+ < \rho < \varkappa_{q+1}^+$. This does not decrease generality, because if necessary the cone K can be replaced by the cone $\widehat{K} = K^{\widehat{\Gamma}}$ satisfying the conditions: 1) $\widehat{\Gamma}$ is a domain in S_1 with smooth boundary; 2) $\widetilde{\Gamma} \Subset \widehat{\Gamma} \Subset \Gamma$; 3) the numbers $\varkappa_j^+(\widehat{\Gamma})$ satisfy the inequality $\varkappa_q^+ < \rho < \varkappa_{q+1}^+$ for some q. We have

$$\begin{aligned}
\int_{(E)_R} |u(x)|\, d\tilde{\sigma}(x) &\leqslant \int_{(\overline{K})_{1,\infty}} d\tau_u(y) \int_{(E)_R} |W_q(x, y)| d\tilde{\sigma}(x) \\
&\leqslant \int_{(\overline{K})_{1,2R}} d|\tau_u(y)| \int_{\substack{E\cap B_R \\ |x|<|y|/2}} |W_q(x, y)| d\tilde{\sigma}(x) \\
&\quad + \int_{(\overline{K})_{1,2R}} d|\tau_u(y)| \int_{\substack{E\cap B_R \\ |y|/2<|x|<2|y|}} |W_q(x, y)| d\tilde{\sigma}(x) \\
&\quad + \int_{(\overline{K})_{1,R/2}} d|\tau_u(y)| \int_{\substack{E\cap B_R \\ |x|<2|y|}} |W_q(x, y)| d\tilde{\sigma}(x) \\
&\quad + \int_{(\overline{K})_{2R,\infty}} d|\tau_u(y)| \int_{(E)_R} |W_q(x, y)| d\tilde{\sigma}(x) \\
&= I_1 + I_2 + I_3 + I_4. \qquad (19)
\end{aligned}$$

Using estimates (1), (5) and (18), we obtain

$$\begin{aligned}
I_1 &\leqslant \int_{(\overline{K})_{1,2R}} d|\tau_u(y)| \int_{(E)_R} c_1 \varphi(x) |x|^{\varkappa_{q+1}^+} |y|^{\varkappa_{q+1}^-}\, d\tilde{\sigma}(x) \\
&\leqslant c_2 \|\varphi\|_\infty \gamma(2R) R^{\varkappa_{q+1}^+ + n-1} \int_1^{2R} t^{\varkappa_{q+1}^-}\, d\|\tau_u\|(t) \\
&= c_2 \|\varphi\|_\infty \gamma(2R) R^{\varkappa_{q+1}^+ + n-1} \left((2R)^{\varkappa_{q+1}^-} \|\tau_u\|(R) - \varkappa_{q+1}^- \int_1^{2R} t^{\varkappa_{q+1}^- - 1} \|\tau_u\|(t)\, dt \right) \\
&\leqslant c_3 \gamma(2R) R^{\varkappa_{q+1}^+ + n-1+\varkappa_{q+1}^- + \rho+n-2} = c_3 R^{\rho+n-1} \gamma(2R). \qquad (20)
\end{aligned}$$

Similarly

$$\begin{aligned}
I_3 &\leqslant c_4 \int_{(\overline{K})_{1,R/2}} d|\tau_u|(y) \int_{\substack{E\cap B_R \\ |x|<2|y|}} \varphi(x) |x|^{\varkappa_q^+} |y|^{\varkappa_q^-}\, d\tilde{\sigma}(x) \\
&\leqslant c_5 \gamma(2R) R^{\rho+n-1}, \qquad (21) \\
I_4 &\leqslant c_1 \int_{(\overline{K})_{2R,\infty}} d|\tau_u|(y) \int_{(E)_R} \varphi(x) |x|^{\varkappa_q^+} |y|^{\varkappa_q^-}\, d\tilde{\sigma}(x) \\
&\leqslant c_6 \gamma(2R) R^{\rho+n-1}. \qquad (22)
\end{aligned}$$

The estimation of I_2 is more complicated. Let us set $d = \min\{\varphi(x) : x \in \widetilde{\Gamma}\}$, $A_1 = \{y : y \in (K)_{1,2R}, \varphi(y) < d/2\}$, and $A_2 = (K)_{1,2R} \setminus A_1$.

Using estimates (13), (14) and (16) we obtain

$$
\begin{aligned}
I_2 &\leqslant c_7 \int_{(\overline{K})_{1,2R}} d|\tau_u|(y) \int_{\substack{E\cap B_R \\ |y|/2<|x|<2|y|}} |x|^{\varkappa_q^+} |y|^{\varkappa_q^-}\, d\tilde{\sigma}(x) \\
&\quad + c_8 \int_{A_1} \frac{d|\tau_u|(y)}{|y|^{n-2}} \int_{\substack{E\cap B_R \\ |y|/2<|x|<2|y|}} d\tilde{\sigma}(x) \\
&\quad + c_9 \int_{A_2} \frac{d|\tau_u|(y)}{\varphi(y)} \int_{\substack{E\cap B_R \\ |y|/2<|x|<2|y|}} \frac{d\tilde{\sigma}(x)}{|x-y|^{n-2}} \\
&= I_{2,1} + I_{2,2} + I_{2,3}.
\end{aligned} \tag{23}
$$

The value $I_{2,1}$ can be estimated in the standard way by using estimates (1) and (18). We have

$$I_{2,1} \leqslant c_{10}\gamma(2R)R^{\rho+n-1}. \tag{24}$$

Similarly

$$I_{2,2} \leqslant c_{11}\gamma(2R)R^{\rho+n-1}. \tag{25}$$

To estimate $I_{2,3}$, let us construct planes T_{x^0} tangent to $\partial\widetilde{K}$ such that each ray $l_{x^0} = \{x = tx^0 : t > 0\}$, $x^0 \in \partial\widetilde{\Gamma}$, be contained in the corresponding plane T_{x^0}. Let us denote by $d\lambda$ the volume element on T_{x^0} induced by the Euclidean metric of the initial space $\mathbb{R}^n$. We also denote by $\pi = \pi_{x^0}$ the operator of orthogonal projection on T_{x^0} and by $\vec{n}_x$, $x \in \partial\widetilde{K}$, the unit vector of the interior normal to $\partial\widetilde{K}$ at the point x. Then let us choose on $\partial\widetilde{K}$ any conical neighborhood $\widetilde{Y} = \widetilde{Y}_{x^0}$ of the ray l_{x^0} such that the inequality $|\alpha(x', x'')| < \delta$, $\forall x' \in \widetilde{Y}$, $x'' \in \widetilde{Y}$, is valid. Here $\delta > 0$ is so small that the restriction $\pi|_{\widetilde{Y}}$ is a diffeomorphism of $\widetilde{Y}$ on $\pi\widetilde{Y} = Y_{x^0} = Y$. Now let us estimate the value

$$J = J_{x^0} = \int_{A_2} \frac{d|\tau_u|(y)}{\varphi(y)} \int_{\substack{\widetilde{Y}\cap E\cap B_R \\ |y|/2<|x|<2|y|}} \frac{d\tilde{\sigma}(x)}{|x-y|^{n-2}}.$$

We have

$$
\begin{aligned}
J &\leqslant \frac{1}{d}\int_{A_2} d|\tau_u|(y) \int_{\widetilde{Y}\cap E\cap B_R} \frac{d\tilde{\sigma}(x)}{|x-y|^{n-2}} \\
&= \frac{1}{d}\int_{A_2} d|\tau_u|(y) \int_{\widetilde{Y}\cap E\cap B_R} \frac{d\tilde{\sigma}(x)}{|\pi x-\pi y|^{n-2}} \\
&\leqslant \frac{1}{d}\int_{A_2} d|\tau_u|(y) \int_{Y\cap\pi E\cap B_R} \frac{d\lambda(\xi)}{|\xi-\pi y|^{n-2}\cos\alpha(x, x^0)} \\
&\leqslant \frac{1}{d\cos\delta}\int_{A_2} d|\tau_u|(y) \int_{Y\cap\pi E\cap B} \frac{d\lambda(\xi)}{|\xi-\pi y|^{n-2}}.
\end{aligned} \tag{26}
$$

Let us estimate the inner integral on the right-hand side. It is obvious that the set $Y\cap\pi E$ can be covered by $(n-1)$-dimensional balls $\pi B^{(j)}$, where the balls $B^{(j)}$ are the ones appearing in the conditions of the lemma. Therefore it follows from the assumption on the radii of the balls $B^{(j)}$ that

$$\int_{B_R\cap Y\cap\pi E} d\lambda(\xi) \leqslant c_{12}\gamma(2R)R^{n-1}. \tag{27}$$

For brevity let us set $E_{R,Y} = \pi y + B_R \cap Y \cap \pi E$. Using Lemma 1, we get

$$\int_{B_R\cap Y\cap\pi E} \frac{d\lambda(\xi)}{|\xi-\pi y|^{n-2}} \leqslant \int_{E_{R,y}} \frac{d\lambda(\xi)}{|\xi|^{n-2}} \leqslant \int_{|\xi|<r} \frac{d\lambda(\xi)}{|\xi|^{n-2}}, \tag{28}$$

where the number r is defined by the equality

$$\int_{|\xi|<r} d\lambda(\xi) = \int_{E_{R,y}} d\lambda(\xi).$$

It follows from (27) that $r \leqslant c_{13}(\gamma(2R))^{1/(n-1)}R$. Taking this into account in inequality (28), we get

$$\int_{B_R\cap Y\cap\pi E} \frac{d\lambda(\xi)}{|\xi-\pi y|^{n-2}} \leqslant c_{14}(\gamma(2R))^{1/(n-1)}R.$$

Therefore

$$\begin{aligned} J &\leqslant c_{15}(\gamma(2R))^{1/(n-1)}R\int_{A_2} d|\tau_u|(y) \\ &\leqslant c_{15}(\gamma(2R))^{1/(n-1)}\|\tau_u\|(2R) \leqslant c_{16}(\gamma(2R))^{1/(n-1)}R^{\rho+n-1}. \end{aligned}$$

Since the surface $\partial\widetilde{K}$ can be covered by a finite number of surfaces $\widetilde{Y}_{x^0}$, this implies

$$I_{2,3} \leqslant c_{17}(\gamma(2R))^{1/(n-1)}R^{\rho+n-1}. \tag{29}$$

Summing estimates (19)–(25), we conclude that the lemma also holds in the case

$$u(x) = \int_{(\overline{K})_{1,\infty}} W_q(x,y)\,d\tau_u(x).$$

This completes the proof of the lemma.

Now let $u(x)$ be the function that appears in the assertion of Theorem 4 and let $B_{r_j}(x^{(j)})$, $j = 1, 2, \dots$, be balls covering the set $E_\varepsilon(u)\cap\widetilde{K}$ in accordance with the definition of c.r.g. in $\widetilde{K}$. As above, we set

$$\gamma(R) = R^{-\rho-n+1}\sum_{|x^{(j)}|<R} r_j^{n-1}.$$

Let us consider a family of domains $\Gamma^{(t)}$, $0<t<1$ on S_1, having smooth boundaries and smoothly approximating the domain $\widetilde{\Gamma}$ from inside. Let us denote by $K^{(t)}$ the corresponding cones $K^{\Gamma^{(t)}}$ and by $d\sigma^{(t)}$ the volume element on $\partial K^{(t)}$. Using Lemma 3 with $K = K^{(t)}$, we obtain

$$\int_{E_\varepsilon(u)\cap(K^{(t)})_{1,R}} |u(x)|\,d\sigma^{(t)}(x) \leqslant c_1(\gamma(2R))^{1/(n-1)}R^{\rho+n-1}.$$

Analyzing the proof of Lemma 3, we see that if $0 < t < \sigma$, where σ is small enough, then the constant c_1 may be considered independent of t. Therefore

$$\begin{aligned}\int_{(\partial K^{(t)})_{1,R}} &|u(x) - \mathfrak{L}_u^*(x)|\, d\sigma^{(t)}(x) \\ &\leqslant \int_{E_\varepsilon(u)\cap(\partial K^{(t)})_{1,R}} (|u(x)| + |\mathfrak{L}_u^*(x)|)\, d\sigma^{(t)}(x) + \varepsilon \int_{(\partial K^{(t)})_{1,R}} |x|^\rho d\sigma^{(t)}(x) \\ &\leqslant (c_2(\gamma(2R))^{1/(n-1)} + c_3\varepsilon)R^{\rho+n-1}, \end{aligned} \tag{30}$$

where the constants c_2 and c_3 are independent of t just as c_1.

It is easy to see that as $u(x)$ and $\mathfrak{L}_u^*(x)$ are subharmonic in the cone K containing $\widetilde{K}$, then the relation

$$\lim_{t\to 0} \int_{(\partial K^{(t)})_{1,R}} |u(x) - \mathfrak{L}_u^*(x)|\, d\sigma^{(t)}(x) \leqslant \int_{(\partial \widetilde{K}^{(t)})_{1,R}} |u(x) - \mathfrak{L}_u^*(x)|\, d\tilde{\sigma}(x)$$

holds.

From this and (30) it follows that

$$\int_{(\partial \widetilde{K}^{(t)})_{1,R}} |u(x) - \mathfrak{L}_u^*(x)|\, d\tilde{\sigma}(x) \leqslant c_2((\gamma(2R))^{1/(n-1)} + \varepsilon c_3)R^{\rho+n-1}$$

and hence

$$\varlimsup_{R\to\infty} R^{-\rho-n+1} \int_{(\partial\widetilde{K})_{1,R}} |u(x) - \mathfrak{L}_u^*(x)|\, d\tilde{\sigma}(x) \leqslant c_3\varepsilon.$$

Therefore

$$\varlimsup_{R\to\infty} R^{-\rho-n+1} \int_{(\partial\widetilde{K})_{1,R}} |u(x) - \mathfrak{L}_u^*(x)|\, d\tilde{\sigma}(x) \leqslant 0.$$

The proof of Theorem 4 is completed.

COROLLARY. *Let* $u \in SH(K, \rho]$ *be a function of c.r.g. in* K. *Then relation* (11) *holds for any cone* $\widetilde{K} = K^{\widetilde{\Gamma}}$ *when the domain* $\widetilde{\Gamma} \Subset \Gamma$ *has a smooth boundary.*

The proof is evident.

THEOREM 5. *Let* $u \in SH(K, \rho]$ *be a function of c.r.g. in* $\overline{K}$ *and let*

$$\sup_{x\in\Gamma} |\mathfrak{L}_u^*(x)| = M < \infty.$$

Then $u(x)$ *satisfies condition* A.

PROOF. In accordance with the assumption of the theorem, the set $E_\varepsilon(u)$ is C_0^{n-1}-set. Hence it can be covered by some system $\mathscr{Q} = \mathscr{Q}(u, \varepsilon)$ of balls $B^{(j)} = B_{r_j}(x^{(j)})$, $r_j = r_j(\mathscr{Q}, \varepsilon)$, $x^{(j)} = x^{(j)}(\mathscr{Q})$, so as to have

$$\lim_{R\to\infty} R^{-n+1}\gamma^{(n-1)}(R; \mathscr{Q}) = 0, \tag{31}$$

where

$$\gamma^m(R;\mathscr{Q}) = \sum_{|x^{(j)}|<R} r_j^m, \qquad m = 1, 2, \dots .$$

We set

$$\mathfrak{S} = \mathfrak{S}(\mathscr{Q}) = \bigcup_j B^{(j)}.$$

Further we shall assume that the cone can be displaced, i.e., that for some ξ^0 and for all $h > 0$ small enough (to be definite, $0 < h < 1$) the inclusion $h\xi^0 + K \subset K$ is valid. It follows from the reasoning carried out in [2] (where the boundary measure $\mu_{u,\partial}$ was defined and its existence was proved) that this assumption in our situation does not violate generality. Then it is obvious that the functions

$$u_h(x) = u(h\xi^0 + x), \qquad 0 < h < 1,$$

are defined in K, belong to the class $SH^*(K)$, and satisfy the condition

$$u_h(x) \leqslant c_1|x|^\rho + c_2, \qquad \forall x \in K,\ 0 < h < 1,$$

where c_1 and c_2 are independent of h. Besides, the set $E_\varepsilon(U_h)$ can be covered by the system $\mathscr{Q}(u_h, \varepsilon)$ of balls $-h\xi^0 + B^{(j)}$ that satisfies the same property (31). Note also that

$$\mathfrak{S}(u_h, \varepsilon) = -h\xi^0 + \mathfrak{S}.$$

According to the definition of the measure $\mu_{u,\partial}$ we have

$$\begin{aligned}
\|\mu_{u,\partial} - \mu_{\mathfrak{L}_u^*,\partial}\|(R) &\leqslant \varlimsup_{h\to 0} \int_{(\partial K)_{1,R}} |u_h(x) - \mathfrak{L}_u^*(h\xi^0 + x)|\, d\sigma(x) \\
&\leqslant \varlimsup_{h\to 0} \int_{(\partial K)_{1,R}\cap\mathfrak{S}(\mathscr{Q}(u_h,\varepsilon))} |u_h(x)|\, d\sigma(x) \\
&\quad + \varlimsup_{h\to 0} \int_{(\partial K)_{1,R}\cap\mathfrak{S}(\mathscr{Q}(u_h,\varepsilon))} |\mathfrak{L}_u^*(h\xi^0 + x)|\, d\sigma(x) \\
&\quad + \varepsilon \int_{(\partial K)_{1,R}} |x|^\rho\, d\sigma(x) = I_1 + I_2 + I_3.
\end{aligned}$$

Let us estimate I_1. This estimation will be cumbersome. First of all just as it was done in the proof of Lemma 3, for the cone $\widetilde{K} = K$ and for an arbitrarily fixed point $x^0 \in \partial\Gamma$ let us construct a tangent hyperplane T_{x^0}, an operator $\rho = \pi_{x^0}$ of projection on T, a cone Y in T and a cone surface $\widetilde{Y} \subset \partial K$ such that $Y = \pi\widetilde{Y}$, where $\pi|_{\widetilde{Y}}$ is a diffeomorphism. We denote by $\vec{n}_x$ the unit interior normal vector to ∂K at the point x, by $\alpha(x', x'')$ the angle between $\vec{n}_{x'}$ and $\vec{n}_{x''}$. We set $\pi_Y = \pi|_{\widetilde{Y}}$, $\widetilde{E} = \pi^{-1}E$, where $E \subset T$. Remark that $|\alpha(x', x'')| < \delta$, $\forall\, x' \in \widetilde{Y}$, $x'' \in \widetilde{Y}$, where δ is an arbitrarily small positive number previously prescribed. It is easy to see that by virtue of the above assumptions

$$\operatorname{diam} \widetilde{E} \leqslant \operatorname{diam} E/\cos\delta, \quad \forall E \subset Y, \tag{32}$$

$$\operatorname{dist}(y, T_x) \leqslant c_3|\pi_x y - x|^2/|x|, \quad \forall\, x \in Y,\ y \in \widetilde{Y},\ |y - x| < |x/2|, \tag{33}$$

where c_3 is independent of both x and y.

Now let $Y' \Subset Y$ be an open cone with vertex at the origin.[5] Let us show that for the set $E_\varepsilon(u_h) \cap Y'$ a cover $\mathscr{Q}$ exists satisfying the condition (31) and having certain additional properties. In particular, the centers of the corresponding balls have to be somewhere near the set E_ε. To show this, let us consider the set $\mathfrak{S}' = \mathfrak{S}'(\mathscr{Q}(u_h, \varepsilon)) = \pi\mathfrak{S}$, where $\mathfrak{S} = \mathfrak{S}(\mathscr{Q}(u_h, \varepsilon))$. Since the radii of the balls $B^{(j)}$ and $\pi B^{(j)}$ coincide, condition (31) implies that the $(n-1)$-dimensional Lebesgue measure $\operatorname{meas}_{n-1}$ on the set $\mathfrak{S}' \cap B_R$ satisfies the condition

$$\lim_{R\to\infty} R^{-n+1} \operatorname{meas}_{n-1}(\mathfrak{S}' \cap B_R) = 0 \tag{33}$$

uniformly in h. Let us set

$$r(x) = r(x; h, \varepsilon) = \sup\{r : \{y \in T : |x - y| < r\} \subset \mathfrak{S}'\}, \qquad x \in \mathfrak{S}',$$
$$D(x) = D(x; h, \varepsilon) = \{y \in T : |y - x| < r(x)\}.$$

It follows from (33) that $\sup\{r(x) : |x| < R\} < \infty$, $\forall R < \infty$. Therefore we can apply the Besikovitch theorem on choosing covering of finite multiplicity (for example, see [11]). This theorem implies that a countable family $\mathscr{Q}'$ of balls $D^{(j)} = D(y^{(j)})$ exists covering the set $\mathfrak{S}' \cap Y'$, its multiplicity being no more than some $p_0 < \infty$. It is clear that

$$\lim_{|y^{(j)}|\to\infty} r(y^{(j)})/|y^{(j)}| = 0.$$

Hence without loss of generality we can assume that $D(y^{(j)}) \subset Y$, $\forall j$. It is obvious also that $\gamma^{(n-1)}(R; \mathscr{Q}') \leqslant c_4 \operatorname{meas}_{n-1}(\mathfrak{S}' \cap B_R)$. Therefore we see that $\lim_{R\to\infty} R^{-n+1}\gamma^{(n-1)}(R; \mathscr{Q}') = 0$ uniformly in h, $0 < h < 1$.

Let us denote by $\xi^{(j)}$ some point from $\overline{D}^{(j)} \cap \partial\mathfrak{S}'$ and let us set $\eta^{(j)} = \xi^{(j)} + r(y^{(j)})\vec{n}_{\xi^{(j)}}$. In view of properties (32) and (33), it is easy to see that the set $\widetilde{D}^{(j)}$ is contained in the intersection of the layer

$$M_j = \left\{x \in \mathbb{R}^n : |x - y| < c_3 \left(\frac{2r(y^{(j)})}{\cos\delta}\right)^2 \frac{1}{|\xi^{(j)}|},\ y \in T_{\xi^{(j)}}\right\}$$

and the ball $\mathscr{Q}_j = B_{R_j}(\eta^{(j)})$, where

$$R_j = 2r(y^{(j)})\left(\sqrt{1 + \frac{1}{\cos^2\delta}} + \frac{2c_3}{\cos^2\delta}\frac{r(y^{(j)})}{|\xi^{(j)}|}\right).$$

---

[5]By definition $Y' \Subset Y$ if $Y' \cap S_1 \Subset Y \cap S_1$.

Note that it follows from the above properties that the intersection $D_1^{(j)} = \partial K \cap B_{2R_j}(\eta^{(j)})$ is contained in the layer

$$M_j^1 = \{x \in \mathbb{R}^n : |x - y| < 4c_3 R_j^2/|\tilde{\xi}^{(j)}|,\ y \in T_{\tilde{\xi}^{(j)}}\}.$$

It can be easily seen that the ratio $|\xi^{(j)}|/|y^{(j)}|$ is bounded away from zero and infinity when $|y^{(j)}| > R_0$, where R_0 is large enough. Therefore the system $\mathscr{Q}''$ of balls $\mathscr{Q}_j$ covering the set $\mathfrak{S} \cap \partial K \cap \widetilde{Y}'$ satisfies condition (31). Besides, the inclusion $\pi\mathscr{Q} \subset Y$ is valid when $y^{(j)} > R_0'$, $R_0' > R_0$ and R_0' can be considered independent of h. For simplicity of notation let us assume that $\pi\mathscr{Q} \subset Y$, $\forall j$. Now let us consider the point $\widehat{\eta}^{(j)} = \tilde{\xi}^{(j)} + r(y^{(j)})\vec{n}_{x^0}$. Since $|\alpha(x^0, \tilde{\xi}^{(j)})| < \delta$, we have

$$|\widehat{\eta}^{(j)} - \eta^{(j)}| \leqslant r(y^{(j)})\delta.$$

At the same time $\pi\widehat{\eta}^{(j)} = \xi^{(j)}$ implies $\widehat{\eta}^{(j)} = E_\varepsilon(u_h)$. Thus in a sense there exists a point near the center of every ball from the system $\mathscr{Q}''$ which is outside of E_ε.

To estimate I_1, it is obviously sufficient to get the corresponding estimate of the value

$$I_1' = \int_{\widetilde{Y}' \cap \mathfrak{S}(\mathscr{Q}''(u_h, \varepsilon))} |u_h(x)|\, d\sigma(x).$$

This estimate must be independent of the choice of x_0. Using the above properties of the covering $\mathscr{Q}''$, we have

$$I_1' = \sum_{|y^{(j)}|<2R} \int_{\widetilde{Y}' \cap \mathscr{Q}_j} |u_h(x)|\, d\sigma(x) \leqslant \sum_{|y^{(j)}|<2R} \int_{\partial K \cap \mathscr{Q}_j} |u_h(x)|\, d\sigma(x). \tag{34}$$

To every ball $\mathscr{Q}_j$ assign the balls $\mathscr{Q}_{j,r} = B_r(\eta^{(j)}/R_j)$. It is obvious that

$$\int_{\partial K \cap \mathscr{Q}_j} |u_h(x)|\, d\sigma(x) = R_j^{n-1} \int_{\partial K \cap \mathscr{Q}_j} |u_h(R_j y)|\, d\sigma(y). \tag{35}$$

Note that by virtue of the above properties of the intersections $D_1^{(j)}$, the following is valid for the intersection $\mathscr{Q}_{j,2} \cap \partial K$:

For any $\delta_1 > 0$ there exists an $R' > 0$ such that for $|\tilde{\xi}^{(j)}| > R'$ the set $\mathscr{Q}_{j,2} \cap \partial K$ is contained in the layer

$$\widehat{M}_j = \{x \in \mathbb{R}^n : |x - y| < \delta_1,\ y \in T_{\tilde{\xi}^{(j)}}\}.$$

The distance from this layer to the center of the ball $\mathscr{Q}_{j,2}$, i.e., to the point $\eta^{(j)}/R_j$, is contained in the interval between the numbers

$$(1 + \cos^{-2}\delta)^{-1/2}/2 \quad \text{and} \quad \delta_2 = (1 + \cos^{-2}\delta)^{-1/2}/2 - \delta_1.$$

Now let us denote by $\Sigma_t = \Sigma_t^{(j)}$ the half-space that contains the point $\eta^{(j)}$, does not contain the point $\tilde{\xi}^{(j)}$ and is bounded by the hyperplane parallel to

$T_{\tilde{\xi}^{(j)}}$ located at a distance of t from the point $\eta^{(j)}/R$. It follows from the above property that $U_{j,t} \subset \mathscr{Q}_{j,2} \cap K$, $\forall t < \delta_2$, where $U_{j,t} \subset \mathscr{Q}_{j,2} \cap \Sigma_t$.

Now since the function $z = \pi_y^{-1}\xi$, $\xi \in T$, is positive homogeneous of degree 1, it follows that its second derivatives are positive homogeneous of degree -1. Hence they tend to zero as $|\xi| \to \infty$. Using this we can show that if $|\tilde{\xi}^{(j)}| > R''$, $R'' > R'$, where R'' is large enough, then for any point $x \in \partial K \cap \mathscr{Q}_{j,1}$ there exists an ellipsoid[6] of revolution $\Theta = \Theta_x$ such that its axis of revolution is directed along $\vec{n}_x$, its half-axes $a_1 = a$, $a_2 = \cdots = a_n = b$ are independent of x and j and $x \in \Theta_x$, $\Theta_x \subset K \cap \mathscr{Q}_{j,2}$, $\forall x \in \partial K \cap \mathscr{Q}_{j,1}$.

Let us denote by Θ_x^* the section of the ellipsoid Θ_x by the diametral hyperplane orthogonal to its axis of revolution. This section divides the ellipsoid into two parts. Let us denote by Θ_x^+ the part that is joined to x. It is clear that the above properties of the ellipsoids Θ_x will not change if the parameter b becomes less and the parameter a becomes greater (but no greater than $\frac{1}{2}\delta_2$). Hence we can choose a, b, δ, δ_1, $\delta_1 < \frac{1}{2}\delta_2$ so as to have the inclusion $\Theta_x^* \subset U_{j,\delta_2-\delta_1} \setminus U_{j,\delta_2-2\delta_1}$, $\forall x \in \partial K \cap \mathscr{Q}_{j,1}$.

Now let us denote by $\omega_{x'}(x)$ the harmonic measure of the set $\Theta_{x'}^*$ with respect to the domain $\Theta_{x'}^+$. In other words, $\omega_{x'}(x)$ is a function that is harmonic in $\Theta_{x'}^*$ and equals to zero on $\partial\Theta_{x'}^+ \setminus \overline{\Theta}_{x'}^+$. It is obvious that the value $\nu = \partial\omega_{x'}(x)/\partial n_x|_{x=x'}$ is independent of the choice of x'. Note also that $\nu > 0$.

Now let us denote by $\mathfrak{G}_j(x, y)$ Green's function of the domain $\mathscr{Q}_{j,2} \cap K$. The above-mentioned properties of the domains $\mathscr{Q}_{j,2} \cap K$ can be used to give a uniform (in j) lower estimate of the function $\partial\mathfrak{G}(\widehat{n}^{(j)}, y)/\partial n_y$ on the set $\mathscr{Q}_{j,1} \cap \partial K$. First of all note that

$$\inf\{\mathfrak{G}_j(x, y) : y \in \mathscr{Q}_{j,3/2} \cap (\Sigma_{\delta_2-\delta_1}^{(j)} \setminus \Sigma_{\delta_2-2\delta_1}^{(j)}), \\ |x - \eta^{(j)}/R_j| < \delta/2,\ |\xi^{(j)}| > R''\} \stackrel{\text{def}}{=} \delta_3 > 0. \quad (36)$$

Indeed, let $\mathfrak{G}_j^*(x, y)$ be the Green function of the domain U_{j,δ_2}. It is obvious that

$$\inf\{\mathfrak{G}_j^*(x, y) : y \in \mathscr{Q}_{j,3/2} \cap (\Sigma_{\delta_2-\delta_1}^{(j)} \setminus \Sigma_{\delta_2-2\delta_1}^{(j)}), \\ |x - \eta^{(j)}/R_j| < \delta/2,\ j \in \mathbb{N}\} = \delta_4 > 0.$$

It is also plain that $\mathfrak{G}_j(x, y) \geqslant \mathfrak{G}_j^*(x, y)$, $\forall x \in U_{j,\delta_2}$, $y \in U_{j,\delta_2}$.

Therefore inequality (36) holds when $\delta_3 \geqslant \delta_4$.

Now let us consider the harmonic function

$$\Phi(y) = \omega_{x'}(y) - \mathfrak{G}(x, y)/\delta_3$$

[6] Here an ellipsoid is not a surface but the corresponding open set.

in $\Theta^+_{x'}$ when x is fixed and $|x - \eta^{(j)}/R_j| \leqslant \delta/2$. It follows from estimate (35) that $\Phi(y) \leqslant 0$, $\forall y \in \Theta^+_{x'}$. Besides, it is obvious that $\Phi(y) \leqslant 0$, $\forall y \in \partial\Theta^+_{x'} \backslash \overline{\Theta}^+_{x'}$ and that $\sup_{\Theta^+_{x'}} \Phi(y) < \infty$. Hence by the maximum principle $\Phi(y) \leqslant 0$, $\forall y \in \partial\Theta^+_{x'}$. This implies $\partial\Phi/\partial n|_{y=x'} \leqslant 0$. Hence

$$\frac{\partial \mathfrak{G}_j(x, y)}{\partial n_y} \geqslant \delta_3 \frac{\partial \omega_{x'}(y)}{\partial n_y}\bigg|_{y=x'} = \nu\delta_3.$$

Now noting that

$$\left|\frac{\widehat{\eta}^{(j)}}{R_j} - \frac{\eta^{(j)}}{R_j}\right| \leqslant \frac{r(y^{(j)})}{R_j} 2\sin\frac{\delta}{2} \leqslant \frac{1}{2}\delta$$

we obtain

$$\partial \mathfrak{G}_j(\widehat{\eta}^{(j)}/R_j, yt)/\partial n_y \geqslant \nu\delta_3, \qquad \forall y \in \mathscr{Q}_{j,1} \cap \partial K, \ |\tilde{\xi}^{(j)}| > R''.$$

Using this inequality, we estimate the integral

$$\int_{\mathscr{Q}_{j,1}\cap\partial K} |u_h(R_j x)|\, d\sigma(x) = \int_{\mathscr{Q}_{j,1}\cap\partial K} u_h^+(R_j x)\, d\sigma(x) + \int_{\mathscr{Q}_{j,1}\cap\partial K} u_h^-(R_j x)\, d\sigma(x).$$

It follows from the Green representation of a subharmonic function that

$$\begin{aligned} u_h(\eta^{(j)}) &\leqslant \int_{\partial(\mathscr{Q}_{j,2}\cap K)} u_h(R_j y) \frac{\partial \mathfrak{G}_j(\eta^{(j)}, y)}{\partial n_y}\, d\sigma(y) \\ &= \int_{\partial(\mathscr{Q}_{j,2}\cap K)} u_h^+(R_j y) \frac{\partial \mathfrak{G}_j(\eta^{(j)}, y)}{\partial n_y}\, d\sigma(y) \\ &\quad - \int_{\partial(\mathscr{Q}_{j,2}\cap K)} u_h^-(R_j y) \frac{\partial \mathfrak{G}_j(\eta^{(j)}, y)}{\partial n_y}\, d\sigma(y). \end{aligned}$$

Therefore,

$$\begin{aligned} \nu\delta_3 \int_{\mathscr{Q}_{j,1}\cap\partial K} |u_h(R_j y)|\, d\sigma(y) &\leqslant \int_{\mathscr{Q}_{j,1}\cap\partial K} |u_h(R_j y)| \frac{\partial \mathfrak{G}_j(\eta^{(j)}, y)}{\partial n_y}\, d\sigma(y) \\ &\leqslant 2\int_{\partial(\mathscr{Q}_{j,1}\cap\partial K)} u_h^+(R_j y) \frac{\partial \mathfrak{G}_j(\eta^{(j)}, y)}{\partial n_y}\, d\sigma(y) - u_h(\eta^{(j)}) \\ &= \sup_{y\in\mathscr{Q}_{j,2}\cap K} u_h^+(R_j y) - u_h(\eta^{(j)}). \end{aligned}$$

Hence from the fact that $\eta^{(j)} \notin E_\varepsilon(u_h)$ and $|\mathfrak{L}_u^*(x)| \leqslant M|x|^\rho$, we have

$$\int_{\mathscr{Q}_{j,1}\cap\partial K} |u_h(R_j y)|\, d\sigma(y) \leqslant \frac{c_1}{\nu\delta_3}(|\eta^{(j)}| + 2R_j)^\rho + \frac{M}{\nu\delta_3}|\eta^{(j)}|^\rho \leqslant c_4|y^{(j)}|^\rho.$$

From this inequality and (34), (35), we obtain

$$I_1' \leqslant \sum_{|y^{(j)}|<2R} R_j^{n-1} c_4 |y^{(j)}|^\rho \leqslant c_5 R^\rho \sum_{|y^{(j)}|<2R} R_j^{n-1} \leqslant c_5 R^\rho \gamma^{(n-1)}(2R, \mathscr{Q}'').$$

Since the constant c_5 and the function $\gamma^{(n-1)}(2R, \mathscr{Q}'')$ may be considered independent of the choice of the point $x^0 \in \partial\Gamma$, the following estimate is also valid: $I_1 \leqslant c_6 R^\rho \gamma^{(n-1)}(2R, \mathscr{Q}'')$. Hence for any $h \in (0, 1)$ the estimate $I_1 \leqslant R^{\rho+n-1}\gamma_1(R)$ holds when $\gamma(R) \downarrow 0$ as $R \to \infty$ and is independent of the choice of h.

The value I_2 can be estimated using the inequality $|\mathfrak{L}_u^*(x)| \leqslant M|x|^\rho$. Namely,

$$\begin{aligned} I_2 &\leqslant (R+h)^\rho \operatorname{meas}_{n-1}((\partial K)_{1,R} \cap \mathfrak{S}(\mathscr{Q}(u_h, \varepsilon))) \\ &\leqslant c_7 (R+h)^\rho \gamma^{(n-1)}(2R; \mathscr{Q}(u_h, \varepsilon)). \end{aligned}$$

Therefore there exists a function $\gamma_2(R)$ independent of h such that $\gamma_2(R) \downarrow 0$ as $R \to \infty$ and for any $h \in (0, 1)$ $I_2 \leqslant \gamma_2(R) R^{\rho+n-1}$. The estimate of I_3 is obvious. Namely, $I_3 \leqslant \varepsilon c_8 R^{\rho+n-1}$. Summing the results obtained, we get

$$\|\mu_{u,\partial} - \mu_{\mathfrak{L}_u^*,\partial}\|(R) \leqslant (\gamma_1(R) + \gamma(R) + \varepsilon c_8) R^{\rho+n-1}.$$

Consequently

$$\lim_{R\to\infty} \frac{1}{R^{-\rho-n+1}} \|\mu_{u,\partial} - \mu_{\mathfrak{L}_u^*,\partial}\|(R) = 0,$$

i.e., the function $u(x)$ satisfies the condition A. This completes the proof of the theorem.

§3. Sufficient conditions of completely regular growth in $\overline{K}$

The main result of this section is the following theorem.

THEOREM 6. *Let a function $u \in SH(K, \rho]$ be of completely regular growth (c.r.g.) in K and satisfy the condition A. Then it is a function of c.r.g. in the closed cone $\overline{K}$.*

To prove this theorem, let us use the ε-normal points method (see [12]) that we shall modify here. We need some lemmas, where as above without special reservation we shall denote by $c_j(\cdot, \cdot, \cdot)$, $j = 1, 2, \dots$, positive constants depending on parameters. Besides we shall denote by $\theta(t)$, $\theta_1(t)$, $\theta_2(t), \dots$ positive functions monotonically tending to zero as $t \to \infty$. Just as in the case of the constants c_j, the numeration of the functions $\theta_j(t)$ in each lemma differs from that in the other ones.

A point $x \in K$ is referred to as an (ε, ρ)-*normal point* of a real measure τ, defined on $\overline{K} \setminus \{0\}$, if the inequality

$$\int_{(\overline{K})_{0,s}} |y|^{2-n-\rho} d|\tau|(y) \leqslant \varepsilon (s/|x|)^{n-1}$$

is valid at this point for any $s < |x|/2$. The set of all (ε, ρ)-normal points of the measure τ is denoted by $\Delta_\varepsilon(\tau, \varepsilon) = \Delta(\tau, \varepsilon)$.

LEMMA 5. *Let a real measure τ defined on $\overline{K}$ satisfy the conditions: $\operatorname{supp}\tau \subset \overline{K}_{1,\infty}$ and*

$$\|\tau\|(t) \leqslant (\sigma + \theta_1(t))t^{\rho+n-2}, \tag{37}$$

where $\rho \notin \{\varkappa_j^+\}_{j=0}^{\infty}$. Further let q be a number defined by the inequality $\varkappa_q^+ < \rho < \varkappa_{q+1}^+$. Then the estimate

$$\left|\int_{\overline{K}} W_q(x, y)\, d\tau(y)\right| \leqslant c_1(\sigma + \varepsilon + \theta_2(|x|))|x|^\rho \tag{38}$$

is valid on the set $\Delta_\rho(\tau, \varepsilon)$ for some c_1 and $\theta_2(t)$ depending on the choice of the function θ_1 and the number q only.

PROOF. Let p_1, p_2, p_3 be numbers such that $0 < p_1 < 1 < p_2$, $0 < p_3 < 1$. We put each $x \in (K)_{1/p_1,\infty}$ into correspondence with a partition of the cone $\overline{K}$ into the sets

$$\Omega_0 = (\overline{K})_{0,1}, \quad \Omega_1 = \Omega_1(x) = \{y \in \overline{K} : |y| > p_2|x|\},$$
$$\Omega_2 = \Omega_2(x) = \{y \in \overline{K} : 1 \leqslant |y| < p_1|x|\},$$
$$\Omega_3 = \Omega_3(x) = \{y \in \overline{K} : p_1|x| < |y| < p_2|x|,\ |x/|x| - y/|y|| \geqslant p_3\},$$
$$\Omega_4 = \Omega_4(x) = \{y \in \overline{K} : p_1|x| < |y| < p_2|x|,\ |x/|x| - y/|y|| \leqslant p_3\}.$$

Using this partition, we get

$$\int_{\overline{K}} W_q(x, y)\, d\tau(y) = \sum_{j=1}^{4} \int_{\Omega_j} W_q(x, y)\, d\tau(y).$$

Let us estimate the integral over Ω_1. Using estimates (5) and (13) and integrating by parts, we obtain

$$\begin{aligned}\int_{\Omega_1} |W_q(x, y)|\, d|\tau|(y) &\leqslant c_1\|\varphi\|_\infty |x|^{\varkappa_{q+1}^+} \int_{\Omega_1} |y|^{\varkappa_{q+1}^-} d|\tau|(y) \\ &\leqslant c_1\|\varphi\|_\infty \int_{p_2|x|}^{\infty} t^{\varkappa_{q+1}^-}\, d\|\tau\|(y) \leqslant c_2(\sigma + \theta_2(|x|))|x|^\rho. \end{aligned} \tag{39}$$

Similarly, using estimates (37) and (12), we get

$$\int_{\Omega_2} |W_q(x, y)|\, d|\tau|(y) \leqslant c_3(\sigma + \theta_3(|x|))|x|^\rho. \tag{40}$$

In view of estimates (37) and (16), it is also obvious that

$$\int_{\Omega_3} |W_q(x, y)|\, d|\tau|(y) \leqslant c_4 \int_{(\overline{K})_{p_1|x|, p_2|x|}} |y|^{2-n}\, d|\tau|(y) \leqslant c_5(\sigma + \theta_4(|x|))|x|^\rho. \tag{41}$$

Finally let us estimate Ω_4. We represent Ω_4 in the form $\Omega_4 = \Omega' + \Omega''$, where $\Omega' = \Omega_4 \cap B_{d/4}$, $d = d(x) = \operatorname{dist}(x, \partial K)$. Now let us represent Ω' and Ω'' in the form

$$\Omega' = \bigcup_{k=1}^{\infty} \Omega'_k, \qquad \Omega'' = \bigcup_{k=1}^{\infty} \Omega''_k,$$

where

$$\Omega'_k = \{y \in \Omega' : 2^{-k-1} d \leqslant |y - x| \leqslant 2^{-k} d\},$$
$$\Omega''_k = \{y \in \Omega'' : 2^{k-2} d \leqslant |y - x| \leqslant 2^{k-1} d\}.$$

It follows from (14) that

$$\int_{\Omega'_k} |W_q(x, y)|\, d|\tau|(y) \leqslant c_6 \int_{\Omega'_k} |x|^{\varkappa_q^+} |y|^{\varkappa_q^-}\, d|\tau|(y) + \int_{\Omega'_k} \frac{d|\tau|(y)}{\varphi(y)|x - y|^{n-2}}$$
$$= c_6 I_k^{(1)} + I_k^{(2)}.$$

Note that

$$\sum_{k=1}^{\infty} I_k^{(1)} = \int_{B_{d/2} \cap \Omega_4} |x|^{\varkappa_q^+} |y|^{\varkappa_q^-}\, d|\tau|(y) \leqslant \int_{p_1|x|}^{p_2|x|} |x|^{\varkappa_q^+} |y|^{\varkappa_q^-}\, d|\tau|(y).$$

Estimating the latter integral in a standard way we obtain

$$\sum_{k=1}^{\infty} I_k^{(1)} \leqslant c_7(\sigma + \theta_5(|x|))|x|^{\rho}. \tag{42}$$

To estimate $I_k^{(2)}$, let us use the fact that, according to the assertion of the lemma, $x \in \Delta_\rho(\tau, \varepsilon)$. Hence

$$\int_{\widetilde{K} \cap B_s(x)} |y|^{2-n} d|\tau|(y) \leqslant \varepsilon s^{n-1} |x|^{1-n}, \qquad \forall s < |x|/2. \tag{43}$$

Now let us take into account the inequalities

$$d(x)/2 \leqslant d(y) \leqslant 3d(x)/2, \qquad \forall y \in B_{d(x)/2}(x),$$

and the fact that, in view of the above assumption on the cone K, the function $\varphi(x)$ satisfies the condition $c_8 d(x) < |x|\varphi(x) < c_9 d(x)$. This condition and (43) implies

$$I_k^{(2)} \leqslant \int_{2^{-k-1} d \leqslant |y-x| \leqslant 2^{-k} d} \frac{d|\tau|(y)}{\varphi(y)|x - y|^{n-2}}$$
$$\leqslant \left(\frac{2^{k+1}}{d}\right)^{n-2} \frac{2(p_2|x|)^{\rho+n-2}}{c_8 d} \int_{2^{-k-1} d \leqslant |y-x| \leqslant 2^{-k} d} |y|^{-\rho-n+2} d|\tau|(y)$$
$$\leqslant (\varepsilon/c_8) 2^{n-2} p_2^{\rho+n-1} 2^{-k} |x|^{\rho} = c_9 \varepsilon 2^{-k} |x|^{\rho}.$$

Hence

$$\sum_{k=1}^{\infty} I_k^{(2)} \leqslant c_9 \varepsilon |x|^{\rho}.$$

From this and (42) we conclude that

$$\int_{\Omega'} |W_q(x, y)|\, d|\tau|(y) \leqslant (c_9\varepsilon + c_6c_7(\sigma + \theta_5(|x|)))|x|^\rho. \tag{44}$$

The integral $\int_{\Omega'}$ can be estimated similarly using inequalities (13) and (43), namely,

$$\begin{aligned}
&\int_{\Omega''} |W_q(x, y)|\, d|\tau|(y) \\
&\leqslant c_{10}\int_{p_1|x|}^{p_2|x|} |x|^{\varkappa_q^+}|y|^{\varkappa_q^-}\, d|\tau|(y) + c_{11}\sum_{k=1}^{\infty}\int_{\Omega_k''} \frac{\varphi(x)|x||y|}{|x-y|^n}\, d|\tau|(y) \\
&\leqslant c_{10}c_7(\sigma + \theta_5(|x|))|x|^\rho \\
&\qquad + c_{11}\sum_{k=1}^{\infty} \frac{p_2^{\rho+n-1}|x|^{\rho+n-1}3d/2}{d^n 2^{(k-2)n}} \int_{(\overline{K})_{2^{k-1}d}} |y|^{-\rho-n+2}\, d|\tau|(y) \\
&\leqslant c_{10}c_7(\sigma + \theta_5(|x|))|x|^\rho + c_{11}|x|^\rho p_2^{\rho+n-1}\frac{3}{2}2^{n+1}\sum_{k=1}^{\infty}\frac{1}{2^k} \\
&= (c_{10}c_7(\sigma + \theta_5(|x|))c_{12}\varepsilon)|x|^\rho.
\end{aligned}$$

Comparing the obtained estimate with (44), we see that for some constants c_{13} and c_{14}

$$\int_{\Omega_4} |W_q(x, y)|\, d|\tau|(y) \leqslant (c_{13}(\sigma + \theta_5(|x|)) + c_{14}\varepsilon)|x|^\rho.$$

This estimate together with (39)–(41) leads us to the inequality

$$\int_{\overline{K}} |W_q(x, y)|\, d|\tau|(y) \leqslant (c_{15}(\varepsilon + \sigma + \theta_6(|x|))\varepsilon)|x|^\rho, \quad \forall x \in \Delta_\rho(\tau, \varepsilon),$$

and consequently to inequality (38).

The proof of the lemma is completed.

Lemma 5 is not true in the case $\rho \in \{\varkappa_j^+\}_{j=1}^\infty$. The following lemma is the corresponding analog of Lemma 5.

LEMMA 6. *Let a real measure* τ *defined on* $\overline{K}$ *satisfy the conditions*: $\operatorname{supp}\tau \subset (\overline{K})_{1,\infty}$ *and*

$$\|t\|(R) \leqslant (\sigma + \theta_1(R))R^{\rho+n-2}, \qquad R \to \infty, \tag{45}$$

where $\rho \in \{\varkappa_j^+\}_{j=1}^\infty$. *Further let the integers* q *and* p *satisfy the inequalities* $\varkappa_{q-1}^+ < \varkappa_q^+ = \cdots = \varkappa_p^+ = \rho < \varkappa_p^+ + 1$. *Then on the set* $\Delta_\rho(\tau, \varepsilon)$ *the inequality*

$$\left|\int_{(\overline{K})_{t_1|x|}} W_{q-1}(x, y)\, d\tau(y) + \int_{(\overline{K})_{t_1|x|,\infty}} W_p(x, y)\, d\tau(y)\right| \leqslant c(\varepsilon + \sigma + \theta_2(|x|))|x|^\rho$$

is valid for some constant c *and a function* $\theta_2(R)$ *independent of* ε.

We omit the proof of this lemma. It is similar to that of Lemma 5.

Now let us estimate the "mass" of the set of points which belong to K and are not (ε, ρ)-normal ones.

LEMMA 7. *Let τ be a real measure defined on $\overline{K}$, with support in $(\overline{K})_{1,\infty}$. Further let ρ be an arbitrary positive number. Then the set $K\setminus\Delta_\rho(\tau, \varepsilon)$ can be covered by a countable system of balls $B^{(j)} = B_{r_j}(x^{(j)})$ in such a way that*

$$\sum_{|x^{(j)}|<R}\left(\frac{r_j}{|x^{(j)}|}\right)^{n-1} \leqslant \frac{c}{\varepsilon}\int_1^{4R}\frac{d\|\tau\|(s)}{s^{\rho+n-2}}, \qquad c = c(n).$$

PROOF. According to the definition of the set $\Delta_\rho(\tau, \varepsilon)$ for any point $x \in K\setminus\Delta_\rho(\tau, \varepsilon)$ a ball $B_r(x)$ with $r = r(x) < |x|/2$ exists such that

$$\int_{\overline{K}\cap B_r(x)}\frac{d|\tau|(y)}{|y|^{\rho+n-2}} > \varepsilon\frac{r^{n-1}}{|x|^{n-1}}.$$

Consider the set of all balls $B_{r(x)}(x)$ whose centers belong to

$$\Delta^k = (K\setminus\Delta_\rho(\tau, \varepsilon))\cap(K)_{4^k, 4^{k+1}}, \qquad k = 1, 2, \dots .$$

Using the above-mentioned theorem of Besikovitch on coverings of finite multiplicity, we choose from this set of balls a countable subset $\mathscr{Q}^k$ forming a covering of Δ^k of the finite multiplicity $\chi = \chi(n)$. Since $r(x) \leqslant |x|/2$, the balls belonging to $\mathscr{Q}^k$ do not intersect the balls belonging to $\mathscr{Q}^{k+2}$. Therefore the union of all such systems $\mathscr{Q}^k$ forms a no more than countable covering of the set $K\setminus\Delta_\rho(\tau, \varepsilon)$ of multiplicity $\leqslant 3d$. Let us denote by $\widetilde{B}^{(j)}$ elements of this covering, i.e., balls from $\bigcup_k \mathscr{Q}^k$, and by $\tilde{x}^{(j)}$ and $\tilde{r}_j$ their centers and radii.

Let $4^k < R < 4^{k+1}$. Then we have

$$\begin{aligned}\sum_{|\tilde{x}^{(j)}|<R}\left(\frac{\tilde{r}_j}{|\tilde{x}^{(j)}|}\right)^{n-1} &\leqslant \sum_{l=1}^{k}\sum_{\tilde{x}_j\in\Delta^l}\left(\frac{\tilde{r}_j}{|\tilde{x}^{(j)}|}\right)^{n-1} \leqslant \frac{1}{\varepsilon}\sum_{l=1}^{k}\sum_{\tilde{x}_j\in\Delta^l}\int_{\overline{K}\cap B^{(j)}}\frac{d|\tau|(y)}{|y|^{\rho+n-2}}\\ &\leqslant \frac{3\chi}{\varepsilon}\sum_{l=1}^{k}\int_{(\overline{K})_{4^{l+1}, 4^{l+2}}}\frac{d|\tau|(y)}{|y|^{\rho+n-2}} \leqslant \frac{9\chi}{\varepsilon}\int_{(\overline{K})_{4^{l+2}}}\frac{d|\tau|(y)}{|y|^{\rho+n-2}}\\ &\leqslant \frac{9\chi}{\varepsilon}\int_1^{4R}\frac{d\|\tau\|(s)}{s^{\rho+n-2}}.\end{aligned}$$

This completes the proof of the lemma.

A slight change of the reasoning used in the proof of Lemma 7 leads us to the following statement.

LEMMA 8. *Let a measure t satisfy the assumption of Lemma 5 with an arbitrary $\rho > 0$. Then the set $K\setminus\Delta_\rho(\tau, \varepsilon)$ can be covered by a system of balls $B^{(j)} = B_{r_j}(x^{(j)})$ such that*

$$R^{-n+1}\sum_{|x^{(j)}|<R} r_j^{n-1} \leqslant c(\sigma + \theta_2(R))/\varepsilon, \qquad c = c(n).$$

PROOF. Let $\widetilde{B}^{(j)}$, $\widetilde{x}^{(j)}$ and $\tilde{r}_j$ be the same as in Lemma 7. Then for $4^k < R \leqslant 4^{k+1}$, $k = k(R)$, the following inequalities are valid:

$$\begin{aligned} R^{-n+1} \sum_{|x^{(j)}<R} \tilde{r}_j^{n-1} &\leqslant \frac{1}{4^{k(n-1)}} \sum_{l=1}^{k} 4^{(l+1)(n-1)} \sum_{\widetilde{x}_j \in \Delta^l} \left(\frac{\tilde{r}_j}{|\widetilde{x}^{(j)}|} \right)^{n-1} \\ &\leqslant \frac{3\chi}{4^{k(n-1)}} \sum_{l=1}^{k} 4^{(l+1)(n-1)} \int_{(\overline{K})_{4^{l-1},4^{l+1}}} |y|^{2-\rho-n} d|\tau|(y) \\ &= \frac{3\chi}{4^{k(n-1)}} \sum_{l=1}^{k} 4^{(l+1)(n-1)} \int_{4^{l-1}}^{4^{l+1}} s^{2-\rho-n} d\|\tau\|(s) \\ &\leqslant \frac{3\chi}{4^{k(n-1)}} \sum_{l=1}^{k} 4^{(l+1)(n-1)} \left\{ \frac{\|\tau\|(4^{l+1})}{4^{(l+1)(\rho+n-1)}} + (\rho+n-1) \int_{4^{l-1}}^{4^{l+1}} \frac{\sigma+\theta_1(s)}{s} ds \right\} \\ &\leqslant \frac{3\chi 4^{n-1}}{4^{k(n-1)}} \sum_{l=1}^{k} 4^{l(n-1)} \{ \sigma(\rho+n) + \theta_1(4^{l+1}) + (\rho+n-1) \\ &\qquad + \ln 16 \cdot (\sigma + \theta_1(4^{l-1})) \}. \end{aligned}$$

Since

$$\sum_{l=1}^{k} 4^{(l-k)(n-1)} (\theta_1(4^{l+1}) + (\rho+n-1) + \ln 16 \cdot \theta_1(4^{l-1})) \to 0,$$

as $k \to \infty$ or, which is the same, as $R \to \infty$, the right-hand side of the inequality obtained is of the form $c(\sigma + \theta_2(R))$.

The proof of the lemma is complete.

The following lemma is the consequence of Lemmas 8 and 5.

LEMMA 9. *Let a measure τ defined on K satisfy the conditions:* $\operatorname{supp}\tau \subset (\overline{K})_{1,\infty}$ *and*

$$\lim_{R\to\infty} R^{2-\rho-n} \|\tau\|(R) = 0, \qquad \overset{+}{\varkappa}_q < \rho < \overset{+}{\varkappa}_{q+1}.$$

Then the set $E_\varepsilon(U)$ constructed over the function

$$U(x) = \int_K W_q(x, y)\, d\tau(y)$$

is a C_0^{n-1}-set for any $\varepsilon > 0$.

Indeed, by Lemma 5, the estimate $|U(x)| \leqslant c_1(\varepsilon + \theta_1(|x|))|x|^\rho$ holds on the set $\Delta(\tau, \varepsilon)$. Hence for R_0 large enough, $E_\varepsilon(U) \backslash B_{R_0} \subset K \backslash \Delta(\tau, \varepsilon/2)$. At the same time by Lemma 8 the set $K \setminus \Delta(\tau, \varepsilon)$ is a C_0^{n-1}-set for any $\varepsilon > 0$. Consequently $E_\varepsilon(U)$ is a C_0^{n-1}-set.

Now let us pass on directly to the proof of Theorem 6. First we consider the case $\overset{+}{\varkappa}_q < \rho < \overset{+}{\varkappa}_{q+1}$. It follows from Theorem 2 that in this case the function

$u(x)$ appearing in the theorem that we are proving can be represented in the form $u(x) = U_1(x) + U_2(x) + U_3(x)$, where

$$U_1(x) = \int_{K\setminus B_1} \mathfrak{S}_q(x, y)\, d\mu_u(y), \quad U_2(x) = -\frac{1}{\theta_n}\int_{\partial K\setminus B_1} \frac{\partial \mathfrak{S}_q(x, y)}{\partial n_y}\, d\mu_{u,\partial}(y),$$

and $U_3(x)$ satisfies the condition

$$\lim_{|x|\to\infty,\, x\in K} |U_3(x)||x|^{-\rho} = 0.$$

From the above representation of $u(x)$ it follows that in this situation to prove growth regularity of the initial function in K it is sufficient to show that each of the functions $U_1(x)$ and $U_2(x)$ separately is of c.r.g. in $\overline{K}$. First let us consider $U_2(x)$. According to the assumption of Theorem 6, the measures $\mu_{u,\partial}^{[R]}$ defined by equality (10) converge in variation to the measure $\widehat{\mu}_{u,\partial} = \mu_{\mathfrak{L}^*_{u,\partial}}$ as $R\to\infty$. This measure is positive homogeneous of degree $\rho+n-1$, i.e.,

$$\widehat{\mu}_{u,\partial}(tE) = t^{\rho+n-1}\widehat{\mu}_{u,\partial}(E), \qquad \forall t>0,\ E\subset\overline{K}.$$

Besides, it is easy to see that $\mathfrak{S}_q(tx, y) = t^{n-2}\mathfrak{S}_q(x, y/t)$. Therefore the potential

$$J_2(x) = -\frac{1}{\theta_n}\int_{\partial K} \frac{\partial \mathfrak{S}_q(x, y)}{\partial n_y}\, d\widehat{\mu}_{u,\partial}(y)$$

is a positive homogeneous function of degree ρ. It is also clear that J_2 is harmonic in K and Theorem 2 implies that its boundary measure coincides with $d\widehat{\mu}_{u,\partial}$. Now let us consider the function

$$\Phi(x) = U_2(x) - J_2(x) = -\frac{1}{\theta_n}\int_{\partial K} \frac{\partial \mathfrak{S}_q(x, y)}{\partial n_y}\, d\nu(y),$$

where $\nu = \mu_{u,\partial} - \widehat{\mu}_{u,\partial}$ on $(\partial K)_{1,\infty}$ and $\nu = -\widehat{\mu}_{u,\partial}$ on $(\partial K)\cap\overline{B}_1$.

It is obvious that $\mu_{\Phi,\partial} = \nu$ and since the measures $\mu_{u,\partial}^{[R]}$ converge in variation to $\widehat{\mu}_{u,\partial}$ as $R\to\infty$, the measure ν satisfies the condition

$$\lim_{R\to\infty} R^{-\rho-n+1}\|\nu\|(R) = 0.$$

Therefore, by Lemma 9, the set $E^*_\varepsilon = \{x : |U_2(x) - J_2(x)| > \varepsilon|x|^\rho\}$ is a C_0^{n-1}-set for any $\varepsilon > 0$. Now it is easy to see that $J_2(x) = \mathfrak{L}^*_{U_2}(x)$ and hence the above-mentioned characteristics of the set E^*_ε implies the growth regularity of the function $U_2(x)$ in $\overline{K}$.

To prove the growth regularity of $U_1(x)$, let us introduce the measure μ_1 defined by the equality $d\mu_1(x) = \varphi(x)\, d\mu_u(x)$. Using this measure, let us construct a family $\mu_1^{[R]}$ in the following way:

$$\mu_1^{[R]}(E) = R^{-\rho-n+2}\mu_1(E), \qquad \forall E\subset K.$$

The convergence of the measures $\mu_{u,\partial}^{[R]}$ to $\widehat{\mu}_{u,\partial}$ and the fact that the measure τ_u has cone-boundary density (see Theorem 3) imply that the measures $\mu_1^{[R]}$ weakly converge in $C((\overline{K})_{0,r})$, $\forall r > 0$, as $R \to \infty$ to the measure $\widehat{\mu}_1$ defined on $\overline{K}$ by the equalities

$$\widehat{\mu}_1(E) = \int_{E\cap K} \varphi(x) d\widehat{\mu}_u, \qquad E \subset \overline{K},$$

where $\widehat{\mu}_u = \mu_{\mathfrak{L}_u^*}$.

Note that $\widehat{\mu}_u(\partial K) = 0$.

Let us set

$$\Omega'_\delta = \{x \in K : \varphi(x) < \delta\}, \qquad \Omega'_{\delta,R} = \Omega'_\delta \cap B_R,$$
$$\Omega''_\delta = \{x \in K : \varphi(x) \geqslant \delta t\}, \qquad \Omega''_{\delta,R} = \Omega''_\delta \cup B_R.$$

It follows from the properties of the measures $\mu_1^{[R]}$ and $\widehat{\mu}_1$ that for any $\varepsilon > 0$ there exist $\delta_0 = \sigma_0(\varepsilon, r)$ and $\theta_1(R) = \theta_1(R, \delta_0, \varepsilon)$ such that

$$\widehat{\mu}_1(\Omega'_{\delta,R}) \leqslant \varepsilon R^\rho, \qquad \mu_1^{[R]}(\Omega'_{\delta,1}) \leqslant \varepsilon + \theta_1(R),$$

for all $R > 0$, $\delta < \delta_0$.

Now let us represent the measure μ_1 in the form $\mu_1 = \mu'_1 + \mu''_1$, where μ'_1 is the restriction of μ_1 to the cone Ω'_δ and μ''_1 is the restriction of μ_1 to the cone Ω''_δ.

In accordance to this, represent the potential U_1 in the form

$$U_1 = U'_1 + U''_1,$$

where

$$U'_1 = \int_{\Omega'_\delta \setminus B_1} \mathfrak{S}_q(x, y)\, d\mu_u = \int_{(K)_{1,\infty}} W_q(x, y)\, d\mu'_1,$$
$$U''_1 = \int_{\Omega''_\delta \setminus B_1} \mathfrak{S}_q(x, y)\, d\mu_u = \int_{(K)_{1,\infty}} W_q(x, y)\, d\mu''_1.$$

Let us consider the potential U'_1. Lemmas 8 and 5 imply that for any $\varepsilon_1 > 0$ the set of the points $x \in K$ in which $|U'_1(x)| \geqslant \varepsilon_1 |x|^\rho$ can be covered by a system $\mathscr{Q}'$ of balls $B_{r_j}(x^{(j)})$ such that for all sufficiently large R we have

$$R^{-n+1}\gamma^{(n-1)}(R; \mathscr{Q}') \leqslant c_1 \varepsilon / \varepsilon_1,$$

where c_1 is a constant independent of ε and ε_1 and $\gamma^{(n-1)}(R; \mathscr{Q}')$ is the same function as above (see (31)).

To study the behavior of U''_1, which is more complicated, let us consider the cone $\widetilde{K} = K^{\widetilde{\Gamma}}$ parallel to $K = K^\Gamma$, $\Gamma \Subset \widetilde{\Gamma}$, $\partial\widetilde{\Gamma} \in C^2$. The numbers $\widetilde{\varkappa}_j^\pm$ and the functions $\widetilde{\varphi}$, $\widetilde{\mathfrak{S}}$, $\widetilde{\mathfrak{S}}_q$, $j = 0, 1, \ldots$, are introduced accordingly.

Let us assume that the domain $\widetilde{\Gamma}$ is so close to Γ that the inequality $\widetilde{\varkappa}_q^+ < \rho < \widetilde{\varkappa}_{q+1}^+$ is valid with the same q as in the condition $\widetilde{\varkappa}_q^+ < \rho < \widetilde{\varkappa}_{q+1}^+$. Now we represent U_1'' in the form $U_1'' = U_{1,1}'' + U_{1,2}''$, where

$$U_{1,1}'' = \int_{\Omega_\delta''\setminus B_1} \{\mathfrak{S}_q(x, y) - \widetilde{\mathfrak{S}}_q(x, y)\}\, d\mu_u(y),$$

$$U_{1,2}'' = \int_{\Omega_\delta''\setminus B_1} \widetilde{\mathfrak{S}}_q(x, y)\, d\mu_u(y).$$

It is obvious that the function $U_{1,1}''$ is harmonic in K. Let us show also that $U_{1,1}'' \in SH(K, \rho]$. Lemma 2 implies the following estimate for $t_1|x| < |y| < t_2|x|$:

$$|\mathfrak{S}_q(x, y) - \widetilde{\mathfrak{S}}_q(x, y)| \leqslant c_2 \max\{|x|^{\varkappa_q^+}|y|^{\varkappa_q^-}, |x|^{\widetilde{\varkappa}_q^+}|y|^{\widetilde{\varkappa}_q^-}\} \leqslant c_3|x|^{-n+2}.$$

From this in the standard way and taking into account the estimate of the function $\|\tau_u\|(R)$ that follows from Theorem 1, we obtain

$$\int_{(\Omega_\delta'')_{t_1|x|, t_2|x|}} |\mathfrak{S}_q(x, y) - \widetilde{\mathfrak{S}}_q(x, y)|\, d\mu_u(y) \leqslant c_4|x|^\rho.$$

This estimate and the estimates of the integrals $\int_{\Omega_1}$ and $\int_{\Omega_2}$ obtained in the proof of Lemma 5 imply $|U_{1,1}''(x)| \leqslant c_5|x|^\rho$. Hence $U_{1,1}'' \in SH(K, \rho]$.

The fact that $U_{1,2}''$ belongs to $SH(K, \rho]$ is obvious. It is plain also that the measure $\mu_{U_{1,2}''}$ coincides with μ_u in Ω_δ'' and vanishes in Ω_δ'. Besides, $\mu_{U_{1,2}'', \partial} = 0$ (on $\partial\widetilde{K}$). Hence, taking into account the fact that μ_u has cone-boundary density in K and using Theorem 3, we conclude that $U_{1,2}''$ is a function of c.r.g. in $\widetilde{K}$ and hence this is true in $\overline{K}$ as well.

To show that $U_{1,1}''$ is also a function of c.r.g. in $\overline{K}$, note that

$$d\mu_{U_{1,1}'', \delta} = -U_{1,2}''\, d\sigma, \qquad \mu_{U_{1,1}''} = 0.$$

Hence (see Theorem 2),

$$U_{1,1}'' = \frac{1}{\theta_n}\int_{(\partial K)_{1,\infty}} \frac{\partial\mathfrak{S}_q(x, y)}{\partial n_y} U_{1,2}''\, d\sigma + \mathscr{Q}(x),$$

where $\mathscr{Q}(x)$ is a f.c.r.g. in $\overline{K}$.

According to the fact proved above, the function $U_{1,2}''$ is of c.r.g. in $\overline{K}$. So by Theorem 4 it satisfies condition A (with respect to K). Hence, just as the function U_2, the potential

$$-\frac{1}{\theta_n}\int_{(\partial K)_{1,\infty}} \frac{\partial\mathfrak{S}_q(x, y)}{\partial n_y} U_{1,2}''\, d\sigma$$

is a function of c.r.g. in $\overline{K}$. Therefore the above representation of $U_{1,1}''$ implies that $U_{1,1}''$ is a function of c.r.g. in $\overline{K}$. Consequently the function

$U_1'' = U_{1,1}'' + U_{1,2}''$ is of c.r.g. in $\overline{K}$, too. Thus the set $E_{\varepsilon_1}(U_1'')$ can be covered by a system $\mathscr{Q}''$ of balls $B_{r_j'}(\xi^{(j)})$ such that

$$\lim_{R\to\infty} R^{-n+1}\gamma^{(n-1)}(R;\mathscr{Q}'') = 0.$$

Besides, it follows from the representation of the indicator of functions of c.r.g. in K (see [4]) that

$$\mathfrak{L}^*_{U_1''} = \int_{\Omega_\delta''} \mathfrak{S}_q(x, y)\, d\mu_{\mathfrak{L}^*_u}(y) = \int_{\Omega_\delta''} W_q(x, y)\, d\widehat{\mu}_1(y).$$

Further it follows from the estimate $\widehat{\mu}_1^{[R]}(\Omega_{\delta,1}') \leqslant \varepsilon R^\rho$ and Lemmas 5 and 8 that for any $\varepsilon_1 > 0$ the set of all points $x \in K$ where

$$\left|\int_{\Omega_\delta'} W_q(x, y)\, d\widehat{\mu}_1(y)\right| > \varepsilon_1 |x|^\rho$$

can be covered by a system of balls $B_{r_j''}(\eta^{(j)})$ so that for any sufficiently large R

$$R^{-n+1}\sum_{|\xi^{(j)}|<R} (r_j'')^{n-1} \leqslant c_1\varepsilon/\varepsilon_1.$$

Now let us sum up all the above concerning the functions U_1' and U_1''. We see that for any prescribed ε and ε_1 outside of some set that can be covered by a system of balls $\mathscr{Q}''' = \{B_{\lambda_j}(\widehat{\xi}^{(j)})\}$, that satisfies the condition

$$R^{-n+1}\gamma^{(n-1)}(R;\mathscr{Q}''') \leqslant 2c_1\varepsilon/\varepsilon_1 + \theta_2(R),$$

we have the following inequality

$$\left|U_1(x) - \int_K W_q(x, y)\, d\widehat{\mu}_1(y)\right| \leqslant |U_1'(x)| + \left|\int_{\Omega_\delta'} W_q(x, y)\, d\widehat{\mu}_1(y)\right| + \left|U_1''(x) - \int_{\Omega_\delta''} W_q(x, y)\, d\widehat{\mu}_1(y)\right| \leqslant 3\varepsilon_1 |x|^\rho.$$

Since $\varepsilon > 0$ is arbitrary, this means that the set of points $x \in K$ such that the inequality

$$\left|U_1(x) - \int_K W_q(x, y)\, d\widehat{\mu}_1(y)\right| \leqslant 3\varepsilon_1 |x|^\rho$$

holds is a C_0^{n-1}-set. This implies

$$\mathfrak{L}^*_{U_1}(x) = \int_K W_q(x, y)\, d\widehat{\mu}_1(y)$$

and that $U_1(x)$ is a function of c.r.g. in $\overline{K}$. Together with the above statement on the c.r.g. of the function U_2, this proves that $U(x)$ is also a function of c.r.g. in $\overline{K}$.

Thus the proof of the theorem is completed for the case $\rho \notin \{\varkappa_j^+\}_{j=1}^\infty$.

Now let us consider the case $\rho \in \{\varkappa_j^+\}_{j=1}^{\infty}$ or, more precisely, the case

$$\varkappa_{p-1}^+ < \varkappa_p^+ = \cdots = \varkappa_q^+ = \rho < \varkappa_{q+1}^+ .$$

We set $\Gamma_\delta = \{x \in S_1 : 0 < \varphi(x) < \delta\}$ and choose $\delta > 0$ so that for some $\lambda \in \mathbb{Z}_+$ the inequality $\varkappa_\lambda^+(\Gamma_\delta) < \rho < \varkappa_{\lambda+1}^+(\Gamma_\delta)$ is valid. This is possible because the eigenvalues $\lambda_j(\Gamma)$ depend continuously on the domain Γ. By the condition of the theorem that we are proving, $u(x)$ is a function of c.r.g. in K and hence according to the corollary of Theorem 4 the function u satisfies condition A on $\partial(K^{\Gamma_\delta})$. Condition A on ∂K is assumed in the theorem. Hence the restriction of the function u to the cone $\Omega'_\delta = \{x \in K : 0 < \varphi(x) < \delta\}$ satisfies condition A with respect to the cone Ω'. It is obvious also that u is of c.r.g. in K. By the above, we conclude that $u(x)$ is a function of c.r.g. in the cone $\overline{\Omega'_\delta}$. The function $u(x)$ is of c.r.g. in the cone $\Omega''_\delta = K \setminus \Omega'_\delta$ by definition of c.r.g. in K. Then it is obvious that $u(x)$ is function of c.r.g. in $\overline{K} = \overline{(K \setminus \Omega'_\delta)} \cup \overline{\Omega'_\delta}$.

This completes the proof of Theorem 6.

§4. Sets of finite view. The case $\rho = \varkappa^+$

The characteristics of exceptional sets E or E_ε appearing in the definition of functions of c.r.g. can be improved in some cases. A set $E \subset \mathbb{R}^n$ is referred to as a *set of finite view* if it can be covered by balls $B_{r_j}(x^{(j)})$, $j = 1, 2, \ldots$, in such a way that

$$\sum_{j=1}^{\infty} \left(\frac{r_j}{|x^{(j)}|} \right)^{n-1} < \infty.$$

It is obvious that every set of finite view is a C_0^{n-1}-set, but the converse is false.

THEOREM 7. *If a function* $u \in SH(K, \rho]$, $\rho \in \{\varkappa_j^+\}_{j=1}^{\infty}$, *satisfies*

$$\int_{(\overline{K})_{1,\infty}} |x|^{2-\rho-n} d|\tau_u|(x) < \infty, \tag{46}$$

then

$$\mathfrak{L}_u^* = |x|^\rho \sum_{j : \varkappa_j^+ = \rho} a_j \varphi_j(x),$$

the function $u(x)$ *is of c.r.g. in* $\overline{K}$ *and, moreover, it satisfies the condition*:

A_1) *there exists a set* E *of finite view such that*

$$\lim_{\substack{x \to \infty \\ x \in K \setminus E}} |x|^{-\rho} |\mathfrak{L}_u^*(x) - u(x)| = 0. \tag{47}$$

PROOF. Let us define the integers q and p by putting

$$1 + q = \min\{j : \varkappa_j^+ = \rho\}, \qquad p = \max\{j : \varkappa_j^+ = \rho\}.$$

Then Theorem 2 yields the representation

$$\begin{aligned} u(x) = &\int_{(\overline{K})_{1,t|x|}} W_q(x, y)\, d\tau_u(y) + \int_{(\overline{K})_{t|x|,\infty}} W_q(x, y)\, d\tau_u(y) \\ &- \sum_{j=q+1}^{\infty} \left(\int_{(K)_{t|x|,\infty}} \frac{\varphi_j(y)}{\varphi(y)}\, d\tau_u(y) \right. \\ &\left. + \int_{(\partial K)_{t|x|,\infty}} \left(\frac{\partial \varphi(y)}{\partial n}\right)^{-1} \left(\frac{\partial \varphi_j(y)}{\partial n}\right) d\tau_u(y) \right) \\ &+ |x|^{\rho} \sum_{j=q+1}^{p} a_j \varphi_j(x) + \sum_{j=1}^{q} a_j \varphi_j(x) |x|^{\varkappa_j^+} + H_\lambda(x) \\ &- \int_{(K)_1} \mathfrak{S}(x, y) d\mu_u(y) = I_1 + I_2 + I_3 + I_4. \end{aligned}$$

It follows from Lemma 6 and (46) that the inequality $|I_1(x)| \leqslant c_1 \varepsilon |x|^{\rho}$ is valid on the set $\Delta_\rho(\tau_u, \varepsilon) \setminus B_R$ for R large enough. At the same time it follows from Lemma 7 and condition (46) that the set $\Delta_\rho(\tau_u, \varepsilon)$ is of finite view. Then it also evidently follows from (46) that

$$\lim_{\substack{|x|\to\infty \\ x\in K}} I_2(x)|x|^{-\rho} = 0.$$

Taking the above into account and using the obvious equalities

$$\lim_{t\to\infty} I_3(tx)t^{-\rho} = |x|^{\rho} \sum_{j=q+1}^{p} a_j \varphi_j(x), \qquad \lim_{\substack{|x|\to\infty \\ x\in K}} I_4(x)|x|^{-\rho} = 0,$$

we conclude that

$$\mathfrak{L}_u^*(x) = |x|^{\rho} \sum_{j=q+1}^{p} a_j \varphi_j(x)$$

and that the set $E_\varepsilon(u)$ is of finite view for every $\varepsilon > 0$. Now the set E which has finite view and satisfies (47) can be constructed over the sets $E_\varepsilon(u)$ in the standard way. Since a set of finite view is a C_0^{n-1}-set, this implies, in particular, that $u(x)$ is a function of c.r.g. in $\overline{K}$.

This completes the proof of the theorem.

In the case $\rho \notin \{\varkappa_j^+\}_{j=1}^{\infty}$ the following theorem holds.

THEOREM 8. *Let a function* $u \in SH(K, \rho]$, $\varkappa_q^+ < \rho < \varkappa_{q+1}^+$, *satisfy*

$$\int_{(\overline{K})_{1,\infty}} |x|^{2-\rho-n} d|\tau_u|(x) < \infty.$$

Then $\mathfrak{L}_u^*(x) \equiv 0$, *the function* $u(x)$ *is of c.r.g. in* $\overline{K}$ *and, moreover, it satisfies condition* A_1 *of Theorem* 7.

We omit the proof of this theorem. It is similar to that of Theorem 7 (one must use Lemma 5 instead of Lemma 6).

The case $\rho = \varkappa^+$ is of special interest. In this case it is possible to give a condition of growth regularity that uses the boundary measure only. Note also that for holomorphic functions in the upper half-plane $\mathbb{C}^+$ this case leads us to functions of exponential type (in $\overline{\mathbb{C}^+}$).

THEOREM 9. *Let a function* $u \in SH(K)$ *satisfy the conditions*

$$u(x) \leqslant c_1|x|^{\varkappa^+} + c_2, \quad \forall x \in K, \tag{48}$$

and

$$\int_{(\partial K)_{1,\infty}} |x|^{-\varkappa^+ - n - 2} \frac{\partial \varphi}{\partial n} \, d\mu^+_{u,\partial} < \infty. \tag{49}$$

Then:

a) *the indicator* $\mathfrak{L}^*_u$ *is proportional to the function* $|x|^{\varkappa^+} \varphi(x)$;

b) *the function* $u(x)$ *satisfies the condition* A_1 *of Theorem* 7 (*and therefore it is of c.r.g. in* $\overline{K}$);

c) *the following inequality is valid*

$$\int_{(K)_{1,\infty}} |x|^{-\varkappa^+ - n + 2} \varphi(x) \, d\mu_u(x) < \infty. \tag{50}$$

PROOF. It follows from Theorem 7 that statements a) and b) of Theorem 9 will be proved if we show that $u \in SH(K, \rho]$ and

$$\int_{(\overline{K})_{1,\infty}} |x|^{2-n-\varkappa^+} \, d|\tau_u|(x) < \infty. \tag{51}$$

First we prove that

$$\int_{(\partial K)_{1,\infty}} |x|^{-\varkappa^+ - n + 2} \frac{\partial \varphi}{\partial n} \, d|\tau_{u,\partial}|(x) < \infty.$$

Let us use the generalized Carleman's formula (see [1, 2]) which yields

$$\begin{aligned} \theta_n \int_{(K)_{1,R}} \gamma_R(|x|)\varphi(x)\, d\mu_u(x) &= -\gamma'_R(R) \int_{\Gamma_R} u(x)\varphi(x)\, d\mu_u(x) \\ &\quad + \int_{(\partial K)_{1,R}} \gamma_R(|x|) \frac{\partial \varphi}{\partial n} \, d\mu_{u,\partial}(x) + O(1), \qquad R \to \infty, \end{aligned}$$

where $\gamma_R(t) = t^{\varkappa^-} - t^{\varkappa^+} R^{\varkappa^- - \varkappa^+}$ and $\lim_{R\to\infty} O(1)$ exists.

Representing the measure μ by standard way in the form $\mu = \mu^+ - \mu^-$ and using the above formula we obtain that

$$\begin{aligned} \int_{(\partial K)_{1,R/2}} |x|^{-\varkappa^+ - n + 2} \frac{\partial \varphi}{\partial n} \, d\mu_{u,\partial} &\leqslant \int_{(\partial K)_{1,R/2}} \frac{|x|^{-\varkappa^+ - n + 1} \gamma_R(|x|)}{\gamma_R(|x|)} \frac{\partial \varphi}{\partial n} \, d\mu^-_{u,\partial} \\ &\leqslant \frac{1}{1 - 2^{\varkappa^- - \varkappa^+}} \left\{ \int_{(\partial K)_{1,R}} \gamma_R(|x|) \frac{\partial \varphi}{\partial n} \, d\mu^+_{u,\partial} \right. \\ &\qquad \left. - \gamma'_R(R) \int_{\Gamma_R} u(x)\varphi(x)\, dS_R(x) \right\} + O(1). \end{aligned}$$

Hence, taking into account (48) and (49) we conclude that

$$\int_{(\partial K)_{1,\infty}} |x|^{-\varkappa^+-n+2} \frac{\partial \varphi}{\partial n} \, d\mu^-_{u,\partial}(x) < \infty.$$

Comparing this inequality with (49) we obtain that

$$\int_{(\partial K)_{1,\infty}} |x|^{-\varkappa^+-n+2} \frac{\partial \varphi}{\partial n} \, d|\mu_{u,\partial}|(x) < \infty. \tag{52}$$

Now let us show that the condition (50), i.e., statement c) of the theorem, is valid.

Using Carleman's formula once more we obtain that

$$\begin{aligned} \int_{(K)_{1,R/2}} |x|^{-\varkappa^+-n+1} \varphi(x) \, d\mu_u(x) &\leqslant \frac{1}{1-2^{\varkappa^- - \varkappa^+}} \int_{(K)_{1,R}} \gamma(|x|)\varphi(x) \, d\mu_u(x) \\ &\leqslant \frac{1}{1-2^{\varkappa^- - \varkappa^+}} \left\{ -\frac{\gamma'_R(R)}{\theta_n} \int_{\Gamma_R} u(x)\varphi(x) \, dS_R(x) \right. \\ &\qquad \left. + \int_{(K)_{1,R}} |x|^{\varkappa^-} \frac{\partial \varphi}{\partial n} \, d|\mu_{u,\partial}|(x) + O(1) \right\}. \end{aligned}$$

Hence, in view of (48) and (52) inequality (50) immediately follows.

Inequality (51) is the obvious consequence of inequalities (50) and (52). Thus to complete the proof of the theorem it remains to prove that $u(x) \in SH(K, \rho]$.

It follows from (48) that

$$\varlimsup_{R\to\infty} R^{-\varkappa^+} \int_\Gamma u^+(xR)\varphi(x) \, dS_1 < \infty. \tag{53}$$

Now we estimate integral of u^-. We have

$$\begin{aligned} &\int_{\Gamma_R} u^-(x)\varphi(x) \, dS_R(x) + \int_{\Gamma_R} u^+(x)\varphi(x) \, dS_R(x) \\ &\quad + \frac{1}{|\gamma'_R(R)|} \int_{(\partial K)_{1,R}} \gamma_R(|x|) \frac{\partial \varphi}{\partial n} \, d|\mu_{u,\partial}|(x) + \frac{1}{|\gamma'_R(R)|} O(1) \\ &\leqslant \int_{\Gamma_R} u^+(x)\varphi(x) \, dS_R(x) + \frac{R^{\varkappa^+ + n-1}}{\varkappa^+ - \varkappa^-} \int_{(\partial K)_{1,R}} |x|^{-\varkappa^+-n+2} \frac{\partial \varphi}{\partial n} \, d|\mu_{u,\partial}|(x) \\ &\quad + R^{\varkappa^+ + n-1} O(1). \end{aligned}$$

From this estimate and estimates (53) and (52) it follows that

$$\varlimsup_{R\to\infty} R^{-\varkappa^+} \int_\Gamma u^-(xR)\varphi(x) \, dS_1(x) < \infty.$$

Comparing this inequality with (53) we conclude that

$$\varlimsup_{R\to\infty} \Phi_u(R) R^{-\varkappa^+} \varlimsup_{R\to\infty} R^{-\varkappa^+} \int_\Gamma |u(xR)|\varphi(x) \, dS_1(x) < \infty.$$

The last inequality together with (48) shows that $u \in SH(K, \varkappa^+]$.

The proof of the theorem is completed.

REMARK 1. The conditions imposed on the cone K in this paper imply $\min_{x\in\partial\Gamma} \partial\varphi/\partial n > 0$. Therefore condition (49) in Theorem 9 is equivalent to the following:

$$\int_{(\partial K)_{1,\infty}} |x|^{-\varkappa^+-n+1} d\mu^+_{u,\partial}(x) < \infty.$$

REMARK 2. Conditions (48) and (49) are obviously satisfied by functions $u \in SH(K)$ that are bounded in K. In this case Theorem 9 coincides with the corresponding theorem of V. Azarin [12]. Note that Azarin's theorem is the generalization and strengthening of one of Hayman's theorems [13] about functions on the half-plane.

A similar theorem for functions in the half-plane was obtained by N. Govorov [5]. Previously, similar theorems concerning the restrictions of entire functions to the half-plane were obtained by M. Cartwright [14] and B. Levin [15]. Note also the paper by M. Essen and H. Jackson [16], where exceptional sets for bounded functions in the half-space were considered.

REMARK 3. In the case $n = 2$, $K = \mathbb{C}^+$ and $u = \ln|f(z)|$, $f \in H(\mathbb{C})$, condition (50) appearing in Theorem 9, can be rewritten in the form

$$\sum_k \sin\theta_k / r_k < \infty,$$

where $z_k = r_k e^{i\theta_k}$ are zeros of the function $f(z)$. Entire functions that satisfy this condition form the so-called class A (Akhiezer's class) described in [5] and [15].

The condition equivalent to (50) is contained in Theorem 10 below. To formulate this theorem we need some notation.

Let $\widehat{\nu}_{u,\partial}$ be the spherical component of the measure $\mu_{\mathfrak{L}^*_u,\partial}$ and $\widehat{\nu}_u$ be the spherical component of the measure $\mu_{\mathfrak{L}^*_u}$. Thus

$$d\mu_{\mathfrak{L}^*_n,\partial} = (\varkappa^+ + n - 2)^{-1} d(r^{\varkappa^+} + n - 2) \otimes d\widehat{\nu}_{u,\partial},$$

$$d\mu_{\mathfrak{L}^*_n} = (\varkappa^+ + n - 2)^{-1} d(r^{\varkappa^+} + n - 2) \otimes d\widehat{\nu}_u.$$

From the Green formula for positive homogeneous functions, which was obtained in [1, 2], it follows that in the case $\rho = \varkappa^+$ we have

$$\int_{(K)_1} |x|^{\varkappa^+} \varphi(x)\, d\mu_{\mathfrak{L}^*_u} = \frac{1}{\theta_n} \int_{(\partial K)_1} |x|^{\varkappa^+} \frac{\partial\varphi}{\partial n}\, d\mu_{\mathfrak{L}^*_u,\partial}.$$

Taking into account the definitions of the measures $\widehat{\nu}_u$ and $\widehat{\nu}_{u,\partial}$ we conclude that the last equality is equivalent to the following one

$$\int_\Gamma \varphi(x)\, d\widehat{\nu}_u = \int_{\partial\Gamma} \frac{\partial\varphi}{\partial n}\, d\widehat{\nu}_u. \tag{54}$$

This implies[7]

$$\int_{\partial\Gamma} \frac{\partial\varphi}{\partial n}\, d\widehat{\nu}_{u,\partial} \geqslant 0, \qquad \forall u \in SH(K, \varkappa^+].$$

Now we can state and prove the above-mentioned theorem.

THEOREM 10. *Let a function* $u \in SH(K, \varkappa^+]$ *be of c.r.g. in* $\overline{K}$ *and*

$$\sup |\mathfrak{L}_u^*(x)| = M < \infty.$$

Then condition (50) *is satisfied if and only if*

$$\exists \lim_{R\to\infty} \int_{(\partial K)_{1,R}} |x|^{-\varkappa^+ - n+2} \frac{\partial\varphi}{\partial n}\, d\mu_{u,\partial}. \tag{55}$$

PROOF. 1) *Necessity.* Since the function $u(x)$ is of c.r.g. in K, by [2], Theorem 4, the following limit exists

$$\lim_{\substack{R\to\infty\\ R\notin\Lambda_u}} R^{-\varkappa^+ - n+1} \int_{\Gamma_R} u(x)\varphi(x)\, dS_R = \int_\Gamma \mathfrak{L}_u^*(x)\varphi(x)\, dS_1 \tag{56}$$

and the following relation

$$\int_\Gamma \varphi(x)\, d\widehat{\nu}_u = \int_{\partial\Gamma} \frac{\partial\varphi}{\partial n}\, d\widehat{\nu}_u$$

holds (see [2]). Now from the existence of limits (55) and (56) it easily follows by Carleman's formula that

$$\exists \lim_{R\to\infty} \int_{(\partial K)_{1,R}} \gamma_R(|x|) \frac{\partial\varphi}{\partial n}\, d\mu_{u,\partial}.$$

Indeed,

$$\int_{(K)_{1,R}} |x|^{-\varkappa^+ - n+2} \varphi(x)\, d\mu_u(x) = R^{-\varkappa^+ - n+2} \int_{(\partial K)_{1,R}} \varphi(x)\, d\mu_u(x) + (\varkappa^+ + n - 2) \int_1^R t^{-\varkappa^+ - n+1} dt \int_{(\partial K)_{1,t}} \varphi(x)\, d\mu_u(x)$$

and therefore (50) implies that

$$\int_1^\infty t^{-\varkappa^+ - n+1} dt \int_{(K)_{1,t}} \varphi(x)\, d\mu_u(x) < \infty.$$

In its turn, this implies

$$\lim_{r\to\infty} R^{-\varkappa^+ - n+2} \int_{(K)_{1,R}} \varphi(x)\, d\mu_u(x) = 0.$$

Consequently,

$$\lim_{R\to\infty} R^{\varkappa^- - \varkappa^+} \int_{(K)_{1,R}} |x|^{\varkappa^+} \varphi(x)\, d\mu_u(x) = 0.$$

---

[7]In the case $n = 2$ and correspondingly $K = \mathbb{C}^+$ this inequality transforms into the well-known inequality $h_u(+0) + h_u(\pi - 0) \geqslant 0$, where $h_u(\theta) = \mathfrak{L}_u(\varepsilon^{i\theta})$.

Comparing this equality with condition (50), we conclude that

$$\exists \lim_{R\to\infty} \int_{(\partial K)_{1,R}} \gamma_R(|x|)\varphi(x)\,d\mu_u(x).$$

Thus in the Carleman formula written for the function under consideration all terms except

$$\int_{(\partial K)_{1,R}} \gamma_R(|x|)\frac{\partial\varphi}{\partial n}\,d\mu_{u,\partial}$$

have limits when $R\to\infty$. Therefore

$$\exists \lim_{R\to\infty} \int_{(\partial K)_{1,R}} \gamma_R(|x|)\frac{\partial\varphi}{\partial n}\,d\mu_{u,\partial}. \tag{57}$$

Then, since $u(x)$ is of c.r.g. in $\overline{K}$, Theorem 6 implies

$$\exists \lim_{R\to\infty} R^{\varkappa^- - \varkappa^+} \int_{(\partial K)_{1,R}} |x|^{\varkappa^+}\frac{\partial\varphi}{\partial n}\,d\mu_{u,\partial} = \int_{(\partial K)_1} |x|^{\varkappa^+}\frac{\partial\varphi}{\partial n}\,d\widehat{\mu}_{u,\partial}.$$

Comparing this equality with (57) and taking into account the relation

$$\gamma_R(t) = t^{-\varkappa^+ - n+2} - R^{\varkappa^- - \varkappa^+} t^{\varkappa^+}$$

we conclude that limit (55) exists. Thus we have proved the necessity part of Theorem 10.

2) *Sufficiency.* First let us note that (55) implies

$$\lim_{r\to\infty} R^{-\varkappa^+ - n+2} \int_{(\partial K)_{1,R}} \frac{\partial\varphi}{\partial n}\,d\mu_{u,\partial} = 0. \tag{58}$$

Indeed, setting

$$I = \lim_{r\to\infty} \int_{(\partial K)_{1,R}} |x|^{-\varkappa^+ - n+2}\frac{\partial\varphi}{\partial n}\,d\mu_{u,\partial} \tag{59}$$

we have

$$\begin{aligned}
&\lim_{R\to\infty} R^{-\varkappa^+ - n+2} \int_{(\partial K)_{1,R}} \frac{|x|^{-\varkappa^+ - n+2}\partial\varphi/\partial n}{|x|^{-\varkappa^+ - n+2}}\,d\mu_{u,\partial} \\
&\quad = \lim_{R\to\infty} R^{-\varkappa^+ - n+2}\Big\{R^{\varkappa^+ + n-2} \int_{(\partial K)_{1,R}} |x|^{-\varkappa^+ - n+2}\frac{\partial\varphi}{\partial n}\,d\mu_{u,\partial} \\
&\qquad\qquad - (\varkappa^+ + n-2)\int_1^R t^{\varkappa^+ + n-3}\,dt \int_{(\partial K)_{1,t}} |x|^{-\varkappa^+ - n+2}\frac{\partial\varphi}{\partial n}\,d\mu_{u,\partial}\Big\} \\
&\quad = I - (\varkappa^+ + n-2)\lim_{R\to\infty}\int_1^R t^{\varkappa^+ + n-3}(I + o(1))\,dt = 0.
\end{aligned}$$

It follows from (58) and (59) that

$$\lim_{r\to\infty} \int_{(\partial K)_{1,R}} \gamma_R(|x|)\frac{\partial\varphi}{\partial n}\,d\mu_{u,\partial} = I.$$

Now applying Carleman's formula and taking into account the existence of limit (56), which follows from the growth regularity of $u(x)$ in K, we obtain

$$\exists \lim_{R\to\infty} \int_{(K)_{1,R}} \gamma_R(|x|)\varphi\, d\mu_u. \tag{60}$$

Finally from the criteria of growth regularity in K and $\overline{K}$ contained in Theorem 3 and in Corollary 2 of Theorem 6 it follows that

$$\lim_{R\to\infty} R^{\varkappa^- - \varkappa^+} \int_{(K)_{1,R}} |x|^{\varkappa^+} \varphi(x)\, d\mu_u(x) = \int_{(K)_1} |x|^{\varkappa^+} \varphi(x)\, d\mu_u(x).$$

Now (60) allows us to conclude that condition (50) is satisfied.

The proof of the theorem is completed.

REMARK. The function $u(x)$ in the theorem proved above satisfies the equality

$$\int_{\partial\Gamma} \frac{\partial\varphi}{\partial n}\, d\widehat{\nu}_{u,\partial} = 0.$$

Indeed, it is easy to see that (50) implies the equality

$$\lim_{R\to\infty} R^{-\varkappa^+ - n + 2} \int_{(K)_{1,R}} \varphi(x)\, d\mu_u = 0.$$

Then, as was already noted, the equality

$$\lim_{R\to\infty} R^{-\varkappa^+ - n + 2} \int_{(K)_{1,R}} \varphi(x)\, d\mu_u = \int_{(K)_1} \varphi\, d\widehat{\mu}_u$$

is valid. Therefore

$$\int_{(K)_1} \varphi(x)\, d\widehat{\mu}_u = 0.$$

Thus $\widehat{\mu}_u = 0$, $\widehat{\nu}_u = 0$ and equality (54) implies that

$$\int_{\partial\Gamma} \frac{\partial\varphi}{\partial n}\, d\widehat{\nu}_{u,\partial} = 0.$$

REFERENCES

1. A. Yu. Rashkovskiĭ and L. I. Ronkin, *Subharmonic functions of completely regular growth in a cone*, Dokl. Akad. Nauk SSSR **297** (1987), no. 2, 298–302; English transl., Soviet Math. Dokl. **36** (1988), no. 3, 473–477.

2. ______, *Subharmonic functions of finite order in a cone,* I, Teor. Funktsiĭ, Funktsional. Anal. i Prilozhen. **54** (1990), 30–42; II, **55** (1990), 50–58; III, **57** (1991), 40–51; English transl. to appear in J. Soviet Math .

3. A. Yu. Rashkovskiĭ, *Growth and decrease of subharmonic functions in a cone*, Ukrain. Mat. Zh. **41** (1989), 1252–1258; English transl. in Ukrainian Math. J. **41** (1989).

4. ______, *Integral representation of subharmonic functions of finite order in a cone*, Sibirsk. Mat. Zh. **30** (1989), no. 3, 109–123; English transl. in Siberian Math. J. **30** (1989).

5. N. V. Govorov, *Riemann boundary value problems with infinite index*, "Nauka", Moscow, 1986. (Russian)

6. P. Lelong and L. Gruman, *Entire functions of several complex variables*, Springer-Verlag, Berlin and New York, 1986.

7. J. Lelong-Ferrand, *Etude des fonctions subharmoniques positives dans un cylindre ou dans un cone*, C. R. Acad. Sci. Paris. **229** (1949), no. 5, 340–341.

8. L. I. Ronkin, *Estimates of subharmonic functions and their associated measures in a cone*, Theory of Operators and Subharmonic Functions, "Naukova Dumka", Kiev, 1990, pp. 15–25. (Russian)

9. A. I. Heific, *Subharmonic functions of completely regular growth in the half-plane*, Dokl. Akad. Nauk SSSR **239** (1978), no. 2, 282–285; English transl. in Soviet Math. Dokl. **19** (1978).

10. A. A. Goldberg and I. V. Ostrovskiĭ, *Distribution of values of meromorphic functions*, "Nauka", Moscow, 1970. (Russian)

11. M. Gussman, *Differentiation of integrals in* $\mathbb{R}$, Springer-Verlag, Berlin, Heidelberg, and New York, 1975.

12. V. S. Azarin, *Generalization of a theorem of Hayman on subharmonic functions in n-dimensional cone*, Mat. Sb. **66** (1965), no. 2, 248–264; English transl. in Amer. Math. Soc. Transl. **80** (1969).

13. W. K. Hayman, *Questions of regularity connected with the Phragmén-Lindelöf principle*, J. Math. Pures Appl. **35** (1956), no. 2, 115–126.

14. M. L. Cartwright, *On functions which are regular and of finite order in an angle*, Proc. London Math. Soc. **38** (1934/35), no. 2, 158–179.

15. B. Ya. Levin, *Distribution of zeros of entire functions*, GITTL, Moscow, 1956; English transl., Amer. Math. Soc., Providence, R.I., 1964.

16. M. Essén and H. L. Jackson, *A comparison between thin sets and generalized Azarin sets*, Canad. Math. Bull. **18** (1975), no. 3, 335–346.

Institute for Low Temperature Physics & Egineering, 47 Lenin pr., Kharkov 310164 Ukraine

Translated by A. L. RONKIN

ADVANCES IN SOVIET MATHEMATICS
Volume 11, 1992

Complete Description of the Pair Indicator-Lower Indicator of an Entire Function

L. R. PODOSHEV

Introduction

Let $A(\rho(r))$ be the class of entire functions $f(z)$ of finite order $\rho > 0$ and of normal type with respect to proximate order $\rho(r)$, $\rho(r) \to \rho$. The indicator of a function $f \in A(\rho(r))$ is defined by the equality [1]:

$$h_f(\varphi) = \overline{\lim_{r\to\infty}} \ln|f(re^{i\varphi})|r^{-\rho(r)}.$$

The lower indicator of a function $f \in A(\rho(r))$ can be defined by one of the following equalities. The first definition is

$$\underline{h}_1(\varphi, f) = \sup_C \left\{ \underline{\lim}_{\substack{re^{i\varphi}\to\infty \\ re^{i\varphi} \notin C}} \ln|f(re^{i\varphi})|r^{-\rho(r)} \right\}, \tag{0.1}$$

where C is a C_0-set [1, p. 120], i.e., a set covered by a union of disks of the form $K_{\delta_j}(z_j) = \{z : |z - z_j| < \delta_j\}$ such that

$$\lim_{R\to\infty} \frac{1}{R} \sum_{|z_j|<R} \delta_j = 0.$$

C_0-sets are introduced to exclude from consideration neighborhoods of roots of $f(z)$ in the case when $\ln|f|$ is close to $-\infty$.

Similarly we define

$$\underline{h}_2(\varphi, f) = \sup_E \left\{ \underline{\lim}_{\substack{r\to\infty \\ r \notin E}} \ln|f(re^{i\varphi})|r^{-\rho(r)} \right\}, \tag{0.2}$$

where E are E_0-sets [1, p. 182], i.e., sets which satisfy the following condition on $[0, \infty)$: $\lim_{r\to\infty} \operatorname{meas}\{E \cap [0, r)\}r^{-1} = 0$.

Before introducing one more definition we recall the concept of limit set [2, p. 148] of a subharmonic function.

1991 *Mathematics Subject Classification*. Primary 30D15, 30D20.

Let $SH(\rho(r))$ be a set of subharmonic functions (in the complex plane) of normal type with respect to proximate order $\rho(r)$, $\rho(r) \to \rho$. D' is the set of distributions over the set D of test functions (infinitely differentiable functions on $\mathbb{C}$ with compact support).

The family of subharmonic functions $u_t(x) \stackrel{\text{def}}{=} u(tx)t^{-\rho(t)}$ is precompact in the D'-topology in the following sense: for each sequence $t_n \to \infty$ there exist a subharmonic function v and a subsequence t_{n_k} such that

$$D'\text{-}\lim_{k\to\infty} u_{t_{n_k}} = v.$$

The set of all functions v which are defined by this formula is called the *limit set* of the subharmonic function u and is denoted by $\mathrm{Fr}[u]$.

If $f \in A(\rho(r))$, then by definition $\mathrm{Fr}[f] \stackrel{\text{def}}{=} \mathrm{Fr}[\ln|f|]$. By means of $\mathrm{Fr}[f]$ the indicator and the lower indicator of a function $f \in A(\rho(r))$ are defined by the equalities

$$\overline{h}_f(\varphi) = \sup\{v(e^{i\varphi}) : v \in \mathrm{Fr}[f]\}, \tag{0.3}$$

$$\underline{h}_f(\varphi) = \inf\{v(e^{i\varphi}) : v \in \mathrm{Fr}[f]\}. \tag{0.4}$$

The lower indicator was first introduced by A. F. Leontiev [3] in the following way

$$\underline{h}(\varphi, f) = \varliminf_{r\to\infty} \ln|f(re^{i\varphi})|r^{-\rho(r)}$$

for functions whose zeros do not lie inside a certain angle containing the ray $\{\arg z = \varphi\}$. The lower indicator was studied by A. A. Goldberg, I. F. Krasichkov-Ternovskiĭ, V. S. Azarin and other authors [2–11]. These investigations found their applications in the works by Yu. F. Korobeĭnik, O. V. Epifanov, V. V. Napalkov [12–15]. The definition (0.1) was introduced in [4] (see also [5, p. 842]), the definitions (0.3), (0.4) were introduced in [2].

The following relations hold [9]:

$$\underline{h}_1(\varphi, f) = \underline{h}_2(\varphi, f) = \underline{h}_f(\varphi), \qquad h_f(\varphi) = \overline{h}_f(\varphi);$$

this enables one to investigate properties of the indicator and the lower indicator by using the definitions (0.3) and (0.4).

The indicator is a ρ-trigonometrically convex (ρ-t.c.) and a 2π-periodic function, and, as is known, ρ-trigonometric convexity and 2π-periodicity are its complete characteristics. This means that for any ρ-t.c. function $h(\varphi)$ and every proximate order $\rho(r) \to \rho$ there exists an $f \in A(\rho(r))$ such that

$$h(\varphi) = h_f(\varphi), \qquad \varphi \in [0, 2\pi). \tag{0.5}$$

The lower indicator is obviously upper semicontinuous as an infimum of upper semicontinuous (USC) functions and is 2π-periodic.

In [9, Theorem 2] a complete description of the class of lower indicators is given. This is a class of USC and 2π-periodic functions, i.e., for any USC

2π-periodic function $g(\varphi)$ and for every proximate order $\rho(r) \to \rho$ there exists $f \in A(\rho(r))$ such that

$$g(\varphi) = \underline{h}_f(\varphi), \qquad \varphi \in [0, 2\pi). \tag{0.6}$$

A natural question arises: what conditions must be imposed on the pair of functions h, g of proximate order $\rho(r) \to \rho$ to ensure the existence of a function $f \in A(\rho(r))$ that satisfies (0.5) and (0.6) simultaneously.

The answer is supplied by Theorem 1 announced in [16] and presenting effectively verifiable necessary and sufficient conditions for existence of a function $f \in A(\rho(r))$ that satisfies (0.5) and (0.6) simultaneously.

Let us examine what necessary conditions are imposed on the functions h and g by known facts.

Let $f \in A(\rho(r))$ exist and satisfy (0.5) and (0.6); let (a, b) be an interval where $h(\varphi)$ is a ρ-trigonometric function (ρ-t.f.), i.e., where it admits the representation $h(\varphi) = A\cos\rho\varphi + B\sin\rho\varphi$, A and B being certain constants.

We recall [1, p. 182] that $f \in A(\rho(r))$ is a function of completely regular growth (CRG) on the ray $l_\varphi = \{z : \arg z = \varphi\}$ if the equality $h_f(\varphi) = \underline{h}_f(\varphi)$ holds. The CRG condition for a function f on the ray l_φ will be written as $f \in A_{\mathrm{reg},\varphi}$.

The CRG condition for a function f in the angle

$$\Gamma(a, b) \overset{\text{def}}{=} \{re^{i\varphi} : \varphi \in (a, b),\ r > 0\}$$

will be written as

$$h(\varphi) = g(\varphi), \qquad \varphi \in (a, b). \tag{1}$$

M. Cartwright's theorem [1], when applied to entire functions, will be written as the implication

$$b - a > \pi/\rho \ \Rightarrow\ (1). \tag{2}$$

B. Levin's theorems [1, pp. 213–214] on the completely regular growth of a holomorphic function with trigonometric indicator inside an angle will be written for entire functions as the implications

$$(\exists\varphi_0 \in (a, b) \text{ such that } h(\varphi_0) = g(\varphi_0)) \ \Rightarrow\ (1), \tag{3}$$

$$(h(a) = g(a)) \wedge (h'_+(a) = h'_-(a)) \ \Rightarrow\ (1), \tag{4}$$

$$(h(b) = g(b)) \wedge (h'_+(b) = h'_-(b)) \ \Rightarrow\ (1), \tag{5}$$

where $h'_\pm(a)$ and $h'_\pm(b)$ are the right and left derivatives of the function h at the points a and b.

The condition of the completely regular growth obtained in [18] will be written as the implications

$$(b - a = \pi/\rho) \wedge (h'_+(a) = h'_-(a)) \ \Rightarrow\ (1), \tag{6}$$

$$(b - a = \pi/\rho) \wedge (h'_+(b) = h'_-(b)) \ \Rightarrow\ (1). \tag{7}$$

We shall also show that the following are necessary conditions

$$\left(\varliminf_{\varphi\to a+0}\frac{h(\varphi)-g(\varphi)}{\varphi-a}=0\right)\Rightarrow(1),\tag{8}$$

$$\left(\varliminf_{\varphi\to b-0}\frac{h(\varphi)-g(\varphi)}{b-\varphi}=0\right)\Rightarrow(1).\tag{9}$$

Theorem 1 states that the conditions of completely regular growth expressed by the implications (2)–(9) are necessary for the existence of a function $f\in A(\rho(r))$ which satisfies (0.5) and (0.6).

Now we shall give an exact formulation.

An interval (a, b) is said to be the *maximal interval of ρ-trigonometricity* of the function h if $h(\varphi)$ is a ρ-t.f. when $\varphi\in(a, b)$, but on any larger interval including (a, b) h is not a ρ-t.f.

A function h is said to be strictly ρ-t.c. if h is a ρ-t.c. function and is not a ρ-t.f. on any interval.

Functions h and g are said to be *concordant* if at least one of the following conditions holds:

1) h is strictly ρ-t.c.;
2) for each maximal interval (a, b) of ρ-trigonometricity of the function h implications (2)–(9) are satisfied.

THEOREM 1. *Let $0<\rho<\infty$, $h(\varphi)$ be a ρ-t.c. 2π-periodic function, $g(\varphi)$ be a USC 2π-periodic function, $h(\varphi)\geqslant g(\varphi)$, $\forall\varphi$, $h\not\equiv g$. A function $f\in A(\rho(r))$ which simultaneously satisfies the identities $h_f\equiv h$, $\underline{h}_f\equiv g$ with an arbitrary proximate order $\rho(r)\to\rho$ exists if and only if the functions h and g are concordant.*

This paper consists of six sections. In §1 the necessity of the conditions of Theorem 1 will be proved, as well as the implications (3)–(7), by methods of the theory of limit sets. In §2 we shall prove that the conditions of the theorem are sufficient. Sections 3 to 6 prove Lemmas 3.1, 4.1, 5.1 and 6.1 which are used in the proof of sufficiency.

§1. Proof of the necessity of the conditions of Theorem 1

Let a function $f\in A(\rho(r))$ exist such that $h_f\equiv h$, $\underline{h}_f\equiv g$. Let us prove that h and g are concordant.

One can assume that h is not a strictly ρ-t.c. function, since otherwise the functions h and g are concordant by the condition of the theorem.

Let (a, b) be a maximal interval of ρ-trigonometricity of the function h.

As was shown in the Introduction, the implications (2) to (7) follow from known theorems, so we must verify the implications (8) and (9). Note that the veracity of the implication (2) (M. Cartwright's theorem) follows from (6) or (7) ([18]). By virtue of the symmetry, it will obviously suffice to prove one of every pair of implications: (4) and (5), (6) and (7), (8) and (9).

For the sake of completeness, we shall prove conditions (3) to (7) by using the theory of limit sets.

Let $U[\rho, \sigma]$ be the class of subharmonic functions $v(z)$ which satisfy the conditions $v(z) \leqslant \sigma|z|^{\rho}$, $z \in \mathbb{C}$, $v(0) = 0$.

We shall need some facts from [2].

We define a transformation $(\cdot)_{\tau}$ for functions $v \in U[\rho, \sigma]$ by putting

$$v_{\tau}(z) = v(\tau z)\tau^{-\rho}. \tag{1.1}$$

The set $\mathrm{Fr}[f]$ possesses the following properties:

Fr1) $\mathrm{Fr}[f] \subset U[\rho, \sigma]$ for a $\sigma > 0$;

Fr2) $v \in \mathrm{Fr}[f] \Rightarrow v_{\tau} \in \mathrm{Fr}[f]$, $\forall \tau > 0$.

The latter property is called *invariance* with respect to the transformation $(\cdot)_{\tau}$. It follows from this property that for any function $v \in \mathrm{Fr}[f]$

$$v(\tau e^{i\varphi}) \leqslant \tau^{\rho} h(\varphi), \qquad \varphi \in [0, 2\pi), \ \tau > 0$$

Since $h(\varphi)$ is ρ-t.f. for $\varphi \in (a, b)$, the function H ...
harmonic in the angle $\Gamma(a, b)$, whence the function ...
and nonpositive in $\Gamma(a, b)$; by virtue of the ma...
$v < H$ in $\Gamma(a, b)$ or $v \equiv H$ in $\Gamma(a, b)$ for each ...
that the condition $v \equiv H$ in $\Gamma(a, b)$ implies ...
virtue of the upper semicontinuity of v.

Let us prove that the implication (3) is va...
exists a $\varphi_0 \in (a, b)$ such that $h(\varphi_0) = g$ ...
for any $v \in \mathrm{Fr}[f]$. Combining the de...
$h(\varphi) = \overline{h}_f(\varphi) = \underline{h}_f(\varphi) = g(\varphi)$, i.e., we...

Let us prove the concordance con...
$g(a)$, $h'_{+}(a) = h'_{-}(a)$, but there ex...
Then there exists a $v \in \mathrm{Fr}[f]$ su...

whence

Without loss of gener...
since in the converse ...
$\max(v, -c)$ inste...

We choose τ ...

we apply t...

LEMMA ...
$\Gamma(0, \cdot$ ...

From Lemma 1 we get

$$\alpha_1 = \varlimsup_{\varphi \to a+0} \frac{h(\varphi) - v_{\tau_1}(e^{i\varphi})}{\varphi - a} > 0$$

and likewise

$$\alpha_2 = \varlimsup_{\varphi \to a+0} \frac{h(\varphi) - v_{\tau_2}(e^{i\varphi})}{\varphi - a} > 0.$$

So a $\Delta > 0$ can be chosen such that $a + \Delta < b$ and the inequalities

$$H(\tau_j e^{i\varphi}) - v(\tau_j e^{i\varphi}) > \alpha \tau_j^{\rho}(\varphi - a), \tag{1.3}$$

where $\alpha = (1/2)\min(\alpha_1, \alpha_2)$, $j = 1, 2$, hold for $\varphi \in [a, a + \Delta]$.

We denote

$$\beta = \min(H(\tau e^{i(a+\Delta)}) - v(\tau e^{i(a+\Delta)})) > 0, \qquad \tau \in [\tau_1, \tau_2].$$

Let us choose $\varepsilon > 0$ so small as to ensure

$$\varepsilon < \min(\alpha, \beta \tau_2^{-\rho} \Delta^{-1}) \tag{1.4}$$

and let us consider

$$h_\varepsilon(\varphi) = \rho^{-1}(h'(a) - \varepsilon)\sin\rho(\varphi - a) + h(a)\cos\rho(\varphi - a), \qquad \varphi \in (a, b).$$

Since $h(\varphi)$ is a ρ-t. function for $\varphi \in (a, b)$, we have

$$(\varphi) = \rho^{-1} h'(a)\sin\rho(\varphi - a) + h(a)\cos\rho(\varphi - a), \qquad \varphi \in (a, b).$$

$$= \rho^{-1}\varepsilon \sin\rho(\varphi - a) \leqslant \varepsilon(\varphi - a), \qquad \varphi \in [a, a + \Delta]. \tag{1.5}$$

5), we get

$$^{\rho}(h(\varphi) - \alpha(\varphi - a)) < \tau_j^{\rho}(h(\varphi) - \varepsilon(\varphi - a))$$

$$\varphi \in [a, a + \Delta], \tag{1.6}$$

$$) + \tau^{\rho}\varepsilon\Delta - \beta$$

$$\tau \in [\tau_1, \tau_2]. \tag{1.7}$$

$$\in (\tau_1, \tau_2)\}. \tag{1.8}$$

nction $v(re^{i\varphi})$ –

ciple we have

$$\tag{1.9}$$

Let us consider the function

$$H_1(re^{i\varphi}) = r^\rho h_1(\varphi), \qquad re^{i\varphi} \in \Gamma(a-\Delta, a+\Delta),$$

where

$$h_1(\varphi) = \begin{cases} h(\varphi), & \varphi \in (a-\Delta, a], \\ h_\varepsilon(\varphi), & \varphi \in [a, a+\Delta). \end{cases}$$

Note that the function H_1 is continuous in $\Gamma(a-\Delta, a+\Delta)$ and subharmonic in the angles $\Gamma(a-\Delta, a)$, $\Gamma(a, a+\Delta)$. Taking into consideration (1.9), for sufficiently small $R > 0$ we get

$$\mathfrak{M}(H_1, e^{ia}, R) \geqslant \mathfrak{M}(v, e^{ia}, R) \geqslant v(e^{ia}) = H_1(e^{ia}),$$

where $\mathfrak{M}(H_1, e^{ia}, R)$ is the mean value of the function H_1 over the disk $\{z : |z - e^{ia}| < R\}$. It is easy to see that for any $k > 0$, $R > 0$,

$$\mathfrak{M}(H_1, ke^{ia}, kR) = k^\rho \mathfrak{M}(H_1, e^{ia}, R).$$

Hence we get

$$H_1(ke^{ia}) = k^\rho H_1(e^{ia}) \leqslant k^\rho \mathfrak{M}(H_1, e^{ia}, R) = \mathfrak{M}(H_1, ke^{ia}, kR).$$

Sending R to zero, for each fixed k we see that the function H_1 is subharmonic in a vicinity of the ray $\{\varphi = a\}$, whence it follows that h_1 is a ρ-t.c.f. for $\varphi \in (a-\Delta, a+\Delta)$. But, according to the construction,

$$(h_1)'_-(a) = h'_-(a) = h'_+(a) = (h_\varepsilon)'_+(a) + \varepsilon = (h_1)'_+(a) + \varepsilon$$

which contradicts the fact that h_1 is a ρ-t.c.f. We have proved the concordance condition (4).

Let us prove the implication (6). Assume the contrary. Then there exists a $v \in \mathrm{Fr}[f]$, $\varphi_0 \in [a, b]$ such that $g(\varphi_0) \leqslant v(e^{i\varphi_0}) < h(\varphi_0)$, whence, by virtue of the maximum principle, $v(\tau e^{i\varphi}) < \tau^\rho h(\varphi)$, $\tau e^{i\varphi} \in \Gamma(a, b)$. If $v(\tau e^{ia}) = H(\tau e^{ia})$ for a $\tau > 0$, then $v_\tau(e^{ia}) = h(a)$, and it will suffice to repeat the arguments used in proving (4) with v_τ instead of v.

So it is sufficient to examine the case $v(\tau e^{ia}) < H(\tau e^{ia})$, $\tau > 0$. We denote

$$T(\varphi) = h'(a)\rho^{-1} \sin\rho(\varphi - a) + h(a)\cos\rho(\varphi - a).$$

It is easy to see that for $\varphi \in (a-\gamma, a)$, either $T(\varphi) < h(\varphi)$ for some $\gamma > 0$, or $T(\varphi) = h(\varphi)$, $\varphi \in (a-\gamma, a)$ (the latter is true if $h(\varphi)$ is a trigonometric function on the interval $(a-\gamma, b)$).

Let us examine the case when $T(\varphi) < h(\varphi)$, $\varphi \in (a-\gamma, a)$. We set

$$h_1(\varphi) = h(\varphi) - T(\varphi), \qquad \varphi \in (a-\gamma, b),$$
$$v_1(re^{i\varphi}) = v(re^{i\varphi}) - r^\rho T(\varphi), \qquad re^{i\varphi} \in \Gamma(a-\gamma, b).$$

Then $h_1(\varphi) = 0$, $\varphi \in [a, b]$, $h_1(\varphi) > 0$, $\varphi \in [a-\gamma, a)$,

$$h_1'(a) = 0, \tag{1.10}$$

$$v_1(e^{i\varphi}) < 0, \qquad \varphi \in [a, b). \tag{1.11}$$

Since v_1 is an USC function, there exists a δ, $0 < \delta < \gamma$, such that

$$v_1(e^{i\varphi}) < 0, \qquad \varphi \in [a - \delta, b). \tag{1.12}$$

Note that if $v_1(e^{ib}) = 0$, then by virtue of Lemma 1

$$\varlimsup_{\varphi \to b-0} \frac{v_1(e^{i\varphi})}{b - \varphi} < 0.$$

So, taking into account (1.11), one can choose $C_2 > 0$ so as to ensure the inequality

$$v_1(e^{i\varphi}) < C_2 \sin \rho(\varphi - b), \qquad \varphi \in (a, b). \tag{1.13}$$

Denote $C_1 = -\max v_1(e^{i\varphi})$, $\varphi \in [a - \delta, a + \pi/(2\rho)]$, $A = (\rho/4)\min(C_1, C_2)$. We set $h_\varepsilon(\varphi) = A(\rho - \varepsilon)^{-1} \sin(\rho - \varepsilon)(\varphi - b)$, where ε, $0 < \varepsilon < \rho/\pi$, will be chosen later on. Then

$$\begin{aligned} h_\varepsilon(\varphi) &\geqslant A(\rho - \varepsilon)^{-1} \sin \rho(\varphi - b) \cos \varepsilon(\varphi - b) \\ &\geqslant C_2 \sin \rho(\varphi - b), && \varphi \in (b - \pi/(2\rho), b), \\ h_\varepsilon(\varphi) &> -A(\rho - \varepsilon)^{-1} \geqslant -C_1 > v_1(e^{i\varphi}), && \varphi \in [a - \delta, b - \pi/(2\rho)], \end{aligned}$$

whence, taking into consideration (1.13), we get

$$h_\varepsilon(\varphi) > v_1(e^{i\varphi}), \qquad \varphi \in [a - \delta, b), \ \varepsilon \in (0, \rho/\pi). \tag{1.14}$$

Taking (1.10) into account, we choose a $\Delta > 0$ so small as to have the inequalities

$$h_1(a - \Delta) < A\pi^{-1}\Delta, \tag{1.15}$$

$$\Delta < \min(\delta, \rho/\pi, 2/\rho). \tag{1.16}$$

Let us find ε from the relation $\pi((\rho - \varepsilon)^{-1} - \rho^{-1}) = \Delta/2$. We get $\varepsilon = \Delta\rho^2(\Delta\rho + 2\pi)^{-1}$ and it follows from (1.16) that $\varepsilon < \rho/\pi$. So, taking (1.15) into consideration, we have

$$h_\varepsilon(a - \Delta/2) = 0 < h_1(a - \Delta/2),$$

$$h_\varepsilon(a - \Delta) = \frac{A}{\rho - \varepsilon} \sin \frac{\pi\varepsilon}{\rho} \geqslant \frac{A}{\rho - \varepsilon} \frac{2\varepsilon}{\rho} = \frac{A}{\pi}\Delta > h_1(a - \Delta).$$

Since the functions h_ε, h_1 are continuous, there exists a point $\theta_0 \in (a-\Delta, a)$ such that $h_\varepsilon(\theta_0) = h_1(\theta_0) > 0$. We have

$$r^{\rho-\varepsilon} h_\varepsilon(\theta_0) > r^\rho h_1(\theta_0) \geqslant v_1(re^{i\theta_0}), \qquad r \in (0, 1),$$

$$v_1(re^{ib}) = v(re^{ib}) - r^\rho T(b) = v(re^{ib}) - r^\rho h(b) \leqslant 0 = r^{\rho-\varepsilon} h_\varepsilon(b).$$

Whence, using (1.14), we see that $r^{\rho-\varepsilon} h_\varepsilon(\varphi) \geqslant v_1(re^{i\varphi})$ for $re^{i\varphi} \in \partial S$, where ∂S is the boundary of the sector

$$S = \{re^{i\varphi} : \varphi \in (\theta_0, b), \ r \in (0, 1)\}.$$

Since the function $r^{\rho-\varepsilon}h_\varepsilon(\varphi)$ is harmonic, $v_1(re^{i\varphi}) \leqslant r^{\rho-\varepsilon}h_\varepsilon(f)$, $re^{i\varphi} \in S$, whence

$$v(re^{i\varphi}) \leqslant r^\rho h(\varphi) + r^{\rho-\varepsilon}h_\varepsilon(\varphi) = H(re^{i\varphi}) + H_\varepsilon(re^{i\varphi}), \qquad re^{i\varphi} \in S, \quad (1.17)$$

where $H_\varepsilon(re^{i\varphi}) = r^{\rho-\varepsilon}h_\varepsilon(\varphi)$. Let $\mathfrak{N}(R, v)$ be the mean value of the function v over $B(0, R)$, the disk of radius R with center at the origin. Taking (1.17) into consideration, we get

$$\mathfrak{N}(R, v) \leqslant \frac{1}{\pi R^2}\int_{B(0,R)\cap S}[H(re^{i\varphi}) + H_\varepsilon(re^{i\varphi})]r\,dr\,d\varphi$$
$$+ \frac{1}{\pi R^2}\int_{B(0,R)\setminus S} H(re^{i\varphi})r\,dr\,d\varphi \leqslant d_1R^\rho - d_2R^{\rho-\varepsilon},$$

where d_1, d_2 are positive constants. So we get $\mathfrak{N}(R, v) < 0 = v(0)$ for sufficiently small $R > 0$, which contradicts the subharmonicity of the function v.

The case when $T(\varphi) = h(\varphi)$, $\varphi \in (a-\gamma, a)$ for $\gamma > 0$ is simpler and can be treated similarly. The author is indebted to M. L. Sodin for the idea of the above proof of condition (6).

Let us prove implication (8). Assume the contrary: suppose

$$\underline{\lim}_{\varphi\to a+0} \frac{h(\varphi) - g(\varphi)}{\varphi - a} = 0,$$

but there exists a $\varphi_0 \in (a, b)$ such that $h(\varphi_0) > g(\varphi_0)$. Taking into account the definitions (0.3) and (0.4), we conclude that there exists a function $v_0 \in \mathrm{Fr}[f]$ such that

$$g(\varphi_0) \leqslant v^0(e^{i\varphi_0}) < h(\varphi_0). \quad (1.18)$$

Then the function $v^1(re^{i\varphi}) = v^0(re^{i\varphi}) - H(re^{i\varphi})$ is subharmonic and nonpositive in $\Gamma(a, b)$; by virtue of the maximum principle, $v^1(re^{i\varphi}) < 0$, $(re^{i\varphi}) \in \Gamma(a, b)$. Besides, since

$$\underline{\lim}_{\varphi\to a+0} \frac{h(\varphi) - g(\varphi)}{\varphi - a} = 0,$$

we obtain

$$0 = \underline{\lim}_{\varphi\to a+0} \frac{h(\varphi) - g(\varphi)}{\varphi - a} \geqslant \underline{\lim}_{\varphi\to a+0} \frac{h(\varphi) - v^0(e^{i\varphi})}{\varphi - a} = -\overline{\lim}_{\varphi\to a+0} \frac{v^1(e^{i\varphi})}{\varphi - a}$$

whence, recollecting that $v^1(e^{i\varphi}) < 0$, $re^{i\varphi} \in \Gamma(a, b)$, we get

$$\overline{\lim}_{\varphi\to a+0} \frac{v^1(e^{i\varphi})}{\varphi - a} = 0.$$

Applying Lemma 1 to the function

$$W(re^{i\varphi}) = v^1(re^{i(\varphi+a)}), \qquad re^{i\varphi} \in \Gamma(0, \gamma),\ \gamma = b - a,$$

we get $v^1 \equiv 0$ in $\Gamma(a, b)$, which leads to a contradiction. Thus we have proved that the conditions of Theorem 1 are necessary.

§2. Proof of the sufficiency of the conditions of Theorem 1

We introduce the notation

$$U[\rho, \infty) \stackrel{\text{def}}{=} \bigcup_{\sigma>0} U[\rho, \sigma]$$

and write $E \Subset U[\rho, \infty)$ if $E \subset U[\rho, \sigma]$ for a $\sigma > 0$.

The proof of sufficiency in Theorem 1 will be based on the following theorem.

THEOREM C. *Let $\Lambda \Subset U[\rho, \infty)$ be a set closed in D' and invariant with respect to the transformation $(\cdot)_\tau$; let $\rho(r)$ be an arbitrary proximate order for an order $\rho > 0$.*

Then there exists an $f \in A(\rho(r))$ for which

$$h_f(\varphi) = \sup\{v(e^{i\varphi}) : v \in \Lambda\}, \qquad \underline{h}_f(\varphi) = \inf\{v(e^{i\varphi}) : v \in \Lambda\}.$$

For a noninteger ρ, Theorem C follows from [19, Theorem 2, p. 156] and is given in [9, p. 11]. For an integer ρ, Theorem C follows from [21].

By virtue of Theorem C, the proof is reduced to constructing a set $\Lambda \subset U[\rho, \sigma]$ closed in D', invariant with respect to the transformation $(\cdot)_\tau$ and such that

$$h(\varphi) = \sup\{v(e^{i\varphi}) : v \in \Lambda\}, \qquad \varphi \in [0, 2\pi), \tag{2.1}$$

$$g(\varphi) = \inf\{v(e^{i\varphi}) : v \in \Lambda\}, \qquad \varphi \in [0, 2\pi). \tag{2.2}$$

Note that for constructing the set Λ it is sufficient to show that for any $\theta \in [0, 2\pi]$ there exists a sequence $\{u_n^\theta\}_{n=1}^\infty$ of functions, subharmonic in the complex plane $\mathbb{C}$, such that

$$r^\rho g(\varphi) \leqslant u_n^\theta(re^{i\varphi}) \leqslant r^\rho h(\varphi), \quad \varphi \in [0, 2\pi), \ r > 0, \ n = 1, 2, \dots, \tag{2.3}$$

$$\lim_{n\to\infty} u_n^\theta(e^{i\theta}) = g(\theta), \tag{2.4}$$

$$u_n^\theta(0) = 0. \tag{2.5}$$

Indeed, if the sequence $\{u_n^\theta\}_{n=1}^\infty$ is already constructed, then we can consider the set

$$\mathring{\Lambda} = \{(u_n^\theta)_\tau : \tau \in (0, \infty), \ \theta \in [0, 2\pi), \ n = 1, 2, \dots\}.$$

Note that the set $\mathring{\Lambda} \Subset U[\rho, \infty)$ and is invariant with respect to the transformation $(\cdot)_\tau$. We set

$$\Lambda = \{\operatorname{clos} \mathring{\Lambda}\} \cup \{h\}, \tag{2.6}$$

where $\operatorname{clos} \mathring{\Lambda}$ is the closure of the set $\mathring{\Lambda}$ in D'. Let $v \in \Lambda$ and $v = \lim_{j\to\infty} v_j$ in D', $v_j \in \mathring{\Lambda}$. Let us show that inequalities (2.3) are fulfilled for the function v. We need the following proposition.

LEMMA A. *Let* $\{u_n\}$ *be a sequence of functions subharmonic in* $\mathbb{C}$, *and* $u_n \to u$ *in* D'. *Then the equality*

$$\left(\overline{\lim_{n\to\infty}}\, u_n(z)\right)^* = u(z)$$

holds, where * *denotes an upper semicontinuous regularization in* z:

$$(\cdot)^*(z) = \lim_{\varepsilon\to 0} \sup_{|z-z'|<\varepsilon} (\cdot)(z').$$

Combining Lemma A and (2.3), we get

$$\tau^\rho h(\varphi) \geqslant v(\tau e^{i\varphi}) = \left(\overline{\lim_{j\to\infty}}\, v_j\right)^*(\tau e^{i\varphi}) \geqslant \tau^\rho g(\varphi), \qquad \varphi \in [0, 2\pi), \ \tau > 0.$$

On the other hand, inf is attained for any φ on any sequence $u_n^\varphi(e^{i\varphi})$ by property (2.4), whereas sup is attained on the function h; thus (2.1) and (2.2) are fulfilled.

Now, we pass to constructing a family $\{u_n^\theta\}_{n=1}^\infty$ satisfying (2.3)–(2.5).

Let $\theta \in [0, 2\pi)$. If $h(\theta) = g(\theta)$, then for any $n = 1, 2, \dots$, one can set

$$u_n^\theta(re^{i\varphi}) \stackrel{\text{def}}{=} r^\rho h(\varphi), \qquad \varphi \in [0, 2\pi), \ r > 0.$$

Here (2.3)–(2.5) are evidently fulfilled. So it is sufficient to carry out the construction for the case $h(\theta) > g(\theta)$.

Further it will be convenient to regard h and g as 2π-periodic functions on the real axis $\mathbb{R}$. We denote $Q \stackrel{\text{def}}{=} \{\varphi \in \mathbb{R} : h(\varphi) = g(\varphi)\}$.

We shall define the interval (α, β) containing the point θ as follows. If $Q = \varnothing$, i.e., $h(\varphi) > g(\varphi)$, $\forall \varphi \in \mathbb{R}$, then we put $\alpha = -5\pi$, $\beta = 5\pi$. If $Q \neq \varnothing$, we use the fact that the set $\mathbb{R} \setminus Q$ is open and we present it as a union of an at most countable set of nonintersecting intervals

$$\mathbb{R} \setminus Q = \bigcup_{i=1}^{\infty} (\alpha_i, \beta_i).$$

By (α, β) we denote the interval (α_i, β_i) which contains the point θ when $h(\alpha) = g(\alpha)$, $h(\beta) = g(\beta)$ and $\theta - \alpha \leqslant 2\pi$, $\beta - \theta \leqslant 2\pi$; since h is a 2π-periodic function, we get $\alpha \geqslant \theta - 2\pi \geqslant -2\pi$, $\beta \leqslant \theta + 2\pi \leqslant 4\pi$. Thus, the interval (α, β) is chosen in such a way that

$$\alpha < \theta < \beta, \tag{2.7}$$

$$h(\varphi) > g(\varphi), \qquad \varphi \in (\alpha, \beta), \tag{2.8}$$

$$\text{if } Q \neq \varnothing, \quad \text{then } h(\alpha) = g(\alpha),\ h(\beta) = g(\beta), \tag{2.9}$$

$$\alpha \geqslant -5\pi, \qquad \beta \leqslant 5\pi, \tag{2.10}$$

$$\text{if } Q = \varnothing, \quad \text{then } \alpha = -5\pi,\ \beta = 5\pi. \tag{2.11}$$

Let Φ be a ρ-t.c. function and $\nu_{\Phi,a,b}$ be the Lebesgue-Stiltjes measure corresponding to the generating function

$$S_\Phi = \Phi'_-(\varphi) + \rho^2 \int_0^\varphi \Phi(t)\, dt.$$

The measure $\nu_{\Phi,a,b}$ is defined on a certain σ-algebra containing all Borel subsets of the interval (a, b).

In what follows we shall use the term "mass distribution of the function Φ" meaning the measure $\nu_{\Phi,a,b}$.

Let us list the following properties of the measure ν:

($\nu1$) $\nu((\varphi_1, \varphi_2)) = 0 \Leftrightarrow \Phi$ is a ρ-t. function, $\varphi \in (\varphi_1, \varphi_2)$.

($\nu2$) $\nu(\{\varphi\}) > 0 \Leftrightarrow \Phi'_+(\varphi) > \Phi'_-(\varphi)$.

We denote

$$\begin{gathered} \mu = \mu_h \overset{\text{def}}{=} \nu_{h,-5\pi,5\pi}, \qquad \mu(\varphi) \overset{\text{def}}{=} \mu(\{\varphi\}), \\ \mu(\varphi_1, \varphi_2) \overset{\text{def}}{=} \mu((\varphi_1, \varphi_2)), \\ \Omega^+(h, g, t) \overset{\text{def}}{=} \lim_{\varphi\to t+0} \frac{h(\varphi) - g(\varphi)}{\varphi - t}, \\ \Omega^-(h, g, t) \overset{\text{def}}{=} \lim_{\varphi\to t-0} \frac{h(\varphi) - g(\varphi)}{t - \varphi}. \end{gathered} \tag{2.12}$$

In constructing the family $\{u_n^\theta\}_{n=1}^\infty$ of subharmonic functions, different techniques are used, depending on the "concentration" of the mass μ in a vicinity of the point θ. The substantial factor here is the position of intervals of trigonometricity of the function h relative to the point θ.

A classification for all possible cases is given by the following lemma.

LEMMA 3.1. *Let $\theta \in [0, 2\pi)$ and let the functions h and g satisfy the conditions of Theorem* 1; *α, β, θ fulfill the conditions* (2.7)–(2.11); *the measure μ is defined by* (2.12).

Then one and only one of the following conditions ($\theta1$)–($\theta8$) *is fulfilled*:

($\theta1$) *either $\rho \geqslant 1$ or $\rho = 1/2$, $\alpha \leqslant \theta_1 < \theta < b \leqslant \beta$, $\mu(\theta_1, \theta) = 0$, $\mu(\theta, b) = 0$, where $\theta_1 = \theta - \pi/\rho$, $b = \theta + \pi/\rho$*;

($\theta2$) *$\rho > 1/2$ and there exists a θ_1 such that*

$$\theta - \pi/\rho < \theta_1 < \theta, \qquad \mu[\theta_1, \theta) > 0, \tag{2.13}$$

$$\theta < b \leqslant \beta, \qquad \mu(\theta, b) = 0, \tag{2.14}$$

$$\theta_1 \geqslant \alpha, \text{ and if } \theta_1 = \alpha, \text{ then } \mu(\theta_1, \theta) = 0 \text{ and } \Omega^+(h, g, \theta_1) > 0, \tag{2.15}$$

$$b - \theta_1 \leqslant 2\pi, \tag{2.16}$$

where $b = \theta + \pi/\rho$;

($\theta3$) *$\rho > 1/2$ and there exists a θ_1 such that*

$$\theta < \theta_1 < \theta + \pi/\rho, \qquad \mu(\theta, \theta_1] > 0, \tag{2.17}$$

$$\alpha \leqslant a < \theta, \qquad \mu(a, \theta) = 0, \tag{2.18}$$

$$\theta_1 \leqslant \beta, \text{ and if } \theta_1 = \beta, \text{ then } \mu(\theta, \theta_1) = 0 \text{ and } \Omega^-(h, g, \theta_1) > 0, \tag{2.19}$$

$$\theta_1 - a \leqslant 2\pi, \tag{2.20}$$

where $a = \theta - \pi/\rho$;

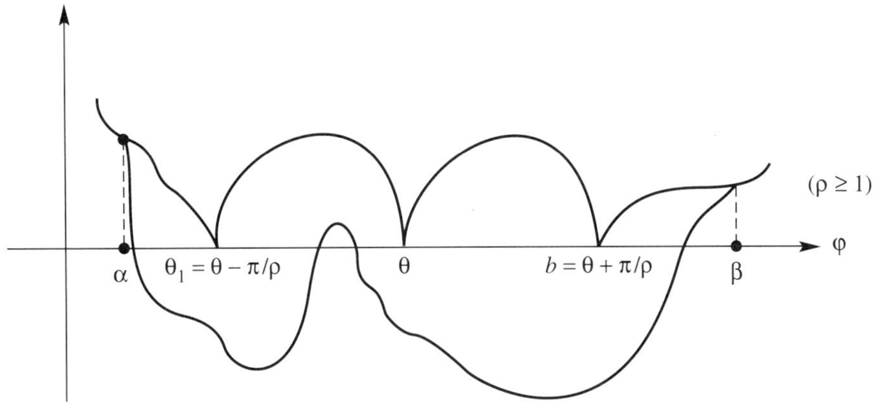

FIGURE 1

($\theta 4$) *$\rho \geqslant 1/2$ and there exist a, b such that $\alpha \leqslant a < \theta < b \leqslant \beta$, $b - a = \pi/\rho$, $\mu(a, b) = 0$;*

($\theta 5$) *$\rho > 0$ and there exist θ_1, θ_2 such that*

$$\alpha \leqslant \theta_1 < \theta < \theta_2 \leqslant \beta, \tag{2.21}$$

$$\theta_2 - \theta_1 < \min(\pi/\rho, 2\pi), \tag{2.22}$$

$$\mu[\theta_1, \theta) > 0, \tag{2.23}$$

$$\mu(\theta, \theta_2] > 0, \tag{2.24}$$

$$\textit{if } \theta_1 = \alpha, \textit{ then } \mu(\theta_1, \theta) = 0 \textit{ and } \Omega^+(h, g, \theta_1) > 0, \tag{2.25}$$

$$\textit{if } \theta_2 = \beta, \textit{ then } \mu(\theta, \theta_2) = 0 \textit{ and } \Omega^-(h, g, \theta_2) > 0; \tag{2.26}$$

($\theta 6$) *$\rho < 1/2$ and $\mu(\theta_1, \theta) = 0$, where $\theta_1 = \theta - 2\pi$;*

($\theta 7$) *$\rho < 1/2$ and there exist a, b such that*

$$a < \theta < b, \qquad b - a = 2\pi, \qquad \mu(a, b) = 0;$$

($\theta 8$) *$\rho > 0$, $\mu(\theta) > 0$ and there exist θ_1, θ_2 satisfying* (2.21), (2.23)–(2.26) *and the conditions*

$$\text{(E1)} \quad \theta - \theta_1 < \min(\pi/\rho, 2\pi), \qquad \text{(E2)} \quad \theta_2 - \theta < \min(\pi/\rho, 2\pi).$$

Note that the relations grouped in ($\theta 1$) give two "adjoining" trigonometricity intervals (θ_1, θ) and (θ, b) of length π/ρ ($\rho \geqslant 1$, Figure 1). When $\rho = 1/2$, these intervals merge into one of length $\pi/\rho = 2\pi$, and θ is its endpoint (Figure 2 on next page).

The relations ($\theta 2$) define one trigonometricity interval of length π/ρ ($\mu(\theta, b) = 0$, $b = \theta + \pi/\rho$), θ being its left endpoint. On the left of the point θ at the distance $< \pi/\rho$ "there is a mass", i.e., there is a point θ_1, $\theta - \pi/\rho < \theta_1 < \theta$, such that $\mu[\theta_1, \theta) > 0$. We assert that if the point θ_1 coincides with α, the left endpoint of the interval (α, β), where

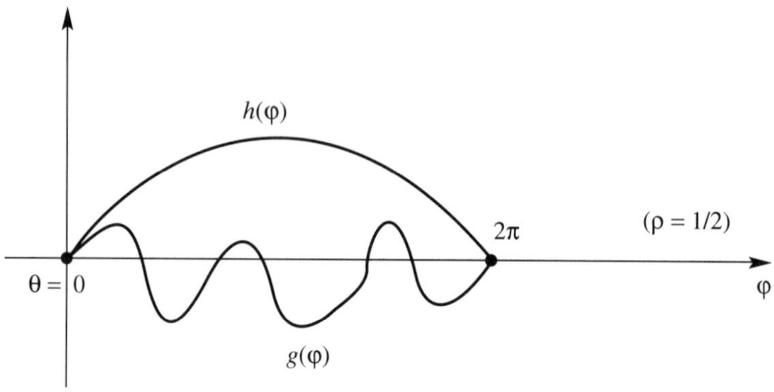

FIGURE 2

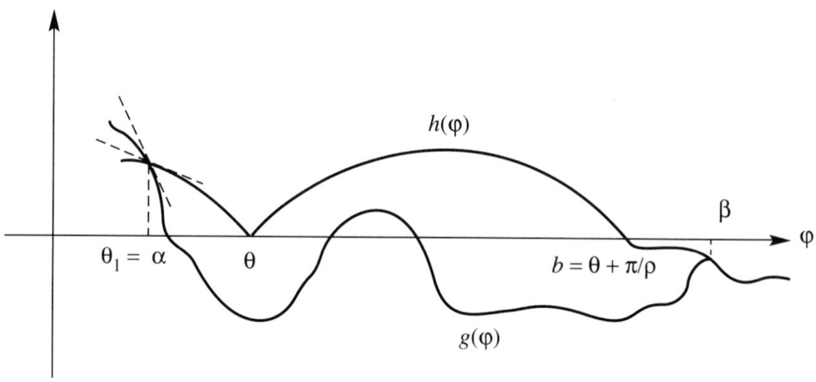

FIGURE 3

$h(\varphi) > g(\varphi)$, then (θ_1, θ) is a trigonometricity interval ($\mu(\theta_1, \theta) = 0$) with mass concentrated at the point θ_1 ($\mu[\theta_1, \theta) > 0$), and the function g "approaches" h at the point $\theta_1 = \alpha$ without "touching" it: $\Omega^+(h, g, \theta_1) > 0$. Note that the inequality $\Omega^+(h, g, \theta_1) > 0$ is implied by the concordance conditions (8) and (9), see Figure 3.

Conditions $(\theta 3)$ are similar to conditions $(\theta 2)$. Conditions $(\theta 4)$ imply the existence of a trigonometricity interval of length π/ρ, but here the point θ is inside the interval.

Conditions $(\theta 5)$ to $(\theta 8)$ imply a "lowering" of the function h (which remains ρ-trigonometrically convex) in the vicinity of the point θ.

The techniques applied for constructing the family $\{u_n^\theta\}$ of subharmonic functions satisfying (2.3)–(2.5) substantially differ depending whether we have the situations $(\theta 1)$–$(\theta 4)$ or $(\theta 5)$–$(\theta 8)$, but in both cases we shall use the same elementary operations which are described below.

The first elementary operation, which we shall call "pasting together", is the following: if $v_1(z)$ is subharmonic in D_1, and $v_2(z)$ is subharmonic in $D_2 \supset \partial D_1$, and if $v_1(z) \geqslant v_2(z)$ in $D_1 \cap D_2$, and $v_2(z) \geqslant v_1(z)$ in $D_2 \setminus D_1$, then the function

$$v_{\text{pasted}}(z) = \begin{cases} v_1(z), & z \in D_1, \\ v_2(z), & z \in D_2 \setminus D_1, \end{cases}$$

is subharmonic.

The second operation is "putting together": if v_1 is harmonic in D_1 and v_2 harmonic in D_2, the complement to D_1, $v_1 = v_2$ on $\partial D_1 = \partial D_2$, and $\partial v_2/\partial \vec{n} - \partial v_1/\partial \vec{n} \geqslant 0$ on ∂D_1 ($\vec{n}$ being the direction of the external normal to D_1), then the function

$$v_{\text{pt}} = \begin{cases} v_1(z), & z \in D_1, \\ v_2(z), & z \in D_2, \end{cases}$$

is subharmonic.

The third operation is constructing a "convex combination". If $v_1(z)$ and $v_2(z)$ are subharmonic in D, then the function

$$v_{\text{con}}(z) = \lambda v_1(z) + (1-\lambda) v_2(z), \qquad \lambda \in (0, 1),$$

is also subharmonic; by a proper choice of λ, the functions v_1 and v_{con} can be made arbitrarily close, together with their normal derivatives on ∂D. All of this is true under the proper assumptions as to smoothness of the boundaries and the functions, of course.

Besides the three above-mentioned operations, which are extended to ρ-t.c. functions as the corresponding operations over subharmonic functions of the form $r^\rho h(\varphi)$, we shall introduce the operation of "continuation" of a ρ-t.c. function $h(\varphi)$ from a point α by means of a trigonometric function. This means that for ρ-t.c. function $h(\varphi)$ we shall consider, at the point α, the following functions

$$\eta^+(h, \alpha, \varphi) \overset{\text{def}}{=} h'_+(\alpha)\rho^{-1} \sin\rho(\varphi - a) + h(\alpha)\cos\rho(\varphi - \alpha),$$

$$\eta^-(h, \alpha, \varphi) \overset{\text{def}}{=} h'_-(\alpha)\rho^{-1} \sin\rho(\varphi - a) + h(\alpha)\cos\rho(\varphi - \alpha),$$

where $h'_\pm(\alpha)$ are the derivatives from the left and from the right of the function h at the point α. The functions $\eta^\pm$ possess the properties

$$\eta^+(h, \alpha, \varphi) \leqslant h(\varphi), \qquad \varphi \in (\alpha, \alpha + \pi/\rho), \ \forall \rho\text{-t.c. } h(\varphi),$$

and

$$\eta^+(h, \alpha, \varphi) < h(\varphi), \qquad \varphi \in (\alpha, \alpha + \pi/\rho),$$

if $h(\varphi)$ is not a ρ-t. function for $\varphi \in (\alpha, \alpha + \pi/\rho)$. The function

$$H(\varphi) = \begin{cases} h(\varphi), & \varphi \in [\alpha - \delta, \alpha], \\ \eta^+(h, \alpha, \varphi), & \varphi \in [\alpha, \alpha + \delta], \end{cases}$$

where $\delta > 0$, is a ρ-t.c.f. The function η^- possesses similar symmetric properties. Note that the functions $\eta^\pm$ were used in [23] for the extrapolation of a ρ-t.c. function from a point α with preservation of the ρ-t.c.

property. The operation of "continuation" will be used for "lowering" the ρ-t.c. function h on some interval.

Further on, in stating and proving lemmas, it will be convenient to denote the right and left derivatives of a function Φ at the point φ as $\Phi'_+(\varphi)$ and $\Phi'_-(\varphi)$, along with $\Phi'(\varphi+0)$ and $\Phi'(\varphi-0)$.

The first operation will be applied in the case of $(\theta 5)$–$(\theta 8)$ and will use the "lowering" of the function h. At first we shall examine this case. Suppose that we have $(\theta 5)$. Then the proof of the theorem is based on the following Lemmas 4.1 and 5.1.

LEMMA 4.1. *Let $H(\varphi)$ be a ρ-t.c. function, $G(\varphi)$ be a USC function, $\varphi \in (-5\pi, 5\pi)$; let H and G be 2π-periodic, and*

$$H(\varphi) \geqslant G(\varphi), \qquad \varphi \in [0, 2\pi), \qquad \omega_0 \in [0, 2\pi);$$

let $\mu = \nu_{H, -5\pi, 5\pi}$ be the mass distribution of the function H, and let there exist ω_1, ω_2, such that

$$\omega_1 < \omega_0 < \omega_2, \tag{2.27}$$

$$H(\varphi) > G(\varphi), \qquad \varphi \in (\omega_1, \omega_2), \tag{2.28}$$

$$\omega_2 - \omega_1 < \min(\pi/\rho, 2\pi), \tag{2.29}$$

$$\mu[\omega_1, \omega_0) > 0, \qquad \mu(\omega_0, \omega_2] > 0, \tag{2.30}$$

satisfying the following implications:

$$(H(\omega_1) = G(\omega_1)) \Rightarrow (\mu(\omega_1) > 0) \wedge (\Omega^+(H, G, \omega_1) > 0), \tag{2.31}$$

$$(H(\omega_2) = G(\omega_2)) \Rightarrow (\mu(\omega_2) > 0) \wedge (\Omega^-(H, G, \omega_2) > 0). \tag{2.32}$$

Then for each $\Delta > 0$ there exist ρ-t.c. functions $f_1(\varphi)$, $f_2(\varphi)$, $\varphi \in (-5\pi, 5\pi)$, and constants ω_1^j, ω_2^j, $j = 1, 2$, ω_3, γ_1, γ_2 such that

$$\omega_1 \leqslant \omega_1^j \leqslant \gamma_1 < \omega_0 < \gamma_2 \leqslant \omega_2^j \leqslant \omega_2, \qquad j = 1, 2, \tag{2.33}$$

$$\omega_1^2 < \omega_3 < \omega_2^2, \tag{2.34}$$

$$G(\varphi) < f_j(\varphi) < H(\varphi), \qquad \varphi \in (\omega_1^j, \omega_2^j), \ j = 1, 2, \tag{2.35}$$

$$f_j(\varphi) = H(\varphi), \qquad \varphi \in (-5\pi, 5\pi) \setminus (\omega_1^j, \omega_2^j), \tag{2.36}$$

$$H'(\omega_1^j + 0) - \Delta < f_j'(\omega_1^j + 0) \leqslant H'(\omega_1^j + 0), \qquad j = 1, 2, \tag{2.37}$$

$$H'(\omega_2^j - 0) \leqslant f_j'(\omega_2^j - 0) < H'(\omega_2^j - 0) + \Delta, \qquad j = 1, 2, \tag{2.38}$$

$$f_1(\varphi) \text{ is strictly } \rho\text{-t.c. for } \varphi \in (\gamma_1, \gamma_2), \tag{2.39}$$

$$f_2(\varphi) \text{ is } \rho\text{-t. for } \varphi \in (\omega_3, \omega_2^2), \tag{2.40}$$

$$f_2'(\omega_3 - 0) < f_2'(\omega_3 + 0), \tag{2.41}$$

$$\mu[\omega_1^j, \omega_0) > 0, \qquad \mu(\omega_0, \omega_2^j] > 0, \tag{2.42}$$

satisfying the following implications:

$$\begin{aligned} &\mu(\omega_i) > 0 \Rightarrow \omega_i = \gamma_i = \omega_i^1 = \omega_i^2, && i = 1, 2, \\ &\mu(\omega_1) > 0 \Rightarrow f_j'(\omega_1 + 0) < H'(\omega_1 + 0), && j = 1, 2, \\ &\mu(\omega_2) > 0 \Rightarrow f_j'(\omega_2 - 0) > H'(\omega_2 - 0), && j = 1, 2. \end{aligned} \tag{2.43}$$

LEMMA 5.1. *Let $\theta \in [0, 2\pi)$ and let the functions h and g satisfy the conditions of Theorem* 1. *Suppose also that there exist a ρ-t.c. 2π-periodic function $f_1(\varphi)$ and constants γ_1, γ_2 such that*

$$\gamma_1 < \theta < \gamma_2, \tag{2.44}$$

$$g(\varphi) \leqslant f_1(\varphi) < h(\varphi), \qquad \varphi \in (\gamma_1, \gamma_2), \tag{2.45}$$

$$g(\varphi) \leqslant f_1(\varphi) \leqslant h(\varphi), \qquad \varphi \in (-5\pi, 5\pi), \tag{2.46}$$

$$\gamma_2 - \gamma_1 < \min(\pi/\rho, 2\pi), \tag{2.47}$$

$$\text{the function } f_1(\varphi) \text{ is strictly } \rho\text{-t.c. for } \varphi \in (\gamma_1, \gamma_2). \tag{2.48}$$

Then there exists a sequence $\{u_n^\theta\}_{n=1}^\infty$ of functions subharmonic in the plane and satisfying (2.3)–(2.5).

Thus, Lemma 4.1 permits us, when there is a mass in the vicinity of the point ω_0, to "lower" the ρ-t.c. function H in this vicinity so that it will be bounded from below by the given function G. These "lowerings" are governed by the functions f_1 and f_2.

The function f_1 has a strictly ρ-t.c. fragment on the interval $(\gamma_1, \gamma_2) \subset (\omega_1, \omega_2)$ and coincides with H outside the interval $(\omega_1^1, \omega_2^1) \subset (\omega_1, \omega_2)$. The right and the left derivatives of the functions f_1 and H at the points ω_1^1, ω_2^1 are, therefore, close enough. If, for example, masses of the initial function H were concentrated at the points ω_1, ω_2, then the intervals (γ_1, γ_2) and (ω_1^1, ω_2^1) coincide with the initial interval (ω_1, ω_2). Condition (2.42) shows that the points ω_1^1, ω_2^1 are chosen so that "masses" on the segment $[\omega_1^1, \omega_2^1]$ are "preserved".

The function f_2 has properties similar to those of f_1. The difference is that the function f_2 is trigonometric on the interval $(\omega_3, \omega_2^2) \subset (\omega_1^2, \omega_2^2)$ and has a concentrated mass at the point ω_3 by virtue of condition (2.41). The function f_2 will be used in some versions of the second operation, in particular, when conditions $(\theta 2)$, $(\theta 3)$ are considered.

The required family of subharmonic functions for the case when such a ρ-t.c. function f_1 with the above-described properties exists is constructed in Lemma 5.1.

Let us show that the case when conditions $(\theta 5)$ hold is reduced to the consecutive application of Lemmas 4.1 and 5.1.

Let Δ be an arbitrary positive number. We set $\omega_1 = \theta_1$, $\omega_2 = \theta_2$, $\omega_0 = \theta$, $H(\varphi) = h(\varphi)$, $G(\varphi) = g(\varphi)$, $\varphi \in (-5\pi, 5\pi)$, where θ_1, θ, θ_2 satisfy (2.21)–(2.26).

First, we shall show that the conditions of Lemma 4.1 are fulfilled. Note that by virtue of (2.7), (2.8), (2.21)–(2.24) the conditions (2.27)–(2.30) are fulfilled. Condition (2.31) follows from (2.23) and (2.25) whereas (2.32) follows from (2.24) and (2.26). First, we apply Lemma 4.1.

Let $f_1(\varphi)$, $\varphi \in (-5\pi, 5\pi)$ be the ρ-t.c. function obtained by virtue of this lemma and fulfilling (2.35)–(2.39). Applying Lemma 5.1, we get the proof of sufficiency in Theorem 1. Thus, the case of $(\theta 5)$ is closed.

Now let us consider the case when either $(\theta 6)$ or $(\theta 7)$ is fulfilled. Note first that it follows from the conditions of the theorem that h cannot be ρ-t. on the entire real axis (circle). Indeed, otherwise ρ must be an integer, and from the concordance condition (2) (see Introduction) it follows that $h = g$, which contradicts the conditions of Theorem 1.

We shall show that in the case when either $(\theta 6)$ or $(\theta 7)$ holds, Lemma 5.1 can be applied, i.e., there exists a function f_1 and constants γ_1, γ_2 fulfilling (2.44)–(2.48).

Note that in the case $(\theta 6)$ we have $h(\varphi) > g(\varphi)$, $\varphi \in [0, 2\pi)$ and $h(\varphi) > c > 0$; in this case any convex combination

$$f_1(\varphi) = (1-\lambda)h(\varphi) + \lambda c, \qquad \varphi \in [0, 2\pi), \ 0 < \lambda < 1, \tag{2.49}$$

can be taken for f_1 with an appropriate choice of $\lambda > 0$.

In the case $(\theta 7)$, we have $h(\varphi) > c = h(a) = h(b)$ and the function f_1 is defined by (2.49). By an argument similar to that used in examining $(\theta 5)$ it can be shown (applying Lemma 4.1) that in the case $(\theta 8)$ there exists an f_1 satisfying (2.44)–(2.48), i.e., we have the conditions of Lemma 5.1.

In the case when one of the conditions $(\theta 1)$–$(\theta 4)$ holds (i.e., the function h has a trigonometricity interval of length π/ρ and the point θ belongs to this interval or is its endpoint) the second operation is applied for constructing the family of subharmonic functions; the technique is based on Lemma 6.1.

LEMMA 6.1. *Let $\theta \in [0, 2\pi)$ and let the functions h and g satisfy the conditions of Theorem* 1, *and one of the conditions* $(\theta 1)$–$(\theta 4)$ *of Lemma* 3.1.

Then there exists a sequence of functions $\{u_n^\theta\}_{n=1}^\infty$ subharmonic in the plane and satisfying (2.3)–(2.5).

Because of the lack of space, in the proof of Lemmas 4.1 and 6.1 we shall confine ourselves to the consideration of the most typical cases. Sometimes we shall only discuss the main ideas, skipping proofs of auxiliary lemmas.

§3. Proof of Lemma 3.1

Note that for the distribution of the mass μ in the vicinity of the point θ the following cases (F1)–(F4) can occur:

(F1) there exist γ_1, γ_2 such that $\gamma_1 < \theta < \gamma_2$, $\mu(\gamma_1, \theta) = 0$, $(\theta, \gamma_2) = 0$;

(F2) there exists a γ_1 such that $\gamma_1 < \theta$, $\mu(\gamma_1, \theta) = 0$ and $\mu(\theta, \gamma_2) > 0$ for each $\gamma_2 > 0$;

(F3) there exists a γ_2 such that $\gamma_2 > \theta$, $\mu(\theta, \gamma_2) = 0$ and $\mu(\gamma_1, \theta) > 0$ for each $\gamma_1 < \theta$;

(F4) for each γ_1, γ_2 such that $\gamma_1 < \theta < \gamma_2$ we have the inequalities $\mu(\gamma_1, \theta) > 0$, $\mu(\theta, \gamma_2) > 0$.

We are going to show that each of the above-listed cases reduces to one of conditions $(\theta 1)$–$(\theta 8)$.

Examine the case (F1). Set

$$\alpha_1 = \inf\{\varphi : \mu(\varphi, \theta) = 0\}, \tag{3.1}$$

$$\beta_1 = \sup\{\varphi : \mu(\theta, \varphi) = 0\}. \tag{3.2}$$

Note that $\alpha_1 \leqslant \gamma_1 < \theta < \gamma_2 \leqslant \beta_1$.

First let us show that the following relations are valid:

$$\alpha_1 \geqslant \alpha, \tag{3.3}$$

$$\beta_1 \leqslant \beta, \tag{3.4}$$

$$\text{if } \alpha_1 = \alpha, \text{ then } \mu(\alpha, \theta) = 0, \ \mu(\alpha) > 0, \ \Omega^+(h, g, \alpha) > 0, \tag{3.5}$$

$$\text{if } \beta_1 = \beta, \text{ then } \mu(\theta, \beta) = 0, \ \mu(\beta) > 0, \ \Omega^-(h, g, \beta) > 0. \tag{3.6}$$

Note that $\theta - \alpha_1 \leqslant 2\pi$, $\beta_1 - \theta \leqslant 2\pi$, since, in the converse case, the function h will be ρ-t. on the entire real axis (circle), whence $h \equiv g$ which contradicts Theorem 1. We have

$$\alpha_1 \geqslant -2\pi, \tag{3.7}$$

$$\beta_1 \leqslant 4\pi. \tag{3.8}$$

Therefore if $Q = \varnothing$ (i.e., $h(\varphi) > g(\varphi)$, $\varphi \in \mathbb{R}$, where $\mathbb{R}$ is the set of real numbers), then, by virtue of (2.11), (3.7), (3.8), the inequalities (3.3) and (3.4) are fulfilled.

Let $Q \neq \varnothing$ (i.e., $\exists \varphi_0 \in \mathbb{R}$ such that $h(\varphi_0) = g(\varphi_0)$). We shall show that $\alpha_1 \geqslant \alpha$. Assume the contrary: $\alpha_1 < \alpha$; then, by virtue of (2.9) $h(\alpha) = g(\alpha)$; and from the concordance condition (3) applied either to the interval (α_1, θ) or to the interval (α_1, β_1) (if $\mu(\theta) = 0$), we get $\mu(\alpha) > 0$, which contradicts the condition $\mu(\alpha_1, \theta) = 0$. This proves (3.3); relation (3.4) is proved in a similar way.

Let us prove (3.5). Assume that $\alpha_1 = \alpha$; then $h(\alpha) = g(\alpha)$. Indeed, by virtue of (2.7)–(2.11), $h(\alpha) > g(\alpha)$ can hold only in the case when $Q = \varnothing$, but then $\alpha = -5\pi$, and $\alpha_1 = \alpha$ implies that $h(\varphi)$ is a ρ-t. function for all $\varphi \in \mathbb{R}$, which contradicts the condition of the theorem.

As $\alpha_1 = \alpha$, then $\mu(\alpha, \theta) = 0$, and from the concordance condition (8) applied to the maximal interval of ρ-trigonometricity of the function h: (α_1, β_1) or (α_1, θ) (if $\mu(\theta) > 0$) we obtain $\Omega^+(h, g, \alpha) > 0$, while from the concordance condition (4), we get $h'(\alpha+0) > h'(\alpha-0)$, which is equivalent to the condition $\mu(\alpha) > 0$. Thus, we have proved (3.5); condition (3.6) is proved in a similar manner.

Now we pass to the direct consideration of (F1).

First, we assume that $\mu(\theta) = 0$, then $\mu(\alpha_1, \beta_1) = 0$, $\beta_1 - \alpha_1 \leqslant 2\pi$ and, by virtue of the concordance condition (2) $\beta_1 - \alpha_1 \leqslant \pi/\rho$, hence $\beta_1 - \alpha_1 \leqslant \min(\pi/\rho, 2\pi)$. If $\beta_1 - \alpha_1 = \min(\pi/\rho, 2\pi)$, we denote $a = \alpha_1$, $b = \beta_1$ and note that for $\rho \geqslant 1/2$ condition $(\theta 4)$ holds, whereas for $\rho < 1/2$ condition $(\theta 7)$ holds. If $\beta_1 - \alpha_1 < \min(\pi/\rho, 2\pi)$, we shall show that $(\theta 5)$ holds. If $\alpha_1 = \alpha$, then we set $\theta_1 = \alpha$, and (2.23), (2.25) follow from (3.5). If simultaneously $\alpha_1 = \alpha$ and $\beta_1 = \beta$, then θ_1 and θ_2 chosen above obviously satisfy (2.21)–(2.26) and condition $(\theta 5)$ holds.

So we are left with one case, when $\beta_1 - \alpha_1 < \min(\pi/\rho, 2\pi)$ and either $\alpha < \alpha_1$ or $\beta_1 < \beta$. We shall examine one case only (the others are treated similarly). Suppose, for example, that $\alpha < \alpha_1$ and $\beta_1 < \beta$. By definition of α_1 and β_1, we have $\mu(\gamma, \alpha_1] > 0$ for every $\gamma < \alpha_1$, $\mu[\beta_1, \eta) > 0$ for every $\eta > \beta_1$. So, granting that $\beta_1 < \beta$ and $\alpha < \alpha_1$, one can choose θ_1 and θ_2 so that $\alpha < \theta_1 \leqslant \alpha_1$, $\beta_1 \leqslant \theta_2 < \beta$, which ensures (2.23)–(2.24). Thus, when $\mu(\theta) = 0$, $(\theta 5)$ holds.

Now examine the case when $\mu(\theta) > 0$. We denote $d = \min(\pi/\rho, 2\pi)$. The following cases are possible:

(G1) $\theta - \alpha_1 = d$, $\beta_1 - \theta = d$,
(G2) $\theta - \alpha_1 < d$, $\beta_1 - \theta = d$,
(G3) $\theta - \alpha_1 < d$, $\beta_1 - \theta < d$,
(G4) $\theta - \alpha_1 = d$, $\beta_1 - \theta < d$.

Examine the case (G1). Let us show that it is possible only for $\rho \geqslant 1$ or $\rho \leqslant 1/2$. Assume the contrary: $1/2 < \rho < 1$, then $d = \pi/\rho < 2\pi$. Since $\mu(\beta_1) > 0$ by virtue of the concordance condition (7) and since the function h is 2π-periodic, we get $\mu(\beta_1 - 2\pi) > 0$. But $1/2 < \rho < 1$, so $\alpha_1 = \beta_1 - 2\pi/\rho < \beta_1 - 2\pi < \beta_1 - \pi/\rho = \theta$ and we have $(\beta_1 - 2\pi) \in (\alpha_1, \theta)$, which contradicts the condition $\mu(\alpha_1, \theta) = 0$. This contradiction proves that in the case (G1) either $\rho \geqslant 1$ or $\rho \geqslant 1/2$.

We denote $\theta_1 = \alpha_1$, $b = \beta_1$ and claim that if $\rho \geqslant 1$ or $\rho = 1/2$, then $(\theta 1)$ holds, and if $\rho < 1/2$, then $(\theta 6)$ holds.

Let us consider the case (G2). Setting $b = \beta_1$, let us prove that in this case the inequality $\rho > 1/2$ holds. Assume the contrary: $\rho \leqslant 1/2$. Then $b - \theta = 2\pi$, $\mu(\theta, b) = 0$ and since h is 2π-periodic, we get $\mu(\theta - 2\pi, \theta) = 0$. Since $0 < \theta - \theta_1 < 2\pi$, we have $[\theta_1, \theta) \subset (\theta - 2\pi, \theta)$, whence $\mu[\theta_1, \theta) = 0$, which contradicts (2.13). Thus, $\rho > 1/2$.

We assert that $b - \alpha_1 \leqslant 2\pi$. Assume the contrary: $b - \alpha_1 > 2\pi$, then $b - 2\pi \in (\alpha_1, \theta)$ since $b - \theta = \pi/\rho < 2\pi$. But $\mu(b) = \mu(b - 2\pi) > 0$ by virtue of the concordance condition (7), whence we deduce a contradiction with the condition $\mu(\alpha_1, \theta) = 0$. So, $b - \alpha_1 \leqslant 2\pi$.

If $b - \alpha_1 = 2\pi$, then we set $\theta_1 = \alpha_1$, and θ_1 satisfies (2.13)–(2.16). If $b - \alpha_1 < 2\pi$, then by an argument similar to the one used above, we conclude that there exists a θ_1 satisfying (2.13)–(2.16).

Examine the case (G4). Setting $a = \alpha_1$ and reasoning as in the case (G2), we conclude that $(\theta 3)$ holds.

In case (G3), by a similar argument, we conclude that condition $(\theta 8)$ is fulfilled for $\mu(\theta) > 0$. For $\mu(\theta) = 0$ (in the case (G3)) $(\theta 5)$ holds provided that $\beta_1 - \alpha_1 < \min(\pi/\rho, 2\pi)$ and either $(\theta 4)$ or $(\theta 7)$ hold under the condition $\beta_1 - \alpha_1 = \min(\pi/\rho, 2\pi)$.

Examine the case (F2). Let α_1 be defined by (3.1). It is easy to see that if $\theta - \alpha_1 < \min(\pi/\rho, 2\pi)$, then $(\theta 5)$ holds, and if $\theta - \alpha_1 = \min(\pi/\rho, 2\pi)$, then $(\theta 3)$ holds.

Examine the case (F3). Let β_1 be defined by (3.2). If we have the inequality $\beta_1 - \theta < \min(\pi/\rho, 2\pi)$, then $(\theta 5)$ holds; if $\beta_1 - \theta = \min(\pi/\rho, 2\pi)$, then $(\theta 2)$ holds. It is obvious that $(\theta 5)$ holds in the case (F4). Lemma 3.1 is proved.

§4. Proof of Lemma 4.1

Let us examine the case when $\mu(\omega_1) > 0$, $\mu(\omega_2) > 0$,

$$H(\omega_1) = G(\omega_1), \qquad H(\omega_2) = G(\omega_2).$$

Denote

$$\eta^+(\zeta, \theta, \varphi) \stackrel{\text{def}}{=} \zeta'(\theta + 0)\rho^{-1} \sin \rho(\varphi - \theta) + \zeta(\theta) \cos \rho(\varphi - \theta),$$
$$\eta^-(\zeta, \theta, \varphi) \stackrel{\text{def}}{=} \zeta'(\theta - 0)\rho^{-1} \sin \rho(\varphi - \theta) + \zeta(\theta) \cos \rho(\varphi - \theta).$$

Set

$$H_1(\varphi) = \eta^-(H, \omega_1, \varphi), \qquad \varphi \in (\omega_1, \omega_2),$$
$$H_2(\varphi) = \eta^+(H, \omega_2, \varphi), \qquad \varphi \in (\omega_1, \omega_2).$$

One can show that $H_1(\varphi) < H(\varphi)$, $\varphi \in (\omega_1, \omega_2]$, $H_2(\varphi) < H(\varphi)$, $\varphi \in [\omega_1, \omega_2)$. We set

$$\gamma_1 = \omega_1^1 = \omega_1, \qquad \gamma_2 = \omega_2^1 = \omega_2, \qquad H_3(\varphi) = \max\{H_1(\varphi), H_2(\varphi)\},$$
$$f_1(\varphi) = \begin{cases} (1 - \lambda)H(\varphi) + \lambda(H_3(\varphi) + J(\varphi)), & \varphi \in (\omega_1, \omega_2), \\ H(\varphi), & \varphi \in (-5\pi, 5\pi) \setminus (\omega_1, \omega_2), \end{cases}$$

where

$$J(\varphi) = \int_{\omega_1}^{\omega_2} G(\psi, \varphi)\, d\psi, \qquad \varphi \in (\omega_1, \omega_2),$$

$G(\psi, \varphi)$ is the Green function of the differential operator $H'' + \rho^2 H$ on the interval (ω_1, ω_2); the function satisfies the boundary conditions $G(\omega_1, \varphi) = G'(\omega_2, \varphi) = 0$.

Choosing λ small enough and taking into consideration (2.31) and (2.32), one can show that f_1 satisfies the conditions of the lemma. The function $f_1(\varphi)$ is strictly ρ-t.c. for $\varphi \in (\omega_1, \omega_2)$, which is guaranteed by the "additional" summand $J(\varphi)$.

Examination of the cases when $\mu(\omega_1) = 0$ or/and $\mu(\omega_2) = 0$, as well as the construction of the function f_2, reduces to a slight modification of the construction described above. We omit the proof.

§5. Proof of Lemma 5.1

By $\mathscr{L}(\theta, L, t)$ we denote the set of functions $f(\varphi)$, $\varphi \in [\theta - t, \theta + t]$, $t > 0$, $L > 0$, satisfying the condition

$$|f(\varphi) - f(\theta)| \leqslant L|\varphi - \theta|, \qquad \varphi \in [\theta - t, \theta + t].$$

Making use of the fact that g is a USC function, we shall consider a monotonically decreasing sequence of functions $\{g_n^\theta\}_{n=1}^\infty$ which converges from above to the function g and satisfies

$$g(\varphi) \leqslant g_n^\theta(\varphi) < f_1(\varphi), \qquad \varphi \in [\theta - \gamma, \theta + \gamma], \tag{5.1}$$

$$g_n^\theta(\varphi) = g(\varphi), \qquad \varphi \in (-5\pi, 5\pi) \setminus [\theta - \gamma, \theta + \gamma], \tag{5.2}$$

$$g_n^\theta \in \mathscr{L}(\theta, L_n, \gamma), \tag{5.3}$$

where $\gamma > 0$ is chosen to ensure the inequality

$$\gamma < (1/2)\min(\theta - \gamma_1, \gamma_2 - \theta, \pi). \tag{5.4}$$

Thus, it is sufficient to construct a sequence of functions $\{u_n^\theta\}_{n=1}^\infty$ subharmonic in the plane that satisfy

$$r^\rho g_n^\theta(\varphi) \leqslant u_n^\theta(re^{i\varphi}) \leqslant r^\rho h(\varphi), \qquad \varphi \in [\theta_0, \theta_0 + 2\pi), \; r > 0, \tag{5.5}$$

$$u_n^\theta(e^{i\theta}) = g_n^\theta(\theta), \tag{5.6}$$

$$u_n^\theta(0) = 0, \tag{5.7}$$

where θ_0 is a point on the segment $[\theta - 2\pi, \theta + 2\pi]$.

Now, we begin constructing the sequence of functions $\{u_n^\theta\}$ satisfying (5.5)–(5.7). Denote $B(\zeta, R) \stackrel{\text{def}}{=} \{z : |z - \zeta| \leqslant R\}$. We fix an arbitrary integer $n > 0$ and denote

$$G(\varphi) = g_n^\theta(\varphi + \theta) - f_1(\varphi + \theta), \qquad \varphi \in (-\gamma, \gamma). \tag{5.8}$$

By virtue of (5.1)

$$G(\varphi) < 0, \qquad \varphi \in (-\gamma, \gamma). \tag{5.9}$$

Since f_1 is a ρ-t.c. function, it has right and left derivatives at each point, in particular, at the point θ, so there exist positive constants b and c such that

$$f_1 \in \mathscr{L}(\theta, b, c).$$

Using (5.3), we get

$$G \in \mathscr{L}(0, L, t), \tag{5.10}$$

where $L = \max(L_n, b)$, $t = \min(\gamma, c)$.

We choose constants θ_j^k, $j = 1, 2, 3$, $k = 1, 2$, satisfying the inequalities

$$\theta - t < \theta_2^1 < \theta_4^1 < \theta_1^1 < \theta < \theta_1^2 < \theta_2^2 < \theta_4^2 < \theta + t. \tag{5.11}$$

Let $i = 1, 2$; we shall consider ρ-trigonometric functions $H_i(\varphi)$, $\varphi \in (\theta_2^i, \theta_4^i)$ such that

$$H_i(\theta_2^i) = f_1(\theta_2^i), \qquad H_i(\theta_4^i) = f_1(\theta_4^i). \tag{5.12}$$

Using (2.47) and (2.48), we get $H_i(\varphi) > f_1(\varphi)$, $\varphi \in (\theta_2^i, \theta_4^i)$. We now set

$$h_i(\varphi) = \lambda_i H_i(\varphi) + (1 - \lambda_i) f_1(\varphi), \qquad \varphi \in (\theta_2^i, \theta_4^i),\ \lambda_i \in (0, 1). \tag{5.13}$$

Using (2.45), choose λ_i, $i = 1, 2$, so small as to ensure the inequalities $h_i(\varphi) < h(\varphi)$, $\varphi \in (\theta_2^i, \theta_4^i)$. We obtain

$$f_1(\varphi) < h_i(\varphi) < h(\varphi), \qquad \varphi \in (\theta_2^i, \theta_4^i), \tag{5.14}$$

$$h_i(\theta_2^i) = f_1(\theta_2^i), \tag{5.15}$$

$$h_i(\theta_4^i) = f_1(\theta_4^i). \tag{5.16}$$

Denote

$$\alpha_1 = \min_{i=1,2}\left(\max_{\varphi \in (\theta_2^i, \theta_4^i)} \{h_i(\varphi) - f_1(\varphi)\}\right),$$
$$K = 2^{\rho+1} L + \rho 2^{|\rho - 1|} |G(0)|, \qquad v^+(z) \overset{\text{def}}{=} \max(v(z), 0). \tag{5.17}$$

By θ_3^i we shall denote the point at which the expression in braces attains its maximum. We choose $\delta > 0$ so small as to ensure

$$\delta < (1/2)\min(t, -G(0)/K, 1),$$
$$B(1, \delta) \subset \Gamma(\theta_1^1 - \theta, \theta_1^2 - \theta) \subset \Gamma(-t, t).$$

We need the following two lemmas.

LEMMA 5.2. *Let t, δ, α be positive constants,*

$$t < \min(\pi/(2\rho), \pi), \qquad B(1, \delta) \subset \Gamma(-t, t).$$

Then there exists a function $v(z)$ subharmonic and continuous in the angle $\Gamma(-t, t)$ such that

$$v(z) \leqslant \alpha |z|^\rho, \qquad z \in \Gamma(-t, t), \tag{5.18}$$

$$v(z) > 0, \qquad z \in \Gamma(-t, t) \setminus B(1, \delta), \tag{5.19}$$

$$v(1) = -\infty, \tag{5.20}$$

$$v(0) = 0. \tag{5.21}$$

LEMMA 5.3. *Suppose $G(\varphi) < 0$, $\varphi \in (-t, t)$, $G \in \mathscr{L}(0, L, t)$, where L, t, δ, K, S are positive constants such that*

$$K = 2^{S+1} L + S 2^{|S-1|} |G(0)|, \tag{5.22}$$

$$B(1, \delta) \Subset \Gamma(-t, t), \qquad \delta < (1/2)\min(t, -G(0)/K, 1). \tag{5.23}$$

Let a function v *be subharmonic and continuous in the angle* $\Gamma(-t, t)$

$$v(z) \geqslant 0, \qquad z \in \Gamma(-t, t) \setminus B(1, \delta), \tag{5.24}$$
$$v(1) = -\infty. \tag{5.25}$$

Then there exists a function $W(z)$ *subharmonic and continuous in the angle* $\Gamma(-t, t)$ *such that*

$$W(1) = G(0), \tag{5.26}$$
$$W(z) = v(z), \qquad z \in \Gamma(-t, t) \setminus B(1, \delta), \tag{5.27}$$
$$r^S G(\varphi) \leqslant W(re^{i\varphi}) \leqslant v^+(re^{i\varphi}), \qquad re^{i\varphi} \in B(1, \delta). \tag{5.28}$$

We shall construct the family as follows. Without loss of generality, we can set $e^{i\theta} = 1$. At first, using Lemma 5.2, we shall prepare a "trunk", i.e., a subharmonic function $v \in U[\rho, \sigma]$ for which $v(1) = -\infty$. For such a function we take $v(z) = \beta(H(z, p) + c|z|^\rho)$, where

$$H(z, p) = \ln|1 - z| + \operatorname{Re}\left(\sum_{k=1}^{p} \frac{z^k}{k}\right)$$

is the logarithm of the modulus of the Weierstrass canonical factor, $p = [\rho]$, β and C are positive constants. The constant β permits us to make the "trunk" as narrow and its type $\alpha = \max v(z)|z|^{-\rho}$ in (5.18) as small as needed, whereas the constant C permits us to make v positive outside the neighborhood $B(1, \delta)$ from (5.19). Further, we need to point the trunk in such a way that its endpoint touches the function $G(\varphi)$ from above. This is done in Lemma 5.3 by "pasting together" with the function $G(0)+K|re^{i\varphi}-1|$, i.e., for W in Lemma 5.3 we take the function

$$W(re^{i\varphi}) = \begin{cases} \max(v(re^{i\varphi}), G(0) + K|re^{i\varphi} - 1|), & re^{i\varphi} \in B(1, \delta), \\ v(re^{i\varphi}), \quad re^{i\varphi} \in \Gamma(-t, t) \setminus B(1, \delta), \end{cases}$$

whose parameters permit us to control both "pointedness" and "height". Further, the function $W(re^{i\varphi}) + r^\rho f_1(\varphi)$ is "pasted" with "lowering" of the indicator of the function f_1 by means of "trigonometric arcs" $r^\rho h_1(\varphi)$, $r^\rho h_2(\varphi)$.

Now we pass to a more rigorous description. Let us choose $\alpha > 0$ so small as to ensure the inequalities: $\alpha < \alpha_1$ and $f_1(\varphi) + \alpha < h(\varphi)$, $\varphi \in (\theta_2^1, \theta_4^1)$. We set $s = \rho$ and consecutively apply Lemmas 5.2 and 5.3; as a result, we obtain functions v and w satisfying (5.18)–(5.21) and (5.26)–(5.28). Let us consider the functions

$$v_1(re^{i\varphi}) = v(re^{i(\varphi-\theta)}), \qquad u(re^{i\varphi}) = W(re^{i(\varphi-\theta)}) + r^\rho f_1(\varphi),$$

with $re^{i\varphi} \in \Gamma(\theta - t, \theta + t)$. Combining (5.27), (5.19) and (5.18), we have

$$u(re^{i\varphi}) = v_1(re^{i\varphi}) + r^\rho f_1(\varphi), \qquad re^{i\varphi} \in \Gamma(\theta - t, \theta + t) \setminus B(e^{i\theta}, \delta), \tag{5.29}$$

$$v_1(re^{i\varphi}) > 0, \quad re^{i\varphi} \in \Gamma(\theta - t, \theta + t) \setminus B(e^{i\theta}, \delta), \tag{5.30}$$

$$v_1(re^{i\varphi}) \leqslant \alpha r^\rho, re^{i\varphi} \in \Gamma(\theta - t, \theta + t). \tag{5.31}$$

From (5.28) and (5.8) it follows that

$$W(re^{i(\varphi-\theta)}) \geqslant r^\rho G(\varphi - \theta) = r^\rho (g_n^\theta(\varphi) - f_1(\varphi)), \qquad re^{i\varphi} \in B(e^{i\theta}, \delta),$$

whence

$$u(re^{i\varphi}) \geqslant r^\rho g_n^\theta(\varphi), \qquad re^{i\varphi} \in B(e^{i\theta}, \delta). \tag{5.32}$$

From (5.26) and (5.8) it follows that

$$u(e^{i\theta}) = g_n^\theta(\theta). \tag{5.33}$$

From (5.29) and (5.30), using (5.1) and (5.2), we get

$$u(re^{i\varphi}) \geqslant r^\rho f_1(\varphi) \geqslant r^\rho g_n^\theta(\varphi), \qquad re^{i\varphi} \in \Gamma(\theta - t, \theta + t) \setminus B(e^{i\theta}, \delta), \tag{5.34}$$

whence, using (5.32), we get

$$u(re^{i\varphi}) \geqslant r^\rho g_n^\theta(\varphi), \qquad re^{i\varphi} \in \Gamma(\theta - t, \theta + t). \tag{5.35}$$

We set

$$W_1(re^{i\varphi}) = \max\{r^\rho h_1(\varphi), u(re^{i\varphi})\}, \qquad re^{i\varphi} \in \Gamma[\theta_3^1, \theta_4^1],$$
$$W_2(re^{i\varphi}) = \max\{r^\rho h_2(\varphi), u(re^{i\varphi})\}, \qquad re^{i\varphi} \in \Gamma[\theta_2^2, \theta_3^2],$$

and

$$u_n^\theta(re^{i\varphi}) = \begin{cases} u(re^{i\varphi}), & re^{i\varphi} \in \Gamma(\theta_4^1, \theta_2^2), \\ W_1(re^{i\varphi}), & re^{i\varphi} \in \Gamma[\theta_3^1, \theta_4^1], \\ r^\rho h_1(\varphi), & re^{i\varphi} \in \Gamma[\theta_2^1, \theta_3^1], \\ W_2(re^{i\varphi}), & re^{i\varphi} \in \Gamma[\theta_2^2, \theta_3^2], \\ r^\rho h_2(\varphi), & re^{i\varphi} \in \Gamma(\theta_3^2, \theta_4^2], \\ r^\rho f_1(\varphi), & re^{i\varphi} \in \Gamma[\theta_0, \theta_0 + 2\pi) \setminus \Gamma[\theta_2^1, \theta_4^2], \end{cases}$$

where θ_0 is chosen so that $[\theta_0, \theta_0 + 2\pi) \Supset [\theta_2^1, \theta_4^2]$;

$$\Gamma[a, b] \overset{\text{def}}{=} \{re^{i\varphi} : \varphi \in [a, b], \ r \in (0, \infty)\},$$
$$\Gamma(a, b] \overset{\text{def}}{=} \{re^{i\varphi} : \varphi \in (a, b], \ r \in (0, \infty)\},$$
$$\Gamma[a, b) \overset{\text{def}}{=} \{re^{i\varphi} : \varphi \in [a, b), \ r \in (0, \infty)\}.$$

Note that (5.35), (5.1) and (5.14) imply

$$r^\rho g_n^\theta(\varphi) \leqslant u_n^\theta(re^{i\varphi}), \qquad \varphi \in [\theta_0, \theta_0 + 2\pi).$$

Taking into account the choice of α and (5.1), (5.18), (5.27), (5.28), we obtain

$$u_n^\theta(re^{i\varphi}) \leqslant r^\rho h(\varphi), \qquad \varphi \in [\theta_0, \theta_0 + 2\pi), \ r \in (0, \infty).$$

Hence (5.5) holds. Now (5.33) implies (5.6) and (5.21), (5.27) imply (5.7).

Let us show that the function u_n^θ is subharmonic. Note that u is subharmonic by construction. The functions $r^\rho h_1(\varphi)$, $r^\rho h_2(\varphi)$ are subharmonic in the angles $\Gamma(\theta_3^1, \theta_4^1)$, $\Gamma(\theta_2^2, \theta_3^2)$ because h_1 and h_2 are ρ-t.c. functions on the intervals (θ_3^1, θ_4^1) and (θ_2^2, θ_3^2), respectively. The functions W_1 and W_2 are also subharmonic, being the maxima of subharmonic functions. Now we show that the function u_n^θ is subharmonic in a neighborhood of the rays $\varphi = \theta_4^1$, $\varphi = \theta_2^2$, $\varphi = \theta_3^1$, $\varphi = \theta_3^2$. From (5.15), (5.16) and (5.34) it follows that

$$W_1(re^{i\varphi}) = u(re^{i\varphi}), \qquad \varphi = \theta_4^1, \ r \in (0, \infty),$$
$$W_2(re^{i\varphi}) = u(re^{i\varphi}), \qquad \varphi = \theta_2^2, \ r \in (0, \infty).$$

From (5.29), (5.31) and (5.17) it follows that

$$W_1(re^{i\varphi}) = r^\rho h_1(\varphi), \qquad \varphi = \theta_3^1, \ r \in (0, \infty),$$
$$W_2(re^{i\varphi}) = r^\rho h_2(\varphi), \qquad \varphi = \theta_3^2, \ r \in (0, \infty).$$

We see that the function u_n^θ is subharmonic in the plane. Lemma 5.1 is proved. The proofs of Lemmas 5.2 and 5.3 are omitted.

§6. Proof of Lemma 6.1

In the cases $(\theta 1)$–$(\theta 4)$ it is impossible to "lower" the function h in such a way that a result of the lowering has a fragment of strict trigonometric convexity in the vicinity of the point θ (so that Lemma 5.1 is not applicable), but the trigonometricity interval is exactly π/ρ long. So, in the role of the "trunk" one can use the Green function $G(\zeta, z)$ for the angle, since this function has the appropriate behavior for $U[\rho, \sigma]$ at 0 and at ∞.

The simplest is the situation $(\theta 4)$, when the point θ is inside the trigonometricity interval (see Lemma 6.4). Here for the "trunk" one can take the convex combination

$$P(re^{i\varphi}) = (1-\lambda) r^\rho h(\varphi) + \lambda G(e^{i\theta}, re^{i\varphi})$$

in the appropriate angle and choose a $\lambda > 0$ so that the function $P(re^{i\varphi})$ is close to $r^\rho h(\varphi)$, i.e., "put it together" with $r^\rho h(\varphi)$ outside the trigonometricity interval. Further the trunk "is pointed" so that the endpoint will touch the function $g(\varphi)$ from above, according to Lemma 5.3.

Situations $(\theta 1)$–$(\theta 3)$ (Lemma 6.2) are more complicated since the pole of the Green function must lie on the angle's side. To this end, for the "trunk" we take the Green function for the angle with local convexity at the point $e^{i\theta}$; this is done to ensure that the point $e^{i\theta}$ will be inside the domain and "put the trunk together" along the domain boundary with a harmonic having zero values and positive in the adjoint angle equal to π/ρ with concavity in the vicinity of the point $e^{i\theta}$. By means of a convex combination one can make the constructed function to have a "trunk" at the point $e^{i\theta}$, but to be near the indicator outside a small angle containing the point $e^{i\theta}$.

These conditions permit us to "put together" the obtained function with the indicator.

For the construction in the case $\rho = 1/2$ (Lemma 6.3), a similar method is used, but here one employs functions of the form $u(\sqrt{z})$, where u is a subharmonic function, and the regular branch is chosen for $\sqrt{z}$.

Let α, β be defined by (2.7)–(2.11), $0 < \gamma < \min\{\theta - \alpha, \beta - \theta, \pi\}$. We consider, as in the proof of Lemma 5.1, a monotonically decreasing sequence of semicontinuous from above functions $\{g_n^\theta\}_{n=1}^\infty$ which converges from above to the function g on the segment $[\theta-\gamma, \theta+\gamma]$ and satisfies the following conditions:

$$g(\varphi) \leqslant g_n^\theta(\varphi) < h(\varphi), \qquad \varphi \in (\alpha, \beta), \tag{6.1}$$

$$g_n^\theta(\varphi) = g(\varphi), \qquad \varphi \in (-5\pi, 5\pi) \setminus [\theta - \gamma, \theta + \gamma], \tag{6.2}$$

$$g_n^\theta \in \mathscr{L}(\theta, L_n, \gamma). \tag{6.3}$$

Note that the functions h and g_n^θ are concordant by virtue of (6.2).

Let us fix a positive integer n.

First we examine the case when $(\theta 1)$ holds. Here either $\rho \geqslant 1$ or $\rho = 1/2$ and there exist θ_1, b such that

$$\theta - \theta_1 = b - \theta = \pi/\rho, \tag{6.4}$$

$$\alpha \leqslant \theta_1, \tag{6.5}$$

$$b \leqslant \beta, \tag{6.6}$$

$$\mu(\theta_1, \theta) = \mu(\theta, b) = 0. \tag{6.7}$$

Taking account of the concordance conditions (6) and (7), one can show that there exists a ρ-t.c. function h_2 such that

$$\begin{aligned} g_n^\theta < h_2(\varphi) < h(\varphi), \qquad & \varphi \in (\theta_1, \theta),\ \varphi \in (\theta, b), \\ h_2(\varphi) = h(\varphi), \qquad & \varphi \in (-5\pi, 5\pi) \setminus \{(\theta_1, \theta) \cup (\theta, b)\}, \\ h_2(\varphi) > g_n^\theta(\varphi), \qquad & \varphi \in (\theta_1, b). \end{aligned}$$

Let us consider

$$\begin{aligned} H(\varphi) = h(\varphi) - h_2(\varphi), \qquad & \varphi \in (-5\pi, 5\pi), \\ G_n(\varphi) = g_n^\theta(\varphi) - h_2(\varphi), \qquad & \varphi \in (-5\pi, 5\pi). \end{aligned}$$

Since, by construction, the function $H(\varphi)$ is ρ-t. and positive for $\varphi \in (\theta_1, \theta)$, $\varphi \in (\theta, b)$, $H(\theta_1) = H(\theta) = H(b) = 0$, it can be written as

$$H(\varphi) = \begin{cases} A_1 \sin \rho(\theta - \varphi), & \varphi \in (\theta_1, \theta), \\ A_2 \sin \rho(\varphi - \theta), & \varphi \in (\theta, b), \\ 0, & \varphi \in (-5\pi, 5\pi) \setminus (\theta_1, b), \end{cases} \tag{6.8}$$

where $A_1 > 0$ and $A_2 > 0$.

By $\mathscr{L}(\theta)$ we denote the set of functions f such that $f \in \mathscr{L}(\theta, L, t)$ for some $L > 0$ and $t > 0$ depending on f.

Note the following properties of the function G_n:

$$G_n(\varphi) < 0, \qquad \varphi \in (\theta_1, b), \tag{6.9}$$

$$G_n \in \mathscr{L}(\theta), \tag{6.10}$$

$$G_n(\varphi) \leqslant 0, \qquad \varphi \in (-5\pi, 5\pi). \tag{6.11}$$

Let us set

$$\theta_0 = \begin{cases} \theta, & \rho = 1/2, \\ \theta_1, & \rho = 1, \\ \theta_1 - \lambda, & \rho > 1, \end{cases}$$

where $\lambda > 0$ is chosen so that $[\theta_1 - \lambda, \theta_1 - \lambda + 2\pi) \Supset [\theta_1, b]$.

Note that to construct a subharmonic function $u_n^\theta(z)$, $z \in \mathbb{C}$, satisfying (2.3)–(2.5), it is sufficient to construct a subharmonic function $v_\theta^n(z)$, $z \in \mathbb{C}$, such that

$$r^\rho G_n(\varphi) \leqslant v_\theta^n(re^{i\varphi}) \leqslant r^\rho H(\varphi), \qquad \varphi \in [\theta_0, \theta_0 + 2\pi), \ r > 0, \tag{6.12}$$

$$v_\theta^n(e^{i\theta}) = G_n(\theta), \tag{6.13}$$

$$v_\theta^n(0) = 0. \tag{6.14}$$

Indeed, the function $u_n^\theta(re^{i\varphi}) = v_\theta^n(re^{i\varphi}) + r^\rho h_2(\varphi)$ satisfies (2.3)–(2.5).

Now we pass to the construction of the function v_θ^n. First, we examine the case when $\rho \geqslant 1$. We shall need the following lemma (the proof is omitted).

LEMMA 6.2. *Let A, ε, Δ be positive constants, $\varepsilon < A$, $\Delta < 1$. The function $G(\varphi)$, $\varphi \in [-\pi, \pi]$, is USC and satisfies*

$$G(\varphi) < 0, \qquad \varphi \in (-\pi, \pi), \tag{6.15}$$

$$G \in \mathscr{L}(0),$$

$$G(-\pi) = G(\pi) \leqslant 0. \tag{6.16}$$

Then there exists a subharmonic function $W(z)$, $z \in \mathbb{C}$, continuous in $\mathbb{C}$ and satisfying

$$rG(\varphi) \leqslant W(re^{i\varphi}) \leqslant Ar \sin|\varphi|, \quad \varphi \in [-\pi, \pi), \ r \in (0, \infty), \tag{6.17}$$

$$W(re^{i\varphi}) \geqslant (A - \varepsilon) r \sin|\varphi|, \quad \varphi \in [-\pi, \pi) \setminus (-\Delta, \Delta), \ r \in (0, \infty), \tag{6.18}$$

$$W(1) = G(0). \tag{6.19}$$

We set $G(\varphi) = G_n(\varphi/\rho + \theta)$, $A = \min(A_1, A_2)$. Let us choose $\varepsilon > 0$ and $\Delta > 0$ so that $\varepsilon < A$, $\Delta < 1$; then, combining (6.9), (6.10) and (6.4), we get (6.15) and (6.16).

Let us apply Lemma 6.2. We denote

$$W_\theta^n(re^{i\varphi}) = W(r^\rho e^{i\rho(\varphi-\theta)}), \qquad \varphi \in (\theta_1, b),\ r \in (0, \infty), \tag{6.20}$$

where W is the function constructed according to Lemma 6.2.

Taking into consideration (6.17) and (6.20), we have

$$r^\rho G_n(\varphi) \leqslant W_\theta^n(re^{i\varphi}) \leqslant Ar^\rho \sin\rho|\varphi-\theta|, \quad \varphi \in (\theta_1, b),\ r \in (0, \infty), \tag{6.21}$$

$$W_\theta^n(e^{i\theta}) = G_n(\theta), \tag{6.22}$$

$$W_\theta^n(re^{i\varphi}) \geqslant (A-\varepsilon)r^\rho \sin\rho|\varphi-\theta|,$$
$$\varphi \in (\theta_1, b) \setminus (\theta - \Delta/\rho, \theta + \Delta/\rho),\ r > 0. \tag{6.23}$$

We set

$$v_\theta^n(re^{i\varphi}) = \begin{cases} W_\theta^n(re^{i\varphi}), & \varphi \in (\theta_1, b),\ r \in (0, \infty), \\ 0, & \varphi \in [\theta_0, \theta_0 + 2\pi) \setminus (\theta_1, b). \end{cases}$$

Note that (6.21), (6.8) and (6.11) imply (6.12). Relation (6.13) follows from (6.22), whereas (6.14) follows from (6.21), (6.23) and the continuity of the function $W_\theta^n(z)$.

Let us show that the function $v_\theta^n(z)$ is subharmonic. We note that the function $W_\theta^n(z)$ is subharmonic and continuous in the angle $\Gamma(\theta_1, b)$, it is equal to zero on the rays $\varphi = \theta_1$, $\varphi = b$. By virtue of (6.21) and (6.23), it follows that $W_\theta^n(z) \geqslant 0$ in the angles $\Gamma(\theta_1, \theta - \Delta/\rho)$, $\Gamma(\theta + \Delta/\rho, b)$, so the function $v_\theta^n(z)$ is subharmonic in the plane.

Examine the case when $\rho = 1/2$. We need the following lemma (the proof is omitted).

LEMMA 6.3. *Let $A > 0$, $G(\varphi)$ ($\varphi \in [0, 2\pi]$) be a USC function, $G(\varphi) < 0$, $\varphi \in [0, 2\pi)$, $G \in \mathscr{L}(0)$, $G(0) = G(2\pi)$.*

Then there exists a function $W(z)$ subharmonic, continuous in the plane and such that

$$r^{1/2} G(\varphi) \leqslant W(re^{i\varphi}) \leqslant Ar^{1/2}\sin(\varphi/2), \qquad \varphi \in [0, 2\pi),\ r > 0, \tag{6.24}$$

$$W(1) = G(0), \tag{6.25}$$

$$W(0) = 0. \tag{6.26}$$

We set $A = A_2$, $G(\varphi) = G_n(\varphi + \theta)$ and use Lemma 6.3. Further we set

$$v_\theta^n(re^{i\varphi}) = W(re^{i(\varphi-\theta)}), \qquad \varphi \in [\theta_0, \theta_0 + 2\pi),\ r \in (0, \infty).$$

We note that (6.12)–(6.14) follow from (6.24)–(6.26). Thus, the case ($\theta 1$) is closed.

Cases ($\theta 2$) or ($\theta 3$) are reduced, by way of certain preparatory constructions, to applications of Lemma 6.2. Since these constructions are somewhat cumbersome, we shall skip the proof.

Let us examine case $(\theta 4)$. Let $\varepsilon > 0$. We set

$$h_1(\varphi) = \begin{cases} (h'(\theta+0) - \varepsilon)\rho^{-1}\sin\rho(\varphi-\theta) + h(\theta)\cos\rho(\varphi-\theta), \\ \qquad \varphi \in (\theta_0, b), \\ h(\varphi), \quad \varphi \in [\theta_0, \theta_0 + 2\pi) \setminus (\theta_0, b). \end{cases}$$

By the same reasoning as in case $(\theta 1)$ we conclude that, by virtue of the concordance conditions (6) and (7), one can choose $\varepsilon > 0$ so small as to ensure the inequalities $g_n^\theta(\varphi) < h_1(\varphi) < h(\varphi)$, $\varphi \in (\theta_0, b)$, and the function $h_1(\varphi)$, $\varphi \in [\theta_0, \theta_0+2\pi)$, to be ρ-t.c. Now, for $\varphi \in [\theta_0, \theta_0+2\pi)$, we consider $H(\varphi) = h(\varphi) - h_1(\varphi)$, $G_n(\varphi) = g_n^\theta(\varphi) - h_1(\varphi)$. We have

$$H(\varphi) = \begin{cases} \varepsilon\rho^{-1}\sin\rho(\varphi-\theta), & \varphi \in (\theta_0, b), \\ 0, & \varphi \in [\theta_0, \theta_0+2\pi) \setminus (\theta_0, b). \end{cases}$$

We shall need the following lemma (the proof is omitted).

LEMMA 6.4. *Let* $\omega \in (0, \pi)$, $A > 0$, $0 < \Delta < \min(\omega, \pi - \omega)$, *let the function* $\Omega(\varphi)$, $\varphi \in [0, \pi]$, *be USC,* $\Omega(\varphi) \leqslant 0$, $\varphi \in [0, \pi]$, $\Omega(\varphi) < 0$, $\varphi \in (0, \pi)$, $\Omega \in \mathscr{L}(\omega)$.

Then there exists a function $P(z)$ *subharmonic in the upper half-plane* $\Pi = \{z : \operatorname{Im} z > 0\}$, *continuous in* $\overline{\Pi} = \{z : \operatorname{Im} z \geqslant 0\}$ *and satisfying*

$$\begin{aligned} r\Omega(\varphi) \leqslant P(re^{i\varphi}) &\leqslant Ar\sin\varphi, && \varphi \in [0, \pi],\ r \in (0, \infty), \\ P(e^{i\omega}) &= \Omega(\omega), \\ P(re^{i\varphi}) &\geqslant 0, && \varphi \in [0, \pi] \setminus (\omega - \Delta, \omega + \Delta),\ r \in (0, \infty). \end{aligned}$$

Using Lemma 6.4 choose a Δ so that $0 < \Delta < \rho\min(\theta - \theta_0, b - \theta)$. We set

$$\begin{aligned} \Omega(\varphi) &= G_n(\theta_0 + \varphi/\rho), & \varphi &\in [0, 2\pi], \\ \omega &= \rho(\theta - \theta_0), & A &= \varepsilon/\rho, \end{aligned}$$

and apply Lemma 6.4. We examine

$$W_1(re^{i\varphi}) = P(r^\rho e^{i\rho(\varphi-\theta_0)}), \qquad \varphi \in (\theta_0, b),\ r \in (0, \infty),$$

where the function P is constructed according to Lemma 6.4. We have

$$\begin{aligned} r^\rho G_n(\varphi) \leqslant W_1(re^{i\varphi}) &\leqslant Ar^\rho \sin\rho(\varphi - \theta_0), && \varphi \in (\theta_0, b),\ r \in (0, \infty), \\ W_1(e^{i\theta}) &= G_n(\theta), \\ W_1(re^{i\varphi}) &\geqslant 0, && \varphi \in [\theta_0, b) \setminus (\theta - d, \theta + d), \end{aligned}$$

where $0 < d < \min(\theta - \theta_0, b - \theta)$. We set

$$u_n^\theta(re^{i\varphi}) = \begin{cases} W_1(re^{i\varphi}) + r^\rho h_1(\varphi), & \varphi \in (\theta_0, b),\ r \in (0, \infty), \\ r^\rho h_1(\varphi), & \varphi \in [\theta_0, \theta_0 + 2\pi) \setminus (\theta_0, b). \end{cases}$$

Then u_n^θ is subharmonic in the plane and satisfies (2.3)–(2.5). Lemma 6.1 is proved.

Acknowledgments. The author is greatly indebted to V. S. Azarin for posing the problem and helpful remarks. The author is also obliged to V. B. Giner, M. L. Sodin and other participants of the seminar on the theory of functions at Kharkov University headed by B. Ya. Levin and I. V. Ostrovskiĭ.

References

1. B. Ya. Levin, *Distribution of zeros of entire functions*, GITTL, Moscow, 1956; English transl., Amer. Math. Soc., Providence, R.I., 1964.

2. V. S. Azarin, *On the asymptotic behavior of subharmonic functions of finite order*, Mat. Sb. **108** (1979), no. 2, 147–167; English transl. in Math. USSR-Sb. **36** (1980), 135–154.

3. A. F. Leontiev, *On the convergence of a sequence of Dirichlet polynomials*, Dokl. Akad. Nauk SSSR **108** (1956), 23–26. (Russian)

4. A. A. Goldberg, *Indicators of entire functions and an integral with respect to a nonadditive measure*, Contemporary Problems in the Theory of Analytic Functions, "Nauka", Moscow, 1966, pp. 88–93. (Russian)

5. I. F. Krasichkov-Ternovskiĭ, *Estimates from below for entire functions of finite order*, Sibirsk. Mat. Zh. **6** (1965), 840–861; English transl. in Amer. Math. Soc. Transl. Ser. 2 **83** (1969).

6. A. A. Kondratyuk, *Entire functions with finite maximal density of zeros* I, Teor. Funktsiĭ, Funktsional. Anal. i Prilozhen. **10** (1970), 57–70; II, **11** (1970), 35–40. (Russian)

7. A. A. Kondratyuk and A. A. Fridman, *The limit value of the lower indicator and estimates from below for entire functions with positive zeros*, Ukrain. Mat. Zh. **24** (1972), 488–494; English transl. in Ukrainian Math. J. **24** (1972).

8. J. Bailette, *Fonctions approchables par des sommes d'exponentielles*, J. Analyse Math. **10** (1962–1963), 91–114.

9. V. S. Azarin and L. R. Podoshev, *Limit sets and indicators of an entire function*, Sibirsk. Mat. Zh. **25** (1984), no. 6, 4–16; English transl. in Siberian Math. J. **25** (1984).

10. V. B. Giner, L. R. Podoshev, and M. L. Sodin, *On addition of lower indicators*, Dokl. Akad. Nauk SSSR **273** (1983), 32–33; English transl. in Soviet Math. Dokl. **28** (1983).

11. ———, *On addition of lower indicators of entire functions*, Teor. Funktsiĭ, Funktsional. Anal. i Prilozhen. **42** (1984), 27–36. (Russian)

12. Yu. F. Korobeĭnik, *On solution of some functional equations in the class of functions analytic in convex domains*, Mat. Sb. **75** (1968), 225–234; English transl. in Math. USSR-Sb. **4** (1968).

13. Yu. F. Korobeĭnik and A. V. Bratishchev, *On some characteristics of growth of subharmonic functions*, Mat. Sb. **106** (1978), 603–625; English transl. in Math. USSR-Sb. **34** (1978).

14. V. V. Napalkov, *On a class of equations of convolution type*, Uspekhi Mat. Nauk **29** (1974), 217–218. (Russian)

15. O. V. Epifanov, *On the convolution epimorphism in a convex set*, Dokl. Akad. Nauk SSSR **217** (1974), no. 3, 18–19; English transl. in Soviet Math. Dokl. **15** (1974).

16. L. R. Podoshev, *A complete description of the indicator-lower indicator pair for an entire function of finite order*, Dokl. Akad. Nauk SSSR **300** (1988), no. 1, 38–40; English transl. in Soviet Math. Dokl. **37** (1988).

17. M. L. Cartwright, *On functions which are regular and of finite order in an angle*, Proc. London Math. Soc. **38** (1934/35), no. 2, 158–179.

18. N. M. Chernykh, *A condition of complete regularity of growth of a function with a trigonometric indicator in an angle*, Manuscript No. 134, deposited at the VINITI, 1980. (Russian)

19. V. S. Azarin, *On rays of completely regular growth of an entire function*, Mat. Sb. **79** (1969), 463–476; English transl. in Math. USSR-Sb. **8** (1969).

20. M. L. Sodin, *On growth in L P metrics of entire functions of finite lower order*, Manuscript No. 420-NKD83, deposited at the UkrNIINTI, 1983. (Russian)

21. R. Sigurdsson, *Growth properties of analytic and plurisubharmonic functions of finite order*, Math. Scand. **59** (1986), 235–304.

22. V. S. Azarin, *Theory of growth of subharmonic functions.* I, Kharkov University, Kharkov, 1978. (Russian)

23. O. V. Epifanov, *A differential operator with polynomial coefficients in classes of entire functions with a given estimate of the indicator*, Mat. Sb. **114** (1981), no. 1, 85–109; English transl. in Math. USSR-Sb. **42** (1982).

KHARKOV STATE UNIVERSITY, 4 INDEPENDENCE SQUARE, KHARKOV 310077 UKRAINE

Translated by V. A. TKACHENKO

ADVANCES IN SOVIET MATHEMATICS
Volume 11, 1992

Solvability Conditions for the Homogeneous Riemann Boundary Problem with an Infinite Index

I. V. OSTROVSKIĬ

§1. Statement of main results

In the sixties N. V. Govorov constructed a theory for the Riemann boundary problem with infinite index of polynomial growth—the theory later described in the monograph [1]. This theory was generalized by a number of authors in a number of directions [2–11]. In this article we shall dwell on the homogeneous problem only. The main difference between previous results and ours lies in much weaker restrictions on the behavior at infinity of the argument of the coefficient of the problem.

We shall consider curves of the form

$$L = \{z : z = re^{i\varphi(r)},\ 2 \leqslant r_0 < r < \infty\}$$

as integration paths, where φ is a real continuously differentiable function such that

$$\lim_{r\to\infty}(r\varphi'(r) - \varphi(r)/\ln r) = 0, \qquad \limsup_{r\to\infty}|\varphi(r)|/\ln r < \infty, \tag{1.1}$$

and, besides, $\varphi' \in \Delta$, where Δ is the class of continuous functions g on $[r_0, \infty)$, whose continuity modulus $\omega_R(r, g)$ on $[r_0, R]$ for any $R > r_0$ satisfies the Dini condition

$$\int_0 \omega_R(\delta, g)\delta^{-1}d\delta < \infty.$$

We shall call such curves *admissible* and characterize them by values

$$C(L) = \limsup_{r\to\infty}|\varphi(r)|/\ln r, \qquad c(L) = \liminf_{r\to\infty}|\varphi(r)|/\ln r.$$

The example of the curve L for which $\varphi(r) = (a + b\sin(\ln\ln r)^{\alpha})\ln r$, where $0 < \alpha < 1$ and $a, b > 0$, shows that admissible curves with any given $c(L)$, $C(L)$, $0 \leqslant c(L) \leqslant C(L) < \infty$, exist.

A plane cut along an admissible curve L will be denoted by $\mathscr{D}$. Let $(\mathscr{D})$ be a set obtained by including into $\mathscr{D}$ both sides of the cut along $L \setminus \{t_0\}$,

---

1991 *Mathematics Subject Classification*. Primary 30E25.

1051-8037/92 $1.00 + $.25 per page

$t_0 = r_0 \exp\{i\varphi(r_0)\}$. We shall consider the problem of finding a function Φ, analytic in $\mathscr{D}$, continuous and bounded in $(\mathscr{D})$ whose values $\Phi^{\pm}$ on the cut's sides satisfy the equation

$$\Phi^{+}(t) = G(t)\Phi^{-}(t), \qquad t \in L \setminus \{t_0\}, \tag{1.2}$$

where G, the coefficient of the problem, is a continuous function not vanishing on L. This problem will be called the *homogeneous Riemann boundary problem.*

We shall assume that $\ln G(re^{i\varphi(r)}) \in \Delta$. To write out the restrictions on the asymptotic behavior of $\ln G$ at infinity, we need the concept of Boutroux proximate order.

We recall that the *Boutroux proximate order* is a function $l(r)$ continuously differentiable on $[0, \infty)$ and satisfying the conditions

$$\lim_{r\to\infty} l'(r) r \ln r = 0, \qquad 0 < \lambda = \liminf_{r\to\infty} l(r) \leqslant \limsup_{r\to\infty} l(r) = \rho < \infty. \tag{1.3}$$

The example $l(r) = \lambda + (1/2)(\rho - \lambda)(1 + \sin\ln\ln(r+b))$, $b > e^e$, shows that for any pair of numbers (λ, ρ), $0 < \lambda \leqslant \rho < \infty$, there exists a Boutroux proximate order satisfying (1.3). We shall use the notation $V(r) = r^{l(r)}$ and assume that the reader is acquainted with the properties of the Boutroux proximate order as they are described in [12, p. 91].

As to the behavior of $\ln G$ at infinity, we shall assume the following. There exists a pair of the Boutroux proximate orders $(l(r), l_1(r))$ such that

$$\begin{aligned} p = [\lambda] < \lambda = \liminf_{r\to\infty} l(r) &\leqslant \limsup_{r\to\infty} l(r) \\ &= \rho < \min\{p+1, (1/2)(1 + c^2(L))\}, \end{aligned} \tag{1.4}$$

$$V_1(r) = o(V(r)/\ln(r)), \qquad r \to \infty \quad (V_1(r) = r^{l_1(r)}), \tag{1.5}$$

and an integer-valued nondecreasing function $n(r) \geqslant 0$, $r_0 \leqslant r < \infty$, which are related to the curve L and to $\ln G$ in the following way. Setting

$$\Lambda(t) = (2\pi)^{-1} \arg G(t) - n(t), \tag{1.6}$$

we shall assume that the following conditions are satisfied:

(A) we have the relation

$$\Lambda(t) = V(|t|) + o(V_1(|t|)), \qquad t \to \infty; \tag{1.7}$$

(B) the inequality

$$\limsup_{r\to\infty} V^{-1}(r) \int_0^{k(r)} (\Lambda(t^{+}(r, u)) - \Lambda(t^{-}(r, u))) u^{-1}\, du \leqslant 0, \tag{1.8}$$

is fulfilled, where $k(r) = \exp\{-V(r)/(V_1(r)\ln r)\}$, and $t^{\pm}(r, k)$ are the points on L such that $|t^{\pm}(r, u) - re^{i\varphi(r)}| = u$, $|t^{+}(r, u)| > r > |t^{-}(r, u)|$ (as we shall show later on, for $0 < u < k(r)$ and r sufficiently large the points $t^{\pm}(r, u)$ are determined uniquely);

(C) we have the estimate

$$|\ln|G(t)||\left\{1+|t|\int_0^{k(|t|/2)}\frac{\omega_{3|t|}(\delta\,,\,\varphi')}{\delta}\,d\delta\right\}=o(V_1(|t|))\,,\qquad t\to\infty. \tag{1.9}$$

Condition (A) means that the function $\arg G(re^{i\varphi(r)})$ may be approximated closely enough by a nondecreasing function; this becomes especially clear when $\arg G(re^{i\varphi(r)})$ grows faster than r^p. Note that if L is the ray $\{t : r_0 \leqslant t < \infty\}$, then $c(L) = 0$ and (1.4) and (1.5) imply that $\rho < 1/2$, $V(r) = o(r^{1/2})$, $V_1(r) = o(r^{1/2}/\ln r)$, $r \to \infty$.

Condition (B) can be regarded as a restriction on regularity of the behavior of $\arg G(re^{i\varphi(r)})$ for large r, which forbids this function to grow too rapidly on small intervals. Note that the condition will be evidently fulfilled if

$$\limsup_{r\to\infty} V^{-1}(r)\int_0^{k(r)}(\arg G(t^+(r\,,\,u))-\arg G(t^-(r\,,\,u)))u^{-1}du\leqslant 0\,, \tag{1.10}$$

but, as we shall see later (see the note at the end of §9), condition (1.10) is much less stringent than (B). Further we shall show (see proof of Theorem 3) that condition (1.10) and, hence, condition (B) will be satisfied if the function $\arg G(re^{i\varphi(r)})$ is continuously differentiable and its derivative has no more than the polynomial growth. Note that when L is a ray, conditions (1.8) and (1.10) respectively take the form

$$\limsup_{r\to\infty} V^{-1}(r)\int_0^{k(r)}(\Lambda(t+u)-\Lambda(t-u))u^{-1}du\leqslant 0\,, \tag{1.11}$$

$$\limsup_{r\to\infty} V^{-1}(r)\int_0^{k(r)}(\arg G(t+u)-\arg G(t-u))u^{-1}du\leqslant 0. \tag{1.12}$$

Condition (C) in the general case includes restrictions on both the growth of $|\ln|G(re^{i\varphi(r)})||$ and the regularity of the behavior of the curve L at infinity. It is trivially fulfilled, of course, when $|G(t)| \equiv 1$; no additional restrictions are imposed on L in this case.

In the general case, condition (C) yields the estimate

$$|\ln|G(t)|| = o\left(V_1(|t|)\right)\,,\qquad t\to\infty\,, \tag{1.13}$$

which is reduced to (C) when L is a ray.

The main result of the present work is the following theorem.

THEOREM 1. *Let L be an admissible curve. If $\ln G \in \Delta$ and conditions (A), (B), (C) are fulfilled, then the problem (1.2) has a nontrivial solution Φ which satisfies the estimate*

$$|\Phi(z)| \leqslant C\exp\{-CV(|z|)\} \tag{1.14}$$

in the domain $\mathscr{D}$.

Henceforth by the letter C, *perhaps with indices, we shall denote positive, generally unequal, constants.*

We shall see from the proof of Theorem 1 (see the note at the end of §6) that the solution Φ has a more regular asymptotic behavior than (1.14) suggests. Let us state the corresponding result for the case when L is a ray.

THEOREM $1'$. *Let* $L = \{t : 2 \leqslant r_0 \leqslant t < \infty\}$. *If* $\ln G \in \Delta$ *and conditions* (1.7), (1.11) *and* (1.13) *hold, then the problem* (1.12) *has a solution* Φ *which, besides* (1.14), *satisfies the condition*: *for each* $\varepsilon > 0$ *the relation*

$$\ln |\Phi(re^{i\theta})| = \left\{ -\frac{\pi}{\sin \pi l(r)} \cos l(r)(\theta - \pi) + o(1) \right\} V(r), \qquad r \to \infty, \qquad (1.15)$$

is fulfilled uniformly with respect to $\theta \in [\varepsilon, 2\pi - \varepsilon]$.

Let us consider also the problem (1.2) under the following conditions which are simpler than (A), (B), (C): there exists an integer-valued function $n(r) \uparrow \infty$ such that the function Λ defined by (1.6) satisfies the conditions:

(A') $\Lambda(t) = O(1)$, $t \to \infty$;

(B') for $q > 0$, we have

$$\limsup_{r\to\infty} (\ln r)^{-1} \int_0^{r^{-q}} (\Lambda(t^+(r, u)) - \Lambda(t^-(r, u)))u^{-1}\, du < \infty.$$

Besides, by (C') we shall denote the condition which is obtained from (C) by replacing $k(|t|)$ by $|t|^{-q}$ and $o(V_1(|t|))$ by $O(1)$ in the right- and left-hand sides.

THEOREM 2. *Let* L *be an admissible curve. If* $\ln G \in \Delta$ *and conditions* (A'), (B'), (C') *are fulfilled, then the problem* (1.2) *has a nontrivial solution* Φ.

As will be seen from the proofs of Theorems 1 and 2, their assertions still hold, if the coefficient G is unimodular, that is $|G(t)| \equiv 1$, for a somewhat wider class than admissible curves, viz. without the condition $\varphi' \in \Delta$.

§2. Relationships with some known results

N. V. Govorov [1, Chapter 5] examined the problem (1.2) in the case when L is a ray $\{t : r_0 \leqslant t < \infty\}$ under the following restrictions on the coefficient G:

(i) the representation

$$(2\pi)^{-1} \arg G(t) = \psi(t)t^{\rho}, \qquad r_0 \leqslant t < \infty, \qquad (2.1)$$

is valid, where $\rho > 0$ and the function ψ satisfies the conditions

$$\psi \in H(\mu_0), \ 1 \geqslant \mu_0 > \left(\frac{2\rho - 1}{2\rho + 1}\right)^+, \ \varphi(\infty) > 0, \ 0 \geqslant \varphi(r_0)r_0^{\rho} > -1; \quad (2.2)$$

(ii) we have $\ln |G| \in H(\mu)$, $0 < \mu \leqslant 1$.

Here $H(\mu)$ denotes the class of functions g on $[r_0, \infty)$ which satisfy the Hölder condition $|\psi(t_1) - \psi(t_2)| \leqslant C|t_1^{-1} - t_2^{-1}|^\mu$, $r_0 \leqslant t_1 < t_2 < \infty$.

Under these restrictions N. V. Govorov established [1, §28] that if a positive number $\sigma < \min(1/2, \rho)$ satisfies $\mu_0 > (\rho - \sigma)/(1 + \rho - \sigma)$, then the problem (1.2) has a nontrivial solution Φ estimated by

$$|\Phi(z)| \leqslant C \exp\{-C|z|^\sigma\}, \qquad z \in D. \tag{2.3}$$

We shall show that this result is contained in Theorem 1.

The fact that $\ln G \in \Delta$ immediately follows from conditions (i) and (ii). As established in [1, p. 166, (28.9)], it follows from condition (i) that there exist a nondecreasing integer-valued function $n(t)$ and a number $\nu \in (0, \sigma)$ such that

$$(2\pi)^{-1} \arg G(t) = n(t) + t^\sigma + O(t^\nu), \qquad t \to \infty. \tag{2.4}$$

Setting $l(t) = \sigma$, $l_1(t) = \nu$, we see that condition (A) is fulfilled when $V(t) = t^\sigma$, $V_1(t) = t^\nu$.

To check condition (B), it is sufficient to verify the validity of (1.12). Since $\psi \in H(\mu_0)$, it follows from (2.1) and (2.2) that for $0 < u \leqslant t/2$ we have

$$\begin{aligned} &|\arg G(t+u) - \arg G(t-u)| \\ &\quad \leqslant |\psi(t+u) - \psi(t-u)|(t+u)^\rho + |\psi(t-u)|((t+u)^\rho - (t-u)^\rho) \\ &\quad \leqslant C_1 u^{\mu_0} t^{\rho - 2\mu_0} + C_2 u t^{\rho - 1}. \end{aligned}$$

Since $k(t) = \exp\{-t^{\sigma - \nu}/\ln t\}$, condition (B) is fulfilled.

Condition (C) is fulfilled since (ii) implies $\ln|G(t)| = O(1)$.

I. E. Sandrigaĭlo [3] considered the Riemann boundary problem for the half-plane, similar to problem (1.2), when L is a ray, under the following restrictions on the coefficient G:

(i′) the representation $(2\pi)^{-1} \arg G(t) = \psi(t) t^{\rho(t)}$, $r_0 \leqslant t < \infty$, is valid, where $\rho(t)$ is a proximate order (a particular case of the Boutroux proximate order with $\lambda = \rho$) and the function ψ satisfies (2.2);

(ii′) $t^{-\beta} \ln|G(t)| \in H(\mu)$, $0 < \mu \leqslant 1$, where $\beta \in (0, \sigma)$ and the number $\sigma < \min(1/2, \rho)$ satisfies $\mu_0 > (\rho - \sigma)/(1 + \rho - \sigma)$.

Judging by [3], it seems probable that I. E. Sandrigaĭlo could prove that under the restrictions indicated the problem (1.2) has a nontrivial solution estimated by (1.14). The latter is also contained in Theorem 1. Indeed, by using arguments similar to those given in [1, pp. 164–166], it is established that condition (i′) implies representation (2.4). So Theorem 1 can be applied as it was applied above (condition (C) is fulfilled since $\ln|G(t)| = O(t^\beta) = o(V_1(t))$, $t \to \infty$).

E. A. Danilov [5] considered the problem (1.2) in the case when L is a logarithmic helix $\{z : z = r\exp(ic\ln r),\ r_0 \leqslant r < \infty\}$, under the following assumptions as to the coefficient G:

(i$''$) the representation $(2\pi)^{-1}\arg G(t) = \psi(t)|t|^\rho$, $t \in L$, $\rho < (1+c^2)/2$, is valid, where $\psi \in H(\mu)$, $0 < \mu \leqslant 1$, $\psi(\infty) > 0$, $\psi(r_0 \exp(ic\ln r_0))r_0^\rho \in (-1, 0)$;

(ii$''$) we have $\ln|G| \in H(\mu)$, $0 < \mu \leqslant 1$.

This result is also contained in Theorem 1. Indeed, $\psi \in H(\mu)$ implies that $\psi(t) = \psi(\infty) + O(|t|^{-\mu})$, $t \to \infty$, whence

$$(2\pi)^{-1}\arg G(t) = \psi(\infty)|t|^\rho + O(|t|^{\rho-\mu}), \quad t \to \infty.$$

Setting $l(r) = \rho + (\ln\psi(\infty))/\ln r$, $l_1(r) = \rho - \mu$, we see that condition (A) with $n(r) \equiv 0$, $V(r) = \psi(\infty)r^\rho$, $V_1(r) = r^{\rho-\mu}$ is fulfilled.

Theorem 1 can be extended so as to cover the result of P. G. Yurov [2] concerning the homogeneous problem (1.2) for the case when $\arg G(t)$ grows as a power of the logarithm (see Theorem 1 in §6).

§3. Results following from the main ones

Theorems 1 and 2 permit one to formulate the conditions of solvability of the problem (1.2), the conditions allowing rather irregular growth of $\arg G(t)$, as $t \to \infty$, as well as arbitrarily fast growth. Theorem 3 treats conditions of the first type, Theorem 4 treats those of the second type. We should remark that the Riemann boundary problem with the argument of the coefficient growing arbitrarily fast but for classes containing unbounded functions Φ, was first considered by S. M. Grudskiĭ [7].

THEOREM 3. *Let L be an admissible curve. Suppose that the coefficient G of the problem* (1.2) *satisfies the conditions*:

(i) *the function* $\arg G(re^{i\varphi(r)})$ *is continuously differentiable, monotonically tends to* $+\infty$ *as* $r \to +\infty$, *and for a certain* $\eta > 0$ *satisfies the estimate*

$$(d/dr)\arg G(re^{i\varphi(r)}) = O(r^\eta), \qquad r \to \infty;$$

(ii) *the function* $\ln|G|$ *belongs to* Δ *and satisfies condition* (B$'$).

Then the problem (1.2) *has a nontrivial solution* Φ.

Note that condition (i) is fulfilled if $\arg G(t) = |t|^{l(|t|)}$, where $l(r)$ is the Boutroux proximate order with any fixed λ and ρ, $0 < \lambda \leqslant \rho < \infty$.

Before formulating Theorem 4, let us denote by F_β, $\beta > 0$, the class of entire functions h, $0 \leqslant h(r) \uparrow \infty$ $(r \uparrow \infty)$ satisfying the conditions:

(i) $|h(z)| \leqslant h(\operatorname{Re} z)$, $z \in \mathbb{C}$;

(ii) $h(r + h^{-\beta}(r)) = O(h(r))$, $r \to \infty$.

Examples of functions belonging to the class F_β for any $\beta > 0$ are $e_n(z)$, $e_n(z^2)$ (where $e_n(z)$ is the nth iteration of e^z), as well as their derivatives $e_n^{(k)}(z)$, $e_n^{(k)}(z^2)$, for any $k \in \mathbb{N}$. It is easy to see that if $h \in F_\beta$, then $h(r) \geqslant Ce^{cr}$ for some $C > 0$. It was established in [13] that the growth of the functions of the class F_β is not limited from above: for any nondecreasing function $0 \leqslant q(r) \uparrow \infty$ and any number $\beta > 0$ there exists a function $h \in F_\beta$ such that $h(r) \geqslant q(r)$, $r \geqslant 0$.

THEOREM 4. *Let L be an admissible curve. Suppose that the coefficient G of the problem* (1.2) *satisfies the conditions*:

(i) *the representation*

$$\frac{1}{2\pi}\arg G(re^{i\varphi(r)}) = \int_{r_0}^{r} h(u)\,du + V(r) - V(r_0), \qquad r \geqslant r_0, \tag{3.1}$$

is valid, where $h \in F_\beta$, $\beta \in (0, 1/2)$, and $V(r) = r^{l(r)}$ is the Boutroux proximate order fulfilling the condition (1.4);

(ii) *the function* $\ln|G|$ *belongs to* Δ *and satisfies condition* (C).

If we have

$$r\int_0^{k(r)} \omega_{2r}(\delta, \varphi')\delta^{-1}\,d\delta = o(V(r)), \qquad r \to \infty, \tag{3.2}$$

then problem (1.2) *has a nontrivial solution* Φ *admitting the estimate* (1.14).

§4. Asymptotics for a special Cauchy type integral

To prove Theorems 1 and 2 we shall need the following fact, which possibly has some interest of itself.

THEOREM 5. *Let $L = \{z : z = re^{i\varphi(r)},\ 2 \leqslant r_0 \leqslant r < \infty\}$ be a curve satisfying the conditions* (1.1), *$l(r)$ be a Boutroux proximate order satisfying the condition* (1.4), *$V(r) = r^{l(r)}$. We set*

$$I(z) = \frac{z^{p+1}}{2\pi i}\int_L \frac{V(|t|)\,dt}{t^{p+1}(t-z)}, \qquad z \in \mathbb{C}\setminus L. \tag{4.1}$$

For $\theta \in (\varphi(r), \varphi(r) + 2\pi)$ the following asymptotics $(r \to \infty)$ is valid:

$$I(re^{i\theta}) = (1 - e^{2\pi i\beta(r)})^{-1}\exp\{\beta(r)(\theta - \varphi(r))\}r^{l(r)} + o(r^{l(r)}), \tag{4.2}$$

uniformly relative to θ with $\beta(r) = (l(r)\ln r)(\ln r + i\varphi(r))^{-1}$.

Note that the integral I converges absolutely. It follows from the conditions (1.4) and from the fact that the length of the admissible curve can be estimated as

$$|dt| = \sqrt{1 + (|t|\varphi'(|t|))^2}\,d|t| \leqslant C\,d|t|.$$

In the case when L is a logarithmic helix $\{z : z = r\exp(ic\ln r)\}$ and $l(r) \equiv \rho$, we have $\beta(r) = \rho/(1+ic)$, so the asymptotics (4.2) can be strengthened:

$$\left(1 - \exp\left(\frac{2\pi i\rho}{1+ic}\right)\right) I(re^{i\theta}) = \exp\left(\frac{\rho(\ln r + i\theta)}{1+ic}\right) + O(r^p). \tag{4.3}$$

Relation (4.3) appears in [14], the proof being as follows. If $\varphi(r) = c\ln r$, then for $t \in L$ we have $|t|^\rho = t^{\rho/(1+ic)}$. Let $\mathscr{D}_R = \{z : r_0 < |z| < R\} \setminus L$; by virtue of the Cauchy theorem we have

$$z^{\rho/(1+ic)} = \frac{z^{p+1}}{2\pi i}\int_{\partial\mathscr{D}_R}\frac{t^{\rho/(1+ic)}\,dt}{t^{p+1}(t-z)}, \qquad z \in \mathscr{D}_R.$$

Letting $R \to \infty$, we get

$$z^{\rho/(1+ic)} = \left(1 - \exp\left(\frac{2\pi i\rho}{1+ic}\right)\right)I(z) - \frac{z^{p+1}}{2\pi i}\int_{|t|=r_0}\frac{t^{\rho/(1+ic)}dt}{t^{p+1}(t-z)},$$

whence (4.3) follows.

In the case when L is a curve of regular rotation and $l(r)$ is the ordinary proximate order (i.e., $\lambda = \rho$ in (1.4)), it was established in [14] that (4.2) holds uniformly relative to $\theta \in (\varphi(r)+\varepsilon, \varphi(r)+2\pi-\varepsilon)$ for any fixed $\varepsilon > 0$. To establish this, the author used an estimate for the difference between the integral $I(z)$ and the integral over the logarithmic helix with parameter $c = \lim_{r\to\infty} r\varphi'(r)$ and $l(r) \equiv \rho$, as well as the equality (4.3) for the latter integral. To prove Theorem 5 we shall use different approach.

Now let us pass to the proof. Note that for $t \in L$ we have $|t|^{l(|t|)} = t^{\beta(|t|)}$. By applying the generalized Cauchy theorem to the domain $\mathscr{D}_R$, we get

$$z^{\beta(|z|)} = \frac{z^{p+1}}{2\pi i}\int_{\partial\mathscr{D}_R}\frac{t^{\beta(|t|)}\,dt}{t^{p+1}(t-z)} + \frac{z^{p+1}}{2\pi i}\iint_{\mathscr{D}_R}\frac{\overline{\partial}t^{\beta(|t|)}}{t^{p+1}(t-z)}dt\wedge d\bar{t}.$$

Noticing that $|z^{\beta(|z|)}| \sim |z|^{l(|z|)}$ and letting $R \to \infty$, we have

$$\begin{aligned} z^{\beta(|z|)} &= \frac{z^{p+1}}{2\pi i}\int_L\frac{(1-e^{2\pi i\beta(|t|)})|t|^{l(|t|)}}{t^{p+1}(t-z)}dt - \frac{z^{p+1}}{2\pi i}\oint_{|t|=r_0}\frac{t^{\beta(|t|)}dt}{t^{p+1}(t-z)} \\ &+ \frac{z^{p+1}}{2\pi i}\iint_{\mathscr{D}_\infty}\frac{\overline{\partial}t^{\beta(|t|)}\,dt\wedge d\bar{t}}{t^{p+1}(t-z)} = I_1 + I_2 + I_3. \end{aligned} \tag{4.4}$$

First, let us estimate the integral I_3. By setting $z = re^{i\theta}$, where $\theta \in (\varphi(r), \varphi(r)+2\pi)$, $r > r_0$, we shall obtain the following estimate uniform with respect to θ

$$I_3(re^{i\theta}) = o(r^{l(r)}), \qquad r \to \infty. \tag{4.5}$$

We have

$$\overline{\partial}t^{\beta(|t|)} = (t^{\beta(|t|)}\ln t)\beta'(|t|)\overline{\partial}|t|, \qquad t = |t|e^{i\tau},\ \tau \in (\varphi(r), \varphi(r)+2\pi).$$

It is easy to see that the first of the conditions (1.1) together with the relation $l'(r)r\ln r = o(1)$, $r \to \infty$, implies the estimate $\beta'(r)r\ln r = o(1)$, $r \to \infty$. Taking into account the fact that

$$|t^{\beta(|t|)}| \sim |t|^{l(|t|)}, \quad |\ln t| = O(\ln|t|), \qquad t \to \infty,$$

(the second of the conditions (1.1) was used), we conclude that

$$|\bar{\partial} t^{\beta(|t|)}| = o(|t|^{l(|t|)-1}) \leqslant \varepsilon(|t|)|t|^{l(|t|)-1},$$

where $\varepsilon(r) > 0$ is some function monotonically tending to 0 as $r \to \infty$.

Hence

$$\begin{aligned} |I_3(re^{i\theta})| &\leqslant \frac{r^{p+1}}{\pi} \iint_{\mathscr{D}_\infty} \frac{\varepsilon(|t|)|t|^{l(|t|)-1}}{|t|^{p+1}|t - re^{i\theta}|} \left(\frac{1}{2i} dt \wedge d\bar{t}\right) \\ &= \frac{r^{p+1}}{\pi} \int_{r_0}^{\infty} \varepsilon(s) s^{l(s)-p-1} ds \int_0^{2\pi} \frac{d\tau}{|se^{i\tau} - re^{i\theta}|} \\ &= 2r^p \int_{r_0}^{\infty} \varepsilon(s) s^{l(s)-p-1} \eta(s/r)\, ds, \end{aligned}$$

where we use the notation

$$\eta(u) = \frac{1}{2\pi} \int_0^{2\pi} |ue^{i\tau} - 1|^{-1} d\tau, \qquad 0 \leqslant u < \infty$$

(the function η is frequently used in the theory of meromorphic functions, see [12, Chapter V]). It is easy to see that for $u \geqslant 2$ we have $\eta(u) \leqslant 2/u$, while for $0 < u \leqslant 2$ we have

$$\begin{aligned} \eta(u) &= \frac{1}{\pi} \left(\int_0^{|u-1|} + \int_{|u-1|}^{\pi/2} + \int_{\pi/2}^{\pi} \right) \frac{d\tau}{|u - e^{i\tau}|} \\ &\leqslant \frac{1}{\pi} \left(1 + \int_{|u-1|}^{\pi/2} \frac{d\tau}{\sin \tau} + \frac{\pi}{2} \right) \leqslant C \ln \frac{4}{|u-1|}. \end{aligned}$$

Therefore,

$$\begin{aligned} |I_3(re^{i\theta})| &\leqslant Cr^p \int_{r_0}^{2r} \varepsilon(s) s^{l(s)-p-1} \ln \frac{4r}{|s-r|} ds + Cr^{p+1} \int_{2r}^{\infty} \varepsilon(s) s^{l(s)-p-2} ds \\ &\leqslant Cr^p \int_{r_0}^{r/2} \varepsilon(s) s^{l(s)-p-1} ds + C\varepsilon(r/2) r^{l(r)-1} \int_{r/2}^{2r} \ln \frac{4r}{|s-1|} ds \\ &\quad + Cr^{p+1} \varepsilon(2r) \int_{2r}^{\infty} s^{l(s)-p-2}\, ds. \end{aligned}$$

By using the properties of the Boutroux proximate order [12, p. 91] we see that estimate (4.5) is valid.

Now let us pass to the integral I_1. We set $Q(z) = I_1(z) - (1 - e^{2\pi i \beta(|z|)}) I(z)$ and show that uniformly relative to $\theta \in (\varphi(r), \varphi(r) + 2\pi)$, we have

$$Q(re^{i\theta}) = o(r^{l(r)}), \qquad r \to \infty. \tag{4.6}$$

Let $\delta \in (0, 1/2)$ be a small number. Considering $r = |z|$ to be sufficiently large, we set

$$\begin{aligned} L_1 &= L \cap \{t : r_0 \leqslant |t| \leqslant \delta r\}, \\ L_2 &= L \cap \{t : \delta r \leqslant |t| \leqslant r/\delta\}, \\ L_3 &= L \cap \{t : |t| \geqslant r/\delta\}. \end{aligned}$$

We have

$$Q(z) = \frac{z^{p+1}}{2\pi i}\left(\int_{L_1} + \int_{L_2} + \int_{L_3}\right) \frac{(e^{2\pi i\beta(r)} - e^{2\pi i\beta(|t|)})|t|^{l(|t|)}\,dt}{t^{p+1}(t-z)}$$
$$= Q_1(z) + Q_2(z) + Q_3(z).$$

Since $|e^{2\pi i\beta(r)}| \leqslant e^{\pi l(r)} \leqslant C$, by using the properties of the Boutroux proximate order [12, p. 91] we get

$$|Q_1(z)| \leqslant Cr^{p+1}\int_{r_0}^{\delta r} \frac{s^{l(s)-p-1}}{r-s}ds \leqslant Cr^p \int_{r_0}^{\delta r} s^{l(s)-p-1}ds \leqslant C\delta^{(\lambda-p)/2} r^{l(r)},$$

$$|Q_3(z)| \leqslant Cr^{p+1}\int_{r/\delta}^{\infty} \frac{s^{l(s)-p-1}}{s-r}ds \leqslant Cr^{p+1}\int_{r/\delta}^{\infty} s^{l(s)-p-2}ds \leqslant C\delta^{(p+1-\rho)/2} r^{l(r)}.$$

To estimate $Q_2(z)$, we shall need the inequality

$$|e^{2\pi i\beta(r)} - e^{2\pi i\beta(s)}| \leqslant C|r-s|\{\delta r\ln(\delta r)\}^{-1}, \qquad s\in[\delta r, r/\delta], \tag{4.7}$$

To check (4.7), it is sufficient to note that

$$|e^{2\pi i\beta(r)} - e^{2\pi i\beta(s)}| \leqslant C|\beta(r)-\beta(s)| \leqslant C|r-s|\max\{|\beta'(u)| : u\in[\delta r, r/\delta]\},$$

and that, as it was already established, $|\beta'(u)| \leqslant C\{u\ln u\}^{-1}$. Using (4.7), we get

$$|Q_2(r)| \leqslant \frac{Cr^p}{\delta\ln(r\delta)}\int_{\delta r}^{r/\delta} s^{l(s)-p-1}\,ds \leqslant \frac{Cr^{l(r)}}{\delta^2\ln(r\delta)} = o(r^{l(r)}), \qquad r\to\infty.$$

Since the choice of δ was arbitrary, we arrive at (4.6).

From (4.6) it follows that

$$I_1(z) = (1-e^{2\pi i\beta(r)})I(z) + o(r^{l(r)}), \qquad r\to\infty.$$

Substituting this equality and (4.5) into (4.4) and taking into account the obvious relation $I_2(z) = O(r^\rho)$, $r\to\infty$, we get the asymptotic relation uniform relative to $\theta\in(\varphi(r), \varphi(r)+2\pi)$

$$(1-e^{2\pi i\beta(r)})I(re^{i\theta}) = \exp\{\beta(r)(\ln r + i\theta)\} + o(r^{l(r)}), \qquad r\to\infty. \tag{4.8}$$

Note that $|\operatorname{Im}\beta(r)| \leqslant l(r)/2$ and by virtue of condition (1.4),

$$0 < \frac{\lambda}{1+C^2(L)} \leqslant \liminf_{r\to\infty}\operatorname{Re}\beta(r) \leqslant \limsup_{r\to\infty}\operatorname{Re}\beta(r) \leqslant \frac{\rho}{1+c^2(L)} < \frac{1}{2}. \tag{4.9}$$

Thus, for sufficiently large r we have

$$|1-e^{2\pi i\beta(r)}| = \{(1-e^{2\pi\operatorname{Im}\beta(r)})^2 + 4e^{-2\pi\operatorname{Im}\beta(r)}\sin^2\pi\operatorname{Re}\beta(r)\}^{1/2}$$
$$\geqslant 2e^{-\pi\operatorname{Im}\beta(r)}\sin\pi\operatorname{Re}\beta(r) \geqslant e^{-\pi\rho/2}\sin\frac{\pi\lambda}{1+C^2(L)} > 0.$$

Therefore, (4.8) is equivalent to (4.2). The theorem is proved.

COROLLARY. *For* $\theta \in (\varphi(r), \varphi(r)+2\pi)$ *the following asymptotics, uniform relative to* θ, *is valid*:

$$\operatorname{Im} I(re^{i\theta}) = Q(r, \theta - \varphi(r))r^{l(r)} + o(r^{l(r)}), \qquad r \to \infty, \tag{4.10}$$

where $Q(r, \alpha)$ *is a bounded function continuous on* $[r_1, \infty) \times [0, 2\pi]$ *for sufficiently large* r_1, *and such that the quantity*

$$A(L, G) = \liminf_{r\to\infty} \min\{Q(r, \alpha) : 0 \leqslant \alpha \leqslant 2\pi\}$$

is strictly positive.

PROOF. Separating the imaginary part in (4.2), we obtain (4.10) with

$$Q(r, \alpha) = \frac{e^{(2\pi-\alpha)\operatorname{Im}\beta(r)} \sin\alpha \operatorname{Re}\beta(r) + e^{-\alpha \operatorname{Im}\beta(r)} \sin(2\pi-\alpha)\operatorname{Re}\beta(r)}{e^{2\pi \operatorname{Im}\beta(r)} + e^{-2\pi\operatorname{Im}\beta(r)} - 2\cos 2\pi \operatorname{Re}\beta(r)}. \tag{4.11}$$

It is obvious that this function is bounded and continuous for sufficiently large r. Taking (4.9) into consideration, we get

$$Q(r, \alpha) \geqslant \frac{2\sin\pi\operatorname{Re}\beta(r)\cos(\pi-\alpha)\operatorname{Re}\beta(r)}{(e^{\pi\operatorname{Im}\beta(r)} + e^{-\pi\operatorname{Im}\beta(r)})^2 e^{2\pi\operatorname{Im}\beta(r)}} \geqslant \frac{\sin 2\pi\operatorname{Re}\beta(r)}{(e^{2\pi|\operatorname{Im}\beta(r)|}+1)^2}.$$

Hence it follows that

$$A(L, G) \geqslant \frac{1}{(e^{\pi\rho}+1)^2} 2\sin\frac{\pi\lambda}{1+C^2(L)} \cos\frac{\pi\rho}{1+c^2(L)} > 0.$$

We shall extend Theorem 5 to the case $p = 0$.

THEOREM 5°. *Let* $L = \{z : z = re^{i\varphi(r)},\ 0 < r_0 \leqslant r < \infty\}$ *where* φ *is a continuously differentiable function such that* $|r\varphi'(r)| \leqslant C$. *Introduce the notation* $V(r) = r^{\rho(r)}$, *where* $\rho(r)$ *is a proximate order with* $\lim_{r\to\infty}\rho(r) = 0$. *We set*

$$I_0(z) = \frac{z}{2\pi i}\int_L \frac{V(|t|)dt}{t(t-z)}, \qquad z \in \mathbb{C}\setminus L.$$

For $\theta \in (\varphi(r), \varphi(r)+2\pi)$ *the asymptotics, uniform relative to* θ,

$$I_0(re^{i\theta}) = -\frac{1}{2\pi i}\int_{L_r} V(|t|)t^{-1}dt + O(V(r)), \qquad r\to\infty,$$

holds, where $L_r = L \cap \{z : |z| \leqslant r\}$.

PROOF. Let us write the integral $I_0(z)$ as

$$\begin{aligned} I_0(z) = &-\frac{1}{2\pi i}\int_{L_{3r/2}} \frac{V(|t|)}{t}dt + \frac{z}{2\pi i}\int_{L\setminus L_{3r/2}} \frac{V(|t|)\,dt}{t(t-z)} \\ &+ \frac{1}{2\pi i}\int_{L_{r/2}} \frac{V(|t|)\,dt}{t-z} + \frac{1}{2\pi i}\int_{L_{3r/2}\setminus L_{r/2}} \frac{V(|t|)-V(r)}{t-z}dt \\ &+ \frac{V(r)}{2\pi i}\int_{L_{3r/2}\setminus L_{r/2}} \frac{dt}{t-z} = I_1 + I_2 + I_3 + I_4 + I_5. \end{aligned} \tag{4.12}$$

It is obvious that (here and further we use well-known properties of the proximate order)

$$I_1 = -\frac{1}{2\pi i}\int_{L_r} V(|t|)t^{-1}dt + O(V(r)), \qquad r \to \infty.$$

Further, we have

$$|I_2| \leqslant Cr\int_{3r/2}^{\infty}\frac{V(u)du}{u(u-r)} \leqslant Cr\int_{3r/2}^{\infty}V(u)u^{-2}du = O(V(r)), \qquad r\to\infty,$$

$$|I_3| \leqslant C\int_{r_0}^{r/2}\frac{V(u)du}{r-u} \leqslant \frac{C}{r}\int_{r_0}^{r/2}V(u)du = O(V(r)), \qquad r\to\infty,$$

$$\begin{aligned}|I_4| &\leqslant C\int_{r/2}^{3r/2}\left|\frac{V(u)-V(r)}{u-r}\right|du \\ &\leqslant Cr\max\{|V'(u)| : r/2 \leqslant u \leqslant 3r/2\} = o(V(r)), \qquad r\to\infty.\end{aligned}$$

Finally, noticing that by virtue of the condition $r|\varphi'(r)| < C$ we have

$$\max\{|\varphi(3r/2)-\theta| + |\varphi(r/2)-\theta| : \varphi(r) \leqslant \theta \leqslant \varphi(r)+2\pi\} \leqslant C,$$

we can conclude that $(z = re^{i\theta})$

$$I_5 = \frac{V(r)}{2\pi i}\ln\left(\frac{(3/2)\exp\{i(\varphi(3r/2)-\theta)\}-1}{(1/2)\exp\{i(\varphi(r/2)-\theta)\}-1}\right) = O(V(r)), \qquad r\to\infty.$$

This completes the proof of the theorem.

Let us set

$$V_0(r) = \int_{r_0}^{r} V(u)u^{-1}\,du. \tag{4.13}$$

It is not difficult to check that $V(r) = o\left(V_0(r)\right)$, $r\to\infty$ and $V_0(r) = r^{\rho_0(r)}$, where $\rho_0(r)$ is a proximate order, $\lim_{r\to\infty}\rho_0(r) = 0$.

Separating the imaginary part in (4.12), we immediately obtain

COROLLARY. *For* $\theta \in (\varphi(r), \varphi(r)+2\pi)$ *we have the asymptotics, uniform relative to* θ, $\operatorname{Im} I_0(re^{i\theta}) = -(2\pi)^{-1}V_0(r) + o(V_0(r))$, $r\to\infty$.

§5. A theorem of the Phragmen-Lindelöf type

To prove Theorems 1 and 2 we shall need the following theorem.

THEOREM 6. *Let* $L_0 = \{z : z = re^{i\varphi(r)}, 0 \leqslant r < \infty\}$, $\varphi(1) = 0$, *be a smooth curve,* $\mathscr{D}_0 = \mathbb{C}\setminus L_0$, Φ *be a function analytic in* $\mathscr{D}_0$ *and bounded in* $\mathscr{D}_0 \cap \{z : |z| < R\}$ *for any* $R > 0$. *If the angular limit values* $\Phi^{\pm}$ *of the function* Φ *on both sides of the cut along* L_0 *satisfy almost everywhere on* L_0 *the condition*

$$|\Phi^{\pm}(\xi)| \leqslant M \tag{5.1}$$

and, besides, for some $\beta > 1$ *we have*

$$\left(\int_{\varphi(r)}^{\varphi(r)+2\pi} (\ln^{+} |\Phi(re^{i\theta})|')^{\beta} d\theta\right)^{1/\beta} = o\left\{(\ln r)^{1/\beta} \exp\left(\frac{\beta-1}{2\beta} \frac{\ln^2 r + \varphi^2(r) - 4\pi|\varphi(r)|}{\ln r}\right)\right\}, \qquad r \to \infty, \tag{5.2}$$

then $|\Phi(z)| \leqslant M$ *for all* $z \in \mathscr{D}_0$.

PROOF. By setting $f(z) = \Phi(e^z)$, $y(x) = \varphi(e^x)$, we see that it is sufficient to prove the following equivalent of Theorem 6.

THEOREM 6$'$. *Let the function* f *be analytic in the curvilinear strip* $S = \{z = x + iy : y(x) < y < y(x) + 2\pi, -\infty < x < \infty\}$, $y(0) = 0$, *and let* $y(x) \in C^1$ *be bounded in* $S_R = S \cap \{z : \operatorname{Re} z < R\}$ *for any* $R > 0$. *If the angular limit values of the function* f *on* ∂S *satisfy the condition* $|f(z)| \leqslant M$ *almost everywhere on* ∂S *and, besides, for some* $\beta > 1$ *as* $x \to +\infty$ *we have*

$$\left(\int_{y(x)}^{y(x)+2\pi} (\ln^{+} |f(x+iy)|)^{\beta} dy\right)^{1/\beta} = o\left\{x^{1/\beta} \exp\left(\frac{\beta-1}{2\beta} \frac{x^2 + y^2(x) - 4\pi|y(x)|}{x}\right)\right\}, \tag{5.3}$$

then $|f(z)| \leqslant M$ *for all* $z \in S$.

PROOF. Without loss of generality, we can assume $M = 1$. Let z_0 be an arbitrary point in S; choose R sufficiently large to ensure $z_0 \in S_R$; denoting the Green function of the domain S_R by g_R, we have

$$\ln^{+} |f(z_0)| \leqslant \frac{1}{2\pi} \int_{\partial S_R} \ln^{+} |f(\xi)| \frac{\partial g_R(z_0, \xi)}{\partial n_\xi} |d\xi|,$$

where the internal normal to ∂S_R at the point ξ is denoted by n_ξ. Since $|f(\xi)| \leqslant 1$ on ∂S, it follows that

$$\ln^{+} |f(z_0)| \leqslant \frac{1}{2\pi} \int_{y(R)}^{y(R)+2\pi} \ln^{+} |f(R+iy)| \frac{\partial g_R(z_0, R+iy)}{\partial n_{R+iy}} dy.$$

Using the Hölder inequality, we get

$$\begin{aligned}\ln^+|f(z_0)| &\leqslant \frac{1}{2\pi}\left\{\int_{y(R)}^{y(R)+2\pi}(\ln^+|f(R+iy)|)^\beta\,dy\right\}^{1/\beta}\\ &\quad\times\left\{\int_{y(R)}^{y(R)+2\pi}\left(\frac{\partial g_R(z_0, R+iy)}{\partial n_{R+iy}}\right)^{\beta/(\beta-1)}dy\right\}^{(\beta-1)/\beta}\\ &\leqslant \frac{1}{2\pi}\left\{\int_{y(R)}^{y(R)+2\pi}(\ln^+|f(R+iy)|)^\beta\,dy\right\}^{1/\beta}\\ &\quad\times\left\{\int_{y(R)}^{y(R)+2\pi}\frac{\partial g_R(z_0, R+iy)}{\partial n_{R+iy}}\,dy\right\}^{(\beta-1)/\beta}\\ &\quad\times\max\left\{\frac{\partial g_R(z_0, R+iy)}{\partial n_{R+iy}} : y(R)\leqslant y\leqslant y(R)+2\pi\right\}^{1/\beta}. \qquad (5.4)\end{aligned}$$

Since the domain S_R is contained in the half-plane $\{z : \operatorname{Re} z \leqslant R\}$, by virtue of Carleman's domain extension principle, we have

$$\frac{\partial g_R(z_0, R+iy)}{\partial n_{R+iy}} \leqslant \operatorname{Re}(R+iy-z_0)^{-1} = O\left(\frac{1}{R}\right), \qquad R\to\infty, \qquad (5.5)$$

uniformly relative to $y\in(y(R), y(R)+2\pi)$. Denoting the harmonic measure relative to the domain G of the set $E\subset\partial G$ by $\omega(z, E, G)$, $z\in G$, and setting $E_R=\{\xi : \operatorname{Re}\xi = R,\ y(R)\leqslant\operatorname{Im}\xi\leqslant y(R)+2\pi\}$, we have

$$\int_{y(R)}^{y(R)+2\pi}\frac{\partial g_R(z_0, R+iy)}{\partial n_{R+iy}}\,dy = \omega(z_0, E_R, S_R). \qquad (5.6)$$

Let us estimate the value of $\omega(z_0, E_R, S_R)$. Denote by $w=w(z)$ the function which conformally maps the curvilinear strip S on the rectilinear one $S_0=\{w : 0<\operatorname{Im} w<1\}$, $w(\pm\infty)=\pm\infty$. By virtue of the conformal invariance of harmonic measure, we have

$$\omega(z_0, E_R, S_R) = \omega(w(z_0), w(E_R), w(S_R)).$$

We set

$$\begin{aligned}\beta_R &= \min\{\operatorname{Re} w(R+iy) : y(R)\leqslant y\leqslant y(R)+2\pi\},\\ \widetilde{E}_R &= \{w : \operatorname{Re} w=\beta_R,\ 0\leqslant\operatorname{Im} w\leqslant 1\},\\ \widetilde{S}_R &= \{w : \operatorname{Re} w<\beta_R,\ 0<\operatorname{Im} w<1\}.\end{aligned}$$

By virtue of the domain expansion principle, we have

$$\omega(w_0, w(E_R), w(S_R)) \leqslant \omega(w_0, \widetilde{E}_R, \widetilde{S}_R).$$

A direct calculation shows that

$$\omega(w_0, \widetilde{E}_R, \widetilde{S}_R) = \frac{1}{\pi}\arg\frac{1+e^{\pi(w_0-\beta_R)}}{1-e^{\pi(w_0-\beta_R)}} = O(e^{-\pi\beta_R}), \qquad R\to\infty,$$

and therefore $\omega(z_0, E_R, S_R) = O(e^{-\pi\beta_R})$, $R\to\infty$.

Taking into account the latter estimate and combining (5.4), (5.5), (5.6) and conditions (5.3), we obtain $\ln^+|f(z_0)| \leqslant H(z_0, R)$, where

$$H(z_0, R) = o\left\{\exp\left(\frac{\beta-1}{2\beta}\frac{R^2+y^2(R)-4\pi|y(R)|}{R} - \pi\beta_R\right)\right\}, \qquad R\to\infty.$$

To complete the proof, one must establish that the quantity on the right-hand side of this relation tends to 0 as $R\to\infty$. This follows from the following lemma which, together with its proof, was kindly communicated to the author by A. A. Goldberg.

LEMMA. *The estimate*

$$\beta_R \geqslant \frac{R^2+y^2(R)-4\pi|y(R)|}{2\pi R} + O(1), \qquad R\to\infty,$$

is valid.

PROOF. We shall use the concept of the extremal length of a family of curves (see, e.g., [15]). Let λ_R be the extremal length of the family Γ_R of curves linking segments E_0 and E_R in S. The length of each curve from Γ_R is not smaller than $\{R^2+((|y(R)|-2\pi)^+)^2\}^{1/2}$, and the area of the domain $S\cap\{z: 0<\operatorname{Re} z<R\}$ is equal to $2\pi R$. So, for one of the admissible functions ρ mentioned in the definition of extremal length, we shall take the function that equals 1 in $S\cap\{z: 0<\operatorname{Re} z<R\}$ and zero for all other values of z; we shall see that

$$\lambda_R \geqslant R^2((|y(R)|-2\pi)^+)^2/(2\pi R). \tag{5.7}$$

On the other hand, as shown in [15, p. 77], $\lambda_R \leqslant 2\Lambda(e^{\pi(\beta_k-\alpha)})$, where $\alpha = \max\{\operatorname{Re} w(z): z\in E_0\}$, $\Lambda(x)$ being the Teichmüller function defined as the modulus of the doubly-connected domain $\mathbb{C}\setminus\{[-1, 0]\cup[x, \infty)\}$. For this function, the estimate $\Lambda(x) \leqslant (1/2\pi)\ln(16(x+1))$ holds [15, p. 76, (4–21)]. So we have

$$\lambda_R \leqslant \pi^{-1}(\ln(e^{\pi(\beta_R-\alpha)}+1)+\ln 16) \leqslant \beta_R + O(1), \qquad R\to\infty. \tag{5.8}$$

The assertion of the lemma follows from (5.7) and (5.8).

NOTE. The lemma proved implies the following theorem of the Phragmen-Lindelöf type which was conjectured by the author and was proved by A. A. Goldberg.

THEOREM 6″. *Let the function f be analytic in the curvilinear strip $S = \{z = x+iy: y(x)<y<y(x)+2\pi\}$, $y(0)=0$, $y(x)\in C(\mathbb{R})$, bounded in $S_R = S\cap\{z: \operatorname{Re} z<R\}$ for any $R>0$ and continuous in the closure of S. If $|f(z)|\leqslant M$ for $z\in\partial S$ and, besides,*

$$\sup\{\ln^+|f(z)| : z\in S_R\} = o\left\{\exp\frac{R^2+y^2(R)-4\pi|y(R)|}{2R}\right\}, \qquad R\to\infty,$$

then $|f(z)|\leqslant M$ for $z\in S$.

Indeed, using the notation of the proof of Theorem $6'$, we shall consider the function $f(z(w))$, where $z = z(w)$ is the mapping inverse to $w = w(z)$. This function is analytic in the strip S_0, continuous in its closure, while the inequality $|f(z(w))| \leqslant M$ holds on ∂S_0. From the conditions of the theorem it follows that

$$\begin{aligned}\sup\{\ln^+ |f(z(w))| : w \in S_0,\ \operatorname{Re} w \leqslant \beta_R\} &\leqslant \sup\{\ln^+ |f(z)| : z \in S_R\} \\ &= o\left(\exp\left(\frac{R^2 + y^2(R) - 4\pi|y(R)|}{2R}\right)\right), \qquad R \to \infty.\end{aligned}$$

By virtue of the lemma we conclude that

$$\sup\{\ln^+ |f(z(w))| : w \in S_0,\ \operatorname{Re} w \leqslant \beta_R\} = o(e^{\pi\beta_R}), \qquad R \to \infty.$$

Applying the ordinary Phragmen-Lindelöf principle for a strip to the function $f(z(w))$, we see that Theorem $6''$ is valid.

§6. Proof of Theorem 1

We set

$$\Phi(z) = \exp\left\{\frac{z^{p+1}}{2\pi i}\int_L \frac{\ln G(t) - 2\pi i n(|t|)}{t^{p+1}(t-z)}\,dt\right\}, \qquad z \in \mathscr{D}. \tag{6.1}$$

Since conditions (A) and (C) imply that

$$|\ln G(t) - 2\pi i n(|t|)| \leqslant \big|\ln |G(t)|\big| + 2\pi|\Lambda(t)| = O(V(|t|)), \qquad t \to \infty,$$

and, because the curve L is admissible, $|dt| \leqslant C d|t|$, taking (1.4) into consideration, we see that the integral in the right-hand side of (6.1) converges absolutely. Since $\ln G \in \Delta$, the function Φ is continuous on the set Φ according to the Plemelj-Privalov theorem.

The Sokhotskiĭ-Plemelj formula shows that the function Φ satisfies condition (1.2). The condition

$$0 \geqslant \arg G(t_0) > -2\pi$$

implies (cf. [1, p. 115]) that the function Φ is bounded in a neighborhood of the point t_0. Thus, the function Φ is bounded in $\mathscr{D} \cap \{z : |z| \leqslant R\}$ for any $R > 0$.

Let us prolong the curve L inside the disk $\{z : |z| < r_0\}$ so that the prolonged curve L_0 has the form stated in Theorem 6. We shall show that the function Φ fulfills conditions (5.1) and (5.2) of Theorem 6.

At first we check condition (5.2). For $\varphi(r) < \theta < \varphi(r) + 2\pi$, we have

$$\begin{aligned}
|\ln \Phi(re^{i\theta})| &\leqslant \frac{r^{p+1}}{2\pi} \int_L \frac{|\ln|G(t)|| + 2\pi|\Lambda(t)|}{|t|^{p+1}|t - re^{i\theta}|}|dt| \\
&\leqslant Cr^{p+1} \int_{r_0}^{\infty} \frac{V(s)\,ds}{s^{p+1}|se^{i\varphi(s)} - re^{i\theta}|} \\
&\leqslant Cr^{p+1}\left\{ \int_{r_0}^{r/2} \frac{V(s)\,ds}{s^{p+1}r} \right. \\
&\qquad \left. + \int_{r/2}^{3r/2} \frac{V(s)\,ds}{s^{p+1}|se^{i\varphi(s)} - re^{i\theta}|} + \int_{3r/2}^{\infty} \frac{V(s)\,ds}{s^{p+2}} \right\} \\
&\leqslant CV(r)\left\{ 1 + \int_{r/2}^{3r/2} \frac{ds}{|se^{i\varphi(s)} - re^{i\theta}|} \right\} \qquad (6.2)
\end{aligned}$$

(we used properties of the Boutroux proximate order [12, p. 91]).

Let us prove the validity of the estimate

$$\int_{r/2}^{3r/2} \frac{ds}{|se^{i\varphi(s)} - re^{i\theta}|} \leqslant C\left(1 + \left|\ln \sin \frac{\theta - \varphi(r)}{2}\right|\right), \quad \varphi(r) < \theta < \varphi(r) + 2\pi. \qquad (6.3)$$

We confine ourselves to the case $\varphi(r) < \theta \leqslant \varphi(r) + \pi$, since the case $\varphi(r) + \pi < \theta < \varphi(r) + 2\pi$ is treated in a similar way.

Let $|s - r| < \delta r|\theta - \varphi(r)|$, where $\delta = \min\{1/(8c(L)),\ 1/(4\pi)\}$. For sufficiently large r, by virtue of (1.1), we have $|\varphi'(r)| \leqslant 2C(L)/r$. Thus

$$\begin{aligned}
|\varphi(r) - \varphi(s)| &\leqslant |r - s| \max\{|\varphi'(t)| : r/2 \leqslant t \leqslant 3r/2\} \\
&\leqslant |r - s| 4C(L)/r \leqslant (1/2)(\theta - \varphi(r))
\end{aligned}$$

and, therefore, $(1/2)(\theta - \varphi(r)) \leqslant \theta - \varphi(s) \leqslant (3/2)(\theta - \varphi(r))$. Since

$$\begin{aligned}
|se^{i\varphi(s)} - re^{i\theta}| &= \{(s - r)^2 + 4sr \sin^2(1/2)(\theta - \varphi(s))\}^{1/2} \\
&\geqslant 2\sqrt{rs}\,|\sin(1/2)(\theta - \varphi(s))|,
\end{aligned}$$

for $\varphi(r) < \theta \leqslant \varphi(r) + \pi/3$ we have

$$|se^{i\varphi(s)} - re^{i\theta}| \geqslant 2\sqrt{rs}(\theta - \varphi(s))/\pi \geqslant C(\theta - \varphi(r)),$$

and for $\varphi(r) + \pi/3 \leqslant \theta \leqslant \varphi(r) + \pi$ we have

$$|se^{i\varphi(s)} - re^{i\theta}| \geqslant 2\sqrt{rs} \sin(\pi/12) \geqslant Cr(\theta - \varphi(r)).$$

Therefore

$$\int_{|s-r| \leqslant \delta r(\theta - \varphi(r))} \frac{ds}{|se^{i\varphi(s)} - re^{i\theta}|} \leqslant C,$$

and so

$$\int_{r/2}^{3r/2} \frac{ds}{|se^{i\varphi(s)} - re^{i\theta}|} \leqslant C + \int_{r/2 \leqslant |s-r| \leqslant \delta r(\theta - \varphi(r))} \frac{ds}{|s - r|} = C + 2\ln \frac{1}{2\delta(\theta - \varphi(r))}.$$

This completes the proof of estimate (6.3).

Combining (6.2) and (6.3), we see that for any $\beta > 1$,

$$\left(\int_{\varphi(r)}^{\varphi(r)+2\pi} |\ln \Phi(re^{i\theta})|^{\beta} d\theta\right)^{1/\beta} \leqslant CV(r)\left\{1 + \left(\int_{\varphi(r)}^{\varphi(r)+2\pi} \left|\ln \sin \frac{\theta - \varphi(r)}{2}\right|^{\beta} d\theta\right)^{1/\beta}\right\} \leqslant CV(r). \quad (6.4)$$

By virtue of condition (1.4), the number β may be taken as large as needed to ensure $\rho < (\beta - 1)(2\beta)^{-1}(1 + c^2(L))$. For such a β we shall have

$$V(r) = o\left\{(\ln r)^{1/\beta} \exp\left(\frac{\beta - 1}{2\beta} \frac{\ln^2 r + \varphi^2(r) - 4\pi|\varphi(r)|}{\ln r}\right)\right\}, \qquad r \to \infty.$$

This proves that the function Φ fulfills condition (5.2).

Now we shall show that the function Φ fulfills condition (5.1) as well.

Since $\ln G \in \Delta$, by virtue of the Sokhotskiĭ-Plemelj formula, we have for all $\xi \in L$, except discontinuity points of the function $n(|\xi|)$, the equality

$$\ln \Phi^{\pm}(\xi) = \pm\frac{1}{2}(\ln G(\xi) - 2\pi i n(|\xi|)) + \frac{\xi^{p+1}}{2\pi i} \int_L \frac{\ln G(t) - 2\pi i n(|t|)}{t^{p+1}(t - \xi)} dt$$

(the integral exists in the sense of the principal value). Separating the real part and setting $|\xi| = r$, $\omega(r) = L \cap \{z : |z - \xi| \leqslant k(r)\}$, $\Omega(r) = L \setminus \omega(r)$, we obtain

$$\begin{aligned} \ln|\Phi^{\pm}(\xi)| &= \pm\frac{1}{2}\ln|G(\xi)| + \operatorname{Re}\frac{\xi^{p+1}}{2\pi i}\int_{\omega(r)} \frac{\ln|G(t)|\,dt}{t^{p+1}(t-\xi)} \\ &\quad + \operatorname{Re}\frac{\xi^{p+1}}{2\pi i}\int_{\Omega(r)} \frac{\ln|G(t)|\,dt}{t^{p+1}(t-\xi)} + \operatorname{Re}\xi^{p+1}\int_{\omega(r)} \frac{\Lambda(t)\,dt}{t^{p+1}(t-\xi)} \\ &\quad + \operatorname{Re}\xi^{p+1}\int_{\Omega(r)} \frac{\Lambda(t)\,dt}{t^{p+1}(t-\xi)} \\ &= \pm\frac{1}{2}\ln|G(\xi)| + \sum_{j=1}^{4}\operatorname{Re} I_j(\xi). \end{aligned} \quad (6.5)$$

We have

$$I_1(\xi) = \frac{1}{2\pi i}\int_{\omega(r)} \frac{\ln|G(t)|\,dt}{t - \xi} + \sum_{k=0}^{p} \frac{\xi^k}{2\pi i}\int_{\omega(r)} \frac{\ln|G(t)|\,dt}{t^{k+1}}.$$

By virtue of condition (C)

$$\left|\frac{\xi^k}{2\pi i}\int_{\omega(r)} \frac{\ln|G(t)|\,dt}{t^{k+1}}\right| \leqslant \frac{r^k}{2\pi}\int_{r-k(r)}^{r+k(r)} \frac{o(V_1(s))\,ds}{s^{k+1}} = o(V_1(r)), \qquad r \to \infty,$$

so

$$\operatorname{Re} I_1(\xi) = \frac{1}{2\pi}\int_{\omega(r)} \ln|G(t)| \operatorname{Im}\left(\frac{dt}{t - \xi}\right) + o(V_1(r)), \qquad r \to \infty. \quad (6.6)$$

At the points $t = se^{i\varphi(s)}$, $\xi = re^{i\varphi(r)}$, we have

$$\operatorname{Im}\left(\frac{dt}{t-\xi}\right) = |t-\xi|^{-2}\{s\varphi'(s)(s - r\cos(\varphi(s)-\varphi(r)) - r\sin(\varphi(s)-\varphi(r)))\}\,ds$$
$$= |t-\xi|^{-2}\{s(\varphi(r)-\varphi(s)-\varphi'(s)(r-s)) + B(r, s)\}\,ds,$$

where $B(r, s)$, for $|s-r| \leqslant r/2$, admits the estimate:

$$|B(r, s)| \leqslant r\sin^2((\varphi(s)-\varphi(r))/2) + r|\sin(\varphi(s)-\varphi(r)) - (\varphi(s)-\varphi(r))|$$
$$+ |r-s|\,|\varphi(s)-\varphi(r)| \leqslant Cr^{-1}(s-r)^2$$

(we have used the finite increments formula and the condition $\varphi'(r) = O(1/r)$, $r \to \infty$). By virtue of the formula for finite increments, we have $|\varphi(r)-\varphi(s)-\varphi'(s)(r-s)| \leqslant \omega_{2r}(|r-s|, \varphi')|r-s|$, $|r-s| \leqslant r/2$, hence

$$\operatorname{Im}\left(\frac{dt}{t-\xi}\right) \leqslant \left(s\frac{\omega_{2r}(|r-s|, \varphi')}{|r-s|} + \frac{C}{r}\right)ds, \qquad |r-s| \leqslant r/2.$$

Therefore, for some $t' \in \omega(r)$ we have

$$\left|\int_{\omega(r)} \ln|G(t)| \operatorname{Im}\left(\frac{dt}{t-\xi}\right)\right|$$
$$\leqslant |\ln|G(t')|| \int_{r-k(r)}^{r+k(r)} \left(s\frac{\omega_{2r}(|s-r|, \varphi')}{|s-r|} + \frac{c}{r}\right) ds$$
$$\leqslant |\ln|G(t')||\left\{4|t'| \int_0^{k(r)} \omega_{2r}(\delta, \varphi')\delta^{-1}\,d\delta + 2k(r)C/r\right\}$$
$$\leqslant |\ln|G(t')||\left\{C|t'| \int_0^{k(|t'|/2)} \omega_{3|t'|}(\delta, \varphi')\delta^{-1}\,d\delta + o(1/r)\right\},$$

and, using condition (C), we conclude from (6.6) that

$$|\operatorname{Re} I_1(\xi)| \leqslant o(V_1(|t'|)) + o\left(V_1(r)\right) = o\left(V_1(r)\right), \qquad r \to \infty. \tag{6.7}$$

Since by virtue of (1.1) we have $|se^{i\varphi(s)} - re^{i\varphi(r)}| \geqslant C_0|s-r|$, we have

$$\Omega(r) \subset \{se^{i\varphi(s)} : |s-r| \geqslant k(r)/C_0\}.$$

Since condition (C) implies $\ln|G(t)| = o\left(V_1(|t|)\right)$, $t \to \infty$, we obtain

$$|I_2(\xi)| \leqslant Cr^{p+1}\left(\int_{r_0}^{r-k(r)/C_0} + \int_{r+k(r)/C_0}^{\infty}\right)\frac{o\left(V_1(s)\right)\,ds}{s^{p+1}|s-r|}$$
$$= Cr^{p+1}\left\{\frac{1}{r}\int_{r_0}^{r/2} o\left(V_1(s)\right)s^{-p-1}\,ds + o(V_1(r))r^{-p-1}\int_{r/2}^{r-k(r)/C_0}\frac{ds}{r-s}\right.$$
$$\left. + o(V_1(r))r^{-p-1}\int_{r+k(r)/C_0}^{3r/2}\frac{ds}{s-r} + \int_{3r/2}^{\infty} o\left(V_1(s)\right)s^{-p-2}\,ds\right\}$$
$$= o\left(V_1(r)\ln\frac{rC_0}{2k(r)}\right) = o(V(r)), \qquad r \to \infty. \tag{6.8}$$

Further

$$I_3(\xi) = \int_{\omega(r)} \frac{\Lambda(t)\,dt}{t-\xi} + \sum_{k=0}^{p} \xi^k \int_{\omega(r)} \frac{\Lambda(t)}{t^{k+1}}\,dt.$$

By virtue of condition (A), we have $\Lambda(t) = O(V(|t|))$, $t \to \infty$, hence

$$\begin{gathered} \left|\xi^k \int_{\omega(r)} \frac{\Lambda(t)\,dt}{t^{k+1}}\right| \leqslant r^k \int_{r-k(r)}^{r+k(r)} \frac{CV(s)}{s^{k+1}}\,ds = o(V(r)), \qquad r \to \infty, \\ \operatorname{Re} I_3(\xi) = \int_{\omega(r)} \Lambda(t) \operatorname{Re}\left(\frac{dt}{t-\xi}\right) + o(V(r)) = A(\xi) + o(V(r)). \end{gathered} \tag{6.9}$$

To estimate the integral $A(\xi)$, at first we shall show that the quantity

$$\sigma = |t-\xi| = \{s^2 + r^2 - 2sr\cos(\varphi(s) - \varphi(r))\}^{1/2}$$

for sufficiently large r is an increasing function of s for $s \in (r, r+k(r))$, and a decreasing function for $s \in (r-k(r), r)$. Indeed, for $|s-r| = o(r)$, $r \to \infty$ we have

$$\begin{aligned} &(s-r)^{-1} d\sigma^2/ds \\ &\quad = \{s - r + 2r\sin^2((\varphi(s) - \varphi(r))/2) + sr\varphi'(s)\sin(\varphi(s) - \varphi(r))\} 2(s-r)^{-1} \\ &\quad = 2 + o(1) + 2sr(\varphi'(s))^2 + \frac{2sr\varphi'(s)}{s-r}\{\varphi(s) - \varphi(r) - \varphi'(s)(s-r)\} + o(1). \end{aligned}$$

The latter expression is positive for sufficiently large r, since the Lagrange theorem and the first condition in (1.1) imply that for some s', s'' lying between r and s we have

$$\begin{aligned} r(s-r)^{-1}\{\varphi(s) - \varphi(r) - \varphi'(s)(s-r)\} &= r\{\varphi'(s') - \varphi'(s)\} \\ &= r\left\{\frac{\varphi(s')}{s'\ln s'} - \frac{\varphi(s)}{s\ln s} + o\left(\frac{1}{r}\right)\right\} \\ &= r(s'-s)\frac{d}{du}\left(\frac{\varphi(u)}{u\ln u}\right)_{u=s''} + o(1) = o(1), \qquad r \to \infty. \end{aligned}$$

It follows from the property of σ that we have just proved that the point $t^{\pm}(r, \sigma)$ mentioned in condition (B) is determined uniquely, and the integral $A(\xi)$ (in the sense of the principal value) can be written as

$$A(\xi) = \int_{\omega(r)} \Lambda(t)\,d\ln|t-s| = \int_0^{k(r)} \{\Lambda(t^+(r,\sigma)) - \Lambda(t^-(r,\sigma))\}\frac{d\sigma}{\sigma}. \tag{6.10}$$

Hence it immediately follows from condition (B) and (6.9) that

$$\limsup_{r\to\infty} V^{-1}(r) \operatorname{Re} I_3(\xi) \leqslant 0. \tag{6.11}$$

The integral $I_4(\xi)$ will be written as

$$\begin{aligned} I_4(\xi) = \xi^{p+1} \int_{\Omega(r)} \frac{\Lambda(t) - V(|t|)}{t^{p+1}(t-\xi)}\,dt - \xi^{p+1} \int_{\omega(r)} \frac{V(|t|)\,dt}{t^{p+1}(t-\xi)} \\ + \xi^{p+1} \int_L \frac{V(|t|)\,dt}{t^{p+1}(t-\xi)} = A_1(\xi) - A_2(\xi) + A_3(\xi). \end{aligned}$$

Using the condition (A) and a calculation similar to that carried out for estimate (6.8) of the integral I_2, we see that $A_1(\xi) = o(V(r))$, $r \to \infty$. Further, we have

$$A_2(\xi) = \int_{\omega(r)} \frac{V(|t|)\,dt}{t-\xi} + \sum_{k=0}^{p} \xi^k \int_{\omega(r)} \frac{V(|t|)\,dt}{t^{k+1}} = \int_{\omega(r)} \frac{V(|t|)\,dt}{t-\xi} + o\,(V(r)). \tag{6.12}$$

Similarly to (6.10), we have the equality

$$\int_{\omega(r)} V(|t|) d \ln |t-\xi| = \int_0^{k(r)} \{V(|t^+(r\,,\sigma)|) - V(|t^-(r\,,\sigma)|)\} \frac{d\sigma}{\sigma}, \tag{6.13}$$

so

$$\operatorname{Re} A_2(\xi) = \int_0^{k(r)} \frac{d\sigma}{\sigma} \int_{|t^-(r\,,\sigma)|}^{|t^+(r\,,\sigma)|} V'(u)\,du + o(V(r)), \qquad r \to \infty.$$

Since $V'(u) \leqslant CV(u)u^{-1}$ and

$$|t^+(r\,,\sigma)| - |t^-(r\,,\sigma)| \leqslant |t^+(r\,,\sigma) - \xi| + |t^-(r\,,\sigma) - \xi| = 2\sigma\,,$$

we have

$$|\operatorname{Re} A_2(\xi)| \leqslant 2CV(r)r^{-1}k(r) + o(V(r)) = o(V(r)), \qquad r \to \infty. \tag{6.14}$$

The integral $A_3(\xi)$ is related to the integral $I(\xi)$ defined by (4.1) via the formula

$$\lim_{\theta \to \varphi(r)+0} 2\pi i I(re^{i\theta}) = \pi i V(|\xi|) + A_3(\xi), \qquad \xi = re^{i\varphi(r)}.$$

Since the estimate (4.10), uniform relative to $\theta \in (\varphi(r)\,,\, \varphi(r)+2\pi)$ holds for $\operatorname{Im} I(re^{i\theta})$, we have

$$\operatorname{Re} A_3(\xi) = -2\pi \lim_{\theta \to \varphi(r)+0} \operatorname{Im} I(re^{i\theta}) \leqslant -CV(r).$$

Thus, for sufficiently large r

$$\operatorname{Re} I_4(\xi) \leqslant -CV(r). \tag{6.15}$$

Substituting the estimates (6.7), (6.8), (6.9) and (6.15) into (6.5) and taking into account condition (C), we conclude that

$$\ln |\Phi^{\pm}(\xi)| \leqslant -CV(r), \qquad r \geqslant r_1 \geqslant r_0. \tag{6.16}$$

Since the function Φ is bounded in $D \cap \{z : |z| \leqslant R\}$ for any $R > 0$, equation (6.16) implies that the condition (5.1) of Theorem 6 is fulfilled. This proves that the function Φ is bounded in D.

It remains to establish the estimate (1.14). To this end, we shall consider the function $\Phi_\varepsilon(z) = \Phi(z)\exp\{-\varepsilon 2\pi i I(z)\}$, where $\varepsilon > 0$ and $I(z)$ is defined by equation (4.1). It is easy to see that the function Φ_ε for any $\varepsilon > 0$ fulfills condition (5.1) of Theorem 6. Relation (4.2) and the fact that the coefficient at $r^{l(r)}$ is bounded imply that the function Φ_ε, for a sufficiently small ε, fulfills the condition (5.2) as well. That is why the function Φ_ε is bounded in D when ε is small enough. Now using the estimate (4.10), we get (1.14). The theorem is proved.

NOTE. The estimate (1.14) can be refined in the following way. For each $\varepsilon > 0$, uniformly relative to $\theta \in [\varphi(r)+\varepsilon, \varphi(r)+2\pi-\varepsilon]$, we have

$$\ln|\Phi(re^{i\theta})| = -2\pi Q(r, \theta-\varphi(r))r^{l(r)} + o(r^{l(r)}), \qquad r\to\infty, \tag{6.17}$$

where $Q(r, \alpha)$ is the quantity mentioned in (4.10).

Indeed, it follows from (6.1) that

$$\begin{aligned}\ln\Phi(z) &= \frac{z^{p+1}}{2\pi i}\int_L \frac{\ln G(t) - 2\pi i V(|t|)}{t^{p+1}(t-z)}\,dt + z^{p+1}\int_L \frac{V(|t|)\,dt}{t^{p+1}(t-z)}\\ &= J(z) + 2\pi i I(z).\end{aligned}$$

By virtue of conditions (A) and (B), we have

$$|J(re^{i\theta})| \leqslant Cr^{p+1}\int_{r_0}^{\infty}\frac{V_1(s)\,ds}{s^{p+1}|se^{i\varphi(s)}-re^{i\theta}|}.$$

Obtaining an estimate similar to (6.1) (substituting V_1 for V) and using (6.3), we see that, uniformly relative to $\theta \in [\varphi(r)+\varepsilon, \varphi(r)+2\pi-\varepsilon]$, the relation $J(re^{i\theta}) = O(V_1(r))$, $r\to\infty$, holds. Taking (4.10) into consideration, we get (6.17).

Now let us derive Theorem $1'$. Let L be a ray, $\varphi(r)\equiv 0$, Φ be a function which solves the problem (1.2) and satisfies inequality (1.14). Relation (1.15) immediately follows from (6.17), (4.11) and the equality $\beta(r) = \operatorname{Re}\beta(r) = l(r)$, which is true in this case.

To conclude the section, we state an analog of Theorem 1 for the case when $l(r) = \rho(r)$, where $\rho(r)$ is the ordinary proximate order such that $\lim_{r\to\infty}\rho(r) = 0$. To this end we shall introduce the function $V_0(r)$ by (4.13) and shall impose the following conditions on the behavior of $\ln G$. There is a proximate order $\rho_1(r)$ such that

$$V_1(r) = r^{\rho_1(r)} = o\left(V_0(r)/\ln r\right), \qquad r\to\infty,$$

and a nondecreasing integer-valued function $n(r)\geqslant 0$, $r\geqslant r_0$ related to the curve L and with $\ln G$ as follows. The function Λ defined by relation (1.6) satisfies condition (1.7) and also condition (1.8) if we replace $V(r)$ by $V_0(r)$ and $k(r)$ by $k_0(r) = \exp\{-V_0(r)/(V_1(r)\ln r)\}$ in the latter. Besides, let condition (1.9) hold with $k_0(r)$ replaced by $k(r)$. Conditions (B) and (C) changed in this way will be called (B°) and (C°).

THEOREM 1°. *Let L be an admissible curve. If $\ln G\in\Delta$ and conditions (A), (B°) and (C°) are satisfied, then the problem (1.2) has a nontrivial solution Φ admitting the following estimate in the domain $\mathscr{D}$*

$$|\Phi(z)| \leqslant C\exp(-CV(|z|)).$$

We shall omit the proof, since it is identical to the proof of Theorem 1 with one difference: instead of Theorem 5 one must apply Theorem 5°.

If $V(r) = \ln^\alpha r$, $\alpha > 0$, then $V_0 = (\alpha+1)^{-1} \ln^{\alpha+1} r + \text{const}$, and Theorem 1° implies one of the results of P. G. Yurov [2].

§7. Proof of Theorem 2

We set

$$\Phi_0(z) = \exp\left\{ \frac{z}{2\pi i} \int_L \frac{\ln G(t) - 2\pi i n(|t|)}{t(t-z)}\, dt \right\}, \qquad z \in \mathscr{D}.$$

By an argument similar to the one used in proving (6.4), but by using conditions (A'), (C') instead of (A), (C), we shall prove that for each $\beta > 1$ we have

$$\left(\int_{\varphi(r)}^{\varphi(r)+2\pi} |\ln \Phi(re^{i\theta})|^\beta \, d\theta \right)^{1/\beta} = O(\ln^2 r), \qquad r \to \infty.$$

Setting

$$\xi \in L, \quad |\xi| = r, \quad \omega_0(r) = L \cap \{z : |z - \xi| \leqslant r^{-q}\}, \quad \Omega_0(r) = L \setminus \omega_0(r),$$

where q is the number from condition (B'), we obtain the following equality similar to (6.5):

$$\begin{aligned} \ln|\Phi_0^\pm(\xi)| &= \pm\frac{1}{2}\ln|G(\xi)| + \operatorname{Re}\frac{\xi}{2\pi i}\int_{\omega_0(r)} \frac{\ln|G(t)|\,dt}{t(t-\xi)} \\ &\quad + \operatorname{Re}\frac{\xi}{2\pi i}\int_{\Omega_0(r)} \frac{\ln|G(t)|\,dt}{t(t-\xi)} + \operatorname{Re}\xi\int_{\omega_0(r)} \frac{\Lambda(t)\,dt}{t(t-\xi)} \\ &\quad + \operatorname{Re}\xi\int_{\Omega_0(r)} \frac{\Lambda(t)\,dt}{t(t-\xi)} = \pm\frac{1}{2}\ln|G(\xi)| + \sum_{j=1}^{4} \operatorname{Re} I_j^0(\xi). \end{aligned}$$

The integrals I_1^0, I_2^0 and I_3^0 are estimated in the same manner as the integrals I_1, I_2 and I_3 in the proof of Theorem 1 (but using conditions (A'), (B'), (C') instead of (A), (B), (C)), while the integral I_4^0 is estimated like I_2, i.e. in a much simpler way than I_4. It follows from these estimates that for sufficiently large r we have $\ln|\Phi_0^\pm(\xi)| \leqslant C \ln r$. Let h be a polynomial whose degree is not smaller than the constant C from the latter equality, while its zeros lie at those points $re^{i\varphi(r)} \in L$ where $n(r)$ is discontinuous. Then the function $\Phi = \Phi_0/h$ satisfies the conditions of Theorem 6 and, therefore, is bounded in $\mathscr{D}$. The function is evidently the required solution of the boundary problem (1.2).

§8. Proof of Theorem 3

Let us check that the coefficient G satisfies the conditions of Theorem 2. Let $n(s) = [(1/2\pi)\arg G(se^{i\varphi(s)})] + 1$. Evidently the function n is nonnegative, integer-valued, and monotonically tends to $+\infty$ as $s \to +\infty$. Since

$\Lambda(t) = (1/2\pi)\arg G(t) - n(|t|) = O(1)$, $t \to \infty$, condition (A') is fulfilled. Since $|t^+(r, \sigma)| > |t^-(r, \sigma)|$, we have

$$n(|t^+(r, \sigma)|) - n(|t^-(r, \sigma)|) > 0$$

and, therefore,

$$\begin{aligned}\Lambda(t^+(r, \sigma)) - \Lambda(t^-(r, \sigma)) &\leqslant \frac{1}{2\pi}\{\arg G(t^+(r, \sigma)) - \arg G(t^-(r, \sigma))\} \\ &= \frac{1}{2\pi}\int_{|t^-(r,\sigma)|}^{|t^+(r,\sigma)|} \frac{d\arg G(se^{i\varphi(s)})}{ds}\, ds \\ &\leqslant C|t^+(r, \sigma)|^{\eta}(|t^+(r, \sigma)| - |t^-(r, \sigma)|).\end{aligned}$$

Since $|t^+(r, \sigma)| - |t^-(r, \sigma)| \leqslant 2\sigma$, $|t^+(r, \sigma)| \leqslant r + \sigma \leqslant 2r$ for $\sigma \leqslant r$, condition (B') is fulfilled for any $q \geqslant \eta$. Since condition (C') occured in Theorem 3, conditions of Theorem 2 are fulfilled.

§9. Proof of Theorem 4

Let us check conditions (A) and (B), with $V(r)$ taken from (3.1) and any $V_1(r) = o(V(r)/\ln r)$, $r \to \infty$, such that $V_1(r) \geqslant V(r)r^{-1/2}$. Then Theorem 4 will follow from Theorem 1.

We set

$$H(s) = \int_{r_0}^{s} h(u)\, du, \qquad s \geqslant r_0,$$

and denote by $\vartheta(x)$, $x \geqslant 0$, the function inverse to H. In the role of $n(s)$, $s \geqslant r_0$, we take the integer-valued nondecreasing function defined by the conditions

$$\begin{gathered} n(\vartheta(m)) = H(\vartheta(m)) = m, \\ \int_{\vartheta(m)}^{\vartheta(m+1)} (H(s) - n(s))\, ds = 0 \qquad (m = 0, 1, \dots). \end{gathered} \tag{9.1}$$

It is clear that

$$|H(s) - n(s)| \leqslant 1, \qquad s \geqslant r_0, \tag{9.2}$$

so the validity of condition (A) is obvious.

By $Z(r)$ we shall denote the integral in the left-hand side of (1.8) in condition (B). Using the notation introduced, it can be rewritten as

$$\begin{aligned} Z(r) = \int_0^{k(r)} &\{(H(|t^+(r, \sigma)|) - H(|t^-(r, \sigma)|)) \\ &- (n(|t^+(r, \sigma)|) - n(|t^-(r, \sigma)|))\}\frac{d\sigma}{\sigma} \\ + \int_0^{k(r)} &(V(|t^+(r, \sigma)|) - V(|t^-(r, \sigma)|))\frac{d\sigma}{\sigma} \\ = Z_1(r) &+ Z_2(r). \end{aligned} \tag{9.3}$$

The integral $Z_2(r)$ was estimated in the proof of Theorem 1 (see the estimate of $A_2(\xi)$, (6.12)–(6.14)), where we established that

$$Z_2(r) = o(V(r)), \qquad r \to \infty. \tag{9.4}$$

Let us estimate the integral $Z_1(r)$. Since $h(r) > Ce^{cr}$ and

$$k(r) = \exp\{-V(r)/(V_1(r)\ln r)\} \geqslant \exp\{-r^{1/2}/\ln r\},$$

for sufficiently large r we have $k(r) > 1/h(r)$. Let us write the integral $Z_1(r)$ as

$$\begin{aligned} Z_1(r) = \int_0^{1/h(r)} &\{(H(|t^+(r,\sigma)|) - H(|t^-(r,\sigma)|)) \\ &- (n(|t^+(r,\sigma)|) - n(|t^-(r,\sigma)|))\}\frac{d\sigma}{\sigma} \\ &+ \int_{1/h(r)}^{k(r)} \{H(|t^+(r,\sigma)|) - n(|t^+(r,\sigma)|)\}\frac{d\sigma}{\sigma} \\ &- \int_{1/h(r)}^{k(r)} \{H(|t^-(r,\sigma)|) - n(|t^-(r,\sigma)|)\}\frac{d\sigma}{\sigma} \\ = W_1(r) &+ W_2^{(1)}(r) - W_2^{(2)}(r). \end{aligned} \tag{9.5}$$

The integral $W_1(r)$ is estimated in the following way

$$\begin{aligned} W_1(r) &\leqslant \int_0^{1/h(r)} (H(|t^+(r,\sigma)|) - H(|t^-(r,\sigma)|))\frac{d\sigma}{\sigma} \\ &\leqslant \int_0^{1/h(r)} h(|t^+(r,\sigma)|)(|t^+(r,\sigma)| - |t^-(r,\sigma)|)\frac{d\sigma}{\sigma} \\ &\leqslant \int_0^{1/h(r)} h(r+\sigma)2\sigma\frac{d\sigma}{\sigma} \leqslant 2h\frac{r+1/h(r)}{h(r)} \leqslant C \end{aligned}$$

by virtue of condition (ii) from the definition of the class F_β.

Now, we shall estimate the integral $W_2^{(1)}(r)$. By $s = s(r,\sigma) > r$ we denote the solution of the equation $|se^{i\varphi(s)} - re^{i\varphi(r)}| = \sigma$. As was established in the proof of Theorem 1, the function $s(r,\sigma) = |t^+(r,\sigma)|$ strictly increases for $\sigma \in [0, k(r)]$ and admits the estimate $r + \sigma \geqslant s(r,\sigma) \geqslant r + \sigma/C_0$. Let $\sigma = \sigma(r,s)$ be the function inverse to $s(r,\sigma)$. Substituting $\sigma \to s$ in the integral $W_2^{(1)}(r)$, we have

$$W_2^{(1)}(r) = \int_{s(r,1/h(r))}^{s(r,k(r))} \{H(s) - n(s)\}\frac{\sigma'(r,s)}{\sigma(r,s)}\,ds.$$

Further we shall need the following lemma, whose proof will be given at the end of this section.

LEMMA. *Let $k(r) > 1/h(r)$, $r > r_1$. By $a(r)$ and $b(r)$ we shall denote arbitrary functions satisfying the inequalities*

$$a(r) \leqslant b(r), \quad 1/C_0 \leqslant a(r)h(r) \leqslant 1, \quad 1/C_0 \leqslant b(r)/k(r) \leqslant 1, \quad r \geqslant r_1,$$

where $C_0 > 1$. *We set*

$$\Theta_1(r) = \int_{r+a(r)}^{r+b(r)} \{H(s) - n(s)\}\frac{ds}{s-r}, \qquad \Theta_2(r) = \int_{r-b(r)}^{r-a(r)} \{H(s) - n(s)\}\frac{ds}{s-r}.$$

Then the estimates $\Theta_1(r) = O(1)$, $\Theta_2(r) = O(1)$, $r \to \infty$, *are valid.*

Using this lemma, we shall continue the proof of the theorem. We set

$$\widetilde{W}_2^{(1)}(r) = \int_{s(r,1/h(r))}^{s(r,k(r))} \{H(s) - n(s)\}\frac{ds}{s-r}.$$

Since $r + \sigma/C_0 \leqslant s(r, \sigma) \leqslant r + \sigma$ for $\sigma \in [0, k(r)]$, the integral $\widetilde{W}_2^{(1)}(r)$ can be regarded as a particular case of the integral $\Theta_1(r)$ from the lemma; by virtue of this lemma, we have $\widetilde{W}_2^{(1)}(r) = O(1)$, $r \to \infty$.

Let us consider the difference

$$W_2^{(1)}(r) - \widetilde{W}_2^{(1)}(r) = \int_{s(r,1/h(r))}^{s(r,k(r))} \{H(s) - n(s)\}\left\{\frac{\sigma'(r,s)}{\sigma(r,s)} - \frac{1}{s-r}\right\}ds.$$

It follows from (9.1) that

$$|W_2^{(1)}(r) - \widetilde{W}_2^{(1)}(r)| \leqslant \int_r^{r+k(r)} \left|\frac{\sigma'(r,s)}{\sigma(r,s)} - \frac{1}{s-r}\right| ds.$$

Let us estimate the integrand

$$\begin{aligned}
\left|\frac{\sigma'(r,s)}{\sigma(r,s)} - \frac{1}{s-r}\right| &= \frac{1}{2}\left|\frac{d}{ds}\ln\frac{\sigma^2(r,s)}{(s-r)^2}\right| \\
&= \frac{1}{2}\left|\frac{d}{ds}\ln\left(1 + \frac{4rs\sin^2(1/2)(\varphi(r) - \varphi(s))}{(s-r)^2}\right)\right| \\
&\leqslant 2r\left|\frac{d}{ds}\frac{s\sin^2(1/2)(\varphi(r) - \varphi(s))}{(s-r)^2}\right| \\
&\leqslant C\left(1 + r\frac{|\varphi(s) - \varphi(r) - \varphi'(s)(s-r)|}{(s-r)^2}\right) \\
&\leqslant C\left(1 + r\frac{\omega_{2r}(|s-r|, \varphi')}{|s-r|}\right).
\end{aligned}$$

Using condition (3.2), we get

$$|W_2^{(1)}(r) - \widetilde{W}_2^{(1)}(r)| \leqslant Ck(r) + Cr\int_0^{k(r)} \omega_{2r}(\delta, \varphi')\frac{d\delta}{\delta} = o(V(r)), \qquad r \to \infty.$$

Thus, we have proved that $W_2^{(1)}(r) = o(V(r))$, $r \to \infty$.

In a similar way, using the estimate for the integral $\Theta_2(r)$ given in the lemma, we obtain the estimate $W_2^{(2)}(r) = o(V(r))$, $r \to \infty$.

Substituting the estimates obtained for W_1, $W_2^{(1)}$ and $W_2^{(2)}$ in (9.5), we conclude that $\limsup_{r\to\infty} V^{-1}(r)Z_1(r) \leqslant 0$. Therefore, it follows from (9.3) and (9.4) that condition (B) is satisfied. This proves Theorem 4.

Now we shall prove the lemma. We set

$$R(t) = \int_{r_0}^{t} (H(u) - n(u))\,du, \qquad t \geqslant r_0.$$

We shall need the estimate

$$|R(t)| \leqslant C/h(t), \qquad t \geqslant r_0. \tag{9.6}$$

In order to get it, we denote by m_t an integer defined by the condition $\vartheta(m_t) \leqslant t \leqslant \vartheta(m_t + 1)$ and note that by virtue of (9.1) and (9.2)

$$\begin{aligned} |R(t)| &= \left| \int_{\vartheta(m_t)}^{t} (H(u) - n(u))du \right| \leqslant \vartheta(m_t + 1) - \vartheta(m_t) \\ &= \int_{m_t}^{m_t+1} \vartheta'(u)\,du = \int_{m_t}^{m_t+1} \frac{1}{h(\vartheta(u))}\,du \leqslant \frac{1}{h(\vartheta(m_t))}. \end{aligned}$$

Now, using condition (ii) from the definition of the class F_β, we can derive the estimate (9.6):

$$|R(t)| \leqslant \frac{1}{h(t)} \frac{h(\vartheta(m_t + 1))}{h(\vartheta(m_t))} \leqslant \frac{1}{h(t)} \frac{h\left(\vartheta(m_t) + 1/h(\vartheta(m_t))\right)}{h(\vartheta(m_t))} \leqslant \frac{C}{h(t)}.$$

The integral Θ_1 is estimated rather simply. Integrating by parts and using the estimate (9.6), we have

$$\begin{aligned} |\Theta_1(t)| &= \left| \frac{R(u)}{u-t}\Big|_{t+a(t)}^{t+b(t)} + \int_{t+a(t)}^{t+b(t)} \frac{R(u)\,du}{(u-t)^2} \right| \\ &\leqslant \frac{C}{a(t)h(t+a(t))} \leqslant \frac{CC_0h(t)}{h(t+1/(C_0h(t)))} \leqslant CC_0. \end{aligned}$$

The estimate of the integral Θ_2 is obtained in a more complicated way. At first, using (9.6), we get

$$\begin{aligned} |\Theta_2(t)| &= \left| \frac{R(u)}{u-t}\Big|_{t-b(t)}^{t-a(t)} - \int_{t-b(t)}^{t-a(t)} \frac{R(u)\,du}{(u-t)^2} \right| \\ &\leqslant \frac{C}{a(t)h(t+a(t))} + \frac{C}{b(t)h(t+b(t))} + C\int_{t-b(t)}^{t-a(t)} \frac{du}{h(u)(t-u)^2}. \end{aligned}$$

By virtue of condition (ii) from the definition of the class F_β, we have

$$h(t) \leqslant h(t - h^{-\beta}(t) + h^{-\beta}(t - h^{-\beta}(t))) \leqslant Ch(t - h^{-\beta}(t)).$$

It follows that for sufficiently large t we have

$$a^{-1}(t) \leqslant C_0h(t) \leqslant C_1h(t - h^{-1}(t)) \leqslant C_1h(t - a(t)).$$

Since $V_1(t) \geqslant V(t)t^{-1/2}$, we have

$$k^{-1}(t) = \exp\left(V(t)/V_1(t)\ln t\right) = o(\exp t^{1/2}), \qquad t \to \infty.$$

Taking into consideration $h(t) \geqslant Ce^{ct}$, $t \geqslant 0$ and $\lim_{t\to\infty} b(t) = 0$, we obtain for sufficiently large t

$$b^{-1}(t) \leqslant C_0 k^{-1}(t) \leqslant Ch(t - b(t)).$$

Hence

$$|\Theta_2(t)| \leqslant O(1) + C \int_{t-b(t)}^{t-a(t)} \frac{du}{h(u)(t-u)^2}.$$

Let us denote the integral in the right-hand side of the latter inequality by I and estimate it. Integrating it by parts twice, we get

$$\begin{aligned} I &= \frac{1}{h(u)(t-u)}\bigg|_{t-b(t)}^{t-a(t)} + \left(\frac{1}{h(u)}\right)' \ln(t-u)\bigg|_{t-b(t)}^{t-a(t)} \\ &\quad - \int_{t-b(t)}^{t-a(t)} \left(\frac{1}{h(u)}\right)'' \ln(t-u)\, du \\ &\leqslant \frac{1}{h(t-a(t))a(t)} + \frac{h'(t-a(t))}{h^2(t-a(t))} |\ln a(t)| \\ &\quad + \max_{u \geqslant t-b(t)} \left\{ \frac{h''(u)}{h^2(u)} + 2\frac{(h'(u))^2}{h^3(u)} \right\} \int_{t-b(t)}^{t} |\ln(t-u)|\, du. \end{aligned}$$

Let us show that the condition $h \in F_\beta$, $\beta \in (0, 1/2)$ implies

$$h'(u) \leqslant Ch^{3/2}(u), \qquad h''(u) \leqslant Ch^3(u). \tag{9.7}$$

According to the Cauchy theorem,

$$h^{(p)}(u) = \frac{p!}{2\pi i} \int_{|\xi - u| = h^{-\beta}(u)} \frac{h(\xi)\, d\xi}{(\xi - u)^{p+1}}, \qquad p = 0, 1, \dots,$$

whence

$$h^{(p)}(u) \leqslant p! h^{p\beta}(u) \max\{|h(\xi)| : |\xi - u| = h^{-\beta}(u)\}.$$

Using the conditions (i) and (ii) from the definition of the class F_β, we see that for $|\xi - u| \leqslant h^{-\beta}(u)$,

$$|h(\xi)| \leqslant h(\operatorname{Re}\xi) \leqslant h(u + h^{-\beta}(u)) \leqslant Ch(u).$$

Therefore $h^{(p)}(u) \leqslant Cp!\, h^{1+p\beta}(u)$, whence the estimates (9.7) follow.

The lemma is proved.

NOTE. From the proof of Theorem 4 one sees that conditions (1.10) and (1.8) are not equivalent. Indeed, it was established that the function (3.1) satisfies condition (1.8). If it satisfied the condition (1.10), we would have

$$\int_0^{k(r)} (H(|t^+(r, \sigma)|) - H(|t^-(r, \sigma)|)) \frac{d\sigma}{\sigma} \leqslant CV(r). \tag{9.8}$$

Since

$$\begin{aligned} H(|t^+(r, \sigma)|) - H(|t^-(r, \sigma)|) &\geqslant h(|t^-(r, \sigma)|)(|t^+(r, \sigma)| - |t^-(r, \sigma)|) \\ &\geqslant h(r - \sigma) 2\sigma / C_0, \end{aligned}$$

inequality (9.8) implies that for sufficiently large r we have

$$Ce^{cr} \leqslant h(r - k(r)) \leqslant CV(r)/k(r) = o(\exp r^{1/2}), \qquad r \to \infty,$$

which is impossible.

REFERENCES

1. N. V. Govorov, *Riemann boundary value problems with infinite index*, "Nauka", Moscow, 1986. (Russian)

2. P. G. Yurov, *Riemann's homogeneous boundary problem with infinite index of logarithmic type*, Izv. Vyssh. Uchebn. Zaved. Mat. **1966**, no. 2, 158–163; English transl. in Soviet Math. (Iz. VUZ) **10** (1966).

3. I. E. Sandrigaĭlo, *On Riemann's boundary problem with infinite index for a half-plane*, Dokl. Akad. Nauk BSSR **19** (1975), 872–875. (Russian)

4. B. A. Kac, *On an exceptional case of Riemann's problem with an oscillating coefficient*, Izv. Vyssh. Uchebn. Zaved. Mat. **1981**, no. 12, 41–50; English transl. in Soviet Math. (Iz. VUZ) **25** (1981).

5. E. A. Danilov, *Riemann's homogeneous problem with infinite index of power order on the logarithmic helix*, Integral and Differential Equations and Approximate Solutions, Izdat. Kalmyk. Gos. Univ., Èlista, 1985. (Russian)

6. S. V. Rogozin, *Boundary problem and singular integral equations with infinite index*, Proceedings of Seminar on Boundary Problems, Izdat. Minsk. Univ., Minsk, 1985. (Russian)

7. S. M. Grudskiĭ, *Singular integral operators with an infinite index and Blaschke products*, Math. Nachr. B. **120** (1986), 313–331.

8. V. N. Monakhov and E. V. Semenko, *Boundary problems with an infinite index in Hardy spaces*, Dokl. Akad. Nauk SSSR **291** (1986), 544–547; English transl. in Soviet Math. Dokl. **34** (1987).

9. A. G. Alekhno, *Riemann's homogeneous boundary problem with an infinite index of arbitrary power order*, Dokl. Akad. Nauk BSSR **32** (1988), 112–115. (Russian)

10. V. B. Dybin, *Well-posed problems for singular integral equations*, Izdat. Rostov. Univ., Rostov-on-Don, 1988. (Russian)

11. S. A. Plaksa, *Riemann's boundary problem with $+\infty$ index on a rectifiable curve*, Ukrain. Mat. Zh. **42** (1990), 1204–1213; English transl. in Ukrainian Math. J. **42** (1990).

12. A. A. Goldberg and I. V. Ostrovskiĭ, *Distribution of values of meromorphic functions*, "Nauka", Moscow, 1970. (Russian)

13. I. P. Kamynin and I. V. Ostrovskiĭ, *On zero sets of entire Hermite-positive functions*, Sibirsk. Mat. Zh. **23** (1982), 66–82; English transl. in Siberian Math. J. **23** (1982).

14. S. K. Balashov, *On entire functions of finite order with radicals on curves of regular rotation*, Izv. Akad. Nauk SSSR Ser. Mat. **37** (1973), 603–629; English transl. in Math. USSR-Izv. **7** (1973).

15. L. Ahlfors, *Conformal invariants*, McGraw-Hill, New York, 1973.

Translated by V. A. TKACHENKO

ADVANCES IN SOVIET MATHEMATICS
Volume 11, 1992

Completeness of a System of Functions on Sets in the Complex Plane

YU. I. LYUBARSKIĬ AND V. A. TKACHENKO

§1. Formulation of the problem and statement of results

The present paper is devoted to the study of completeness problems of the system of functions

$$\mathfrak{H}_\alpha^+ = \{e^{\alpha n z} \sin nz\}_{n=1}^{\infty}$$

in the Hilbert space of functions on a piecewise-Lyapunov set in the complex plane.

For every fixed α, the set $\mathfrak{H}_\alpha = \{e^{\alpha n z} \sin nz\}_{n=-\infty}^{\infty}$ is the system of all eigenfunctions of the quadratic pencil

$$-y'' + 2\alpha\lambda y' + (\alpha^2 + 1)\lambda^2 y = 0, \qquad y(0) = y(\pi) = 0.$$

It is easy to show that the system $\mathfrak{H}_\alpha$ has the property of two-fold Keldysh completeness [1, 2] in the space $L^2(0, \pi)$ and is not therefore minimal. Thus, it is natural to expect that "half" of the system $\mathfrak{H}_\alpha$, namely, the system $\mathfrak{H}_\alpha^+$, will be complete in $L^2(0, \pi)$.

The problem of one-fold completeness of the system $\mathfrak{H}_\alpha^+$ seems to have been first noted by M. G. Zhavadov [3]. Papers by M. G. Gasymov and M. G. Zhavadov [4–6] and G. V. Radzievskiĭ [7] presented conditions of multiple completeness of a certain part of the eigenfunctions and adjoint functions of differential operator pencils of arbitrary order which polynomially depend on the parameter; these conditions in particular imply the completeness of $\mathfrak{H}_\alpha^+$ for real α.

In 1969, A. G. Kostyuchenko asked if the completeness of the system $\mathfrak{H}_\alpha^+$ could be proved by methods of the theory of functions. This was done for real α by B. Ya. Levin [8, pp. 49–53].

The system

$$\left\{ \sum_{k=1}^{r} \{\alpha_k \varphi_k^n(x) + \beta_k \overline{\varphi}_k^n(x)\} \right\}_{n=0}^{\infty} \tag{1.1}$$

1991 *Mathematics Subject Classification.* Primary 30E05; Secondary 34L10.

is of more general form than $\mathfrak{H}_\alpha^+$; it is determined by the set of functions $\{\varphi_k(x)\}_{k=1}^r$ and by the numbers $\{\alpha_k\}_1^r$, $\{\beta_k\}_1^r$. If $r = 1$, $\varphi(x) = \exp(ix)$, then (1.1) coincides with the system $\{A\sin nx + B\cos nx\}_{n=0}^\infty$ whose completeness in the spaces $L^p(0, \pi)$, $C(0, \pi)$ was studied by R. Feinerman and D. J. Newman [9]. The case of an arbitrary $r \geqslant 1$, with some restrictions on $\{\varphi_k\}$, $\{\alpha_k\}$, and $\{\beta_k\}$ was considered by A. A. Shkalikov [10]. From the results of his paper it follows, in particular, that the system $\mathfrak{H}_\alpha^+$ is minimal in $L^2(0, \pi)$.

Note also that the closure of the linear hulls of the systems of functions $\{\operatorname{Im} W^n(t)\}_{n=1}^\infty$, $\{\operatorname{Re} W^n(t)\}_{n=1}^\infty$ and some systems of a more general form in the spaces $C(0, l)$ and $L^p(0, l)$, under certain restrictions on W, was described by Yu. A. Kazmin [11–13], A. N. Barmenkov and Yu. A. Kazmin [14], and A. G. Tumarkin [15, 16]. It is evident that for $\operatorname{Im}\alpha = 0$ the system $\mathfrak{H}_\alpha^+$ coincides with the former system when $W(t) = \exp((\alpha + i)t)$.

It is well known that the problem of completeness of the system of exponentials $\{e^{\lambda_n z}\}$ on plane curves may be reduced to that of the existence of an entire function of known growth which vanishes on the set $\Lambda = \{\lambda_n\}$. In the present situation, the problem of finding an entire function φ of exponential type for which the linear combination $\varphi((\alpha + i)\lambda) - \varphi((\alpha - i)\lambda)$ vanishes at the points $1, 2, \dots, n, \dots$ arises. The corresponding uniqueness theorem is proved in a nontraditional manner, because it is not possible to use here the classical Weierstrass product method.

We shall reduce the problem of completeness to a Riemann boundary problem with a Carleman shift on the boundary of the corresponding domain. In this case, the trivial solvability of the homogeneous problem is equivalent to completeness. The boundary problems with shift that arise are somewhat different from those known in the literature [17, 18] and require special studies to be carried out. In [28] the system $\{a\varphi^k(t) + b\psi^k(t)\}_1^\infty$ is considered as well as certain more general systems.

Most of the results below are briefly presented in [19].

We are pleased to express our sincere gratitude to B. Ya. Levin for his interest in and attention to this work at every stage.

Before formulating the results, let us agree on some definitions and notation. By a *simple Lyapunov curve* in the plane, we mean a set l admitting the parametrization $l = \{s \in \mathbb{C} : z = f(s),\ 0 \leqslant s \leqslant 1\}$ with a complex-valued function $f \in C^1(0, 1)$ such that $f'(s) \neq 0$ for all $s \in [0, 1]$, $f(s_1) \neq f(s_2)$ for $s_1 \neq s_2$, and f' satisfies the Hölder condition with a positive exponent. The points $f(0)$ and $f(1)$ will be called the *ends* of the curve l. The union of a finite number of simple Lyapunov curves that may intersect only at their ends, if at all, will be called a *piecewise Lyapunov set*. The point $t \in K$ will be called an *end*, a *simple point*, or a *node*, if for any of its sufficiently small neighborhoods U_t, the set $U_t \setminus K$ is connected, consists of two connected components, or of more than two connected components, respectively. In this case, the *multiplicity* of the point $t \in K$ will be the number of connected components of the set $U_t \setminus K$ minus 1. It is clear that only a finite number of points of the set K may be ends or nodes.

To each piecewise Lyapunov set K one can associate the Hilbert space $L^2(K)$ of measurable complex-valued functions on K with the scalar product

$$(F, G) = \int_K F(t)\overline{G(t)}\,|dt|, \tag{1.2}$$

where $|dt|$ is an element of arc length.

Let us denote by $\mathfrak{R}_\alpha$ a closed rhombus whose vertices are at the points 0, π, $\pi(\alpha+i)/2\alpha$, and $\pi(\alpha-i)/2\alpha$ and by $I_1^\pm$ the closed intervals with the ends 0 and $\pi(\alpha \pm i)/2\alpha$.

Everywhere below it is assumed that $\alpha > 0$. If $\alpha < 0$, then the transformation $t \mapsto \pi - t$ maps the rhombus $\mathfrak{R}_\alpha$ into itself and transforms the problem of completeness of the system $\mathfrak{H}_\alpha^+$ into a similar one for the system $\mathfrak{H}_\alpha^-$. Therefore, the results for the case of $\alpha < 0$ can be obtained by obvious reformulations of the corresponding statements for $\alpha > 0$.

THEOREM 1.1. *Let the piecewise Lyapunov set K with connected complement lie in $\mathfrak{R}_\alpha$ and either $\pi \notin K$ or K contain no pair of complex conjugate points belonging to the half-open intervals $I_1^+ \setminus \{0\}$ and $I_1^- \setminus \{0\}$. Then the system $\mathfrak{H}_\alpha^+$ is complete in the space $L^2(K)$.*

The geometric restrictions on the set K contained in Theorem 1.1 are essential. This becomes apparent when studying the completeness of the system $\mathfrak{H}_\alpha^+$ in the space $L^2(K_\gamma)$, where $\gamma \in (0, 1/2]$ and K_γ is the union of the two segments $[\gamma\pi(1+i/\alpha), \pi]$ and $[\gamma\pi(1-i/\alpha), \pi]$.

THEOREM 1.2. *The system $\mathfrak{H}_\alpha^+$ is not complete in the space $L^2(K_\gamma)$ and, moreover, has infinite defect.*

Thus the rhombus $\mathfrak{R}_\alpha$ is a critical set outside of which the system $\mathfrak{H}_\alpha^+$ may lose the property of completeness.

§2. Reduction to the boundary problem

Below we shall need the following auxiliary statements.

LEMMA 2.1. *Let m be a piecewise Lyapunov set lying in the left half-plane and $l_{\theta,w} = \{z \in \mathbb{C} : z = re^{i\theta} + w,\ r > 0\}$ be a ray determined by the parameters $w \in \mathbb{C}$ and $\theta \in [-\pi/2, \pi/2]$. Set*

$$\alpha_+ = \max_{\xi \in m} \operatorname{Im} \xi, \qquad \alpha_- = \min_{\xi \in m} \operatorname{Im} \xi.$$

Then the operator $T_{\theta,w}\colon L^2(m) \to L^2(l_{\theta,w})$ defined by the relation

$$T_{\theta,w}\varphi = \begin{cases} e^{-i\alpha_- z} \int_m e^{\xi z}\varphi(\xi)\,d\xi, & \theta \in [0, \pi/2], \\ e^{-i\alpha_+ z} \int_m e^{\xi z}\varphi(\xi)\,d\xi, & \theta \in [-\pi/2, 0), \end{cases}$$

is bounded, and

$$\|T_{\theta, w}\| \leqslant \begin{cases} C \max_{\xi \in m} |e^{w(\xi - i\alpha_-)}|, & \theta \in [0, \pi/2], \\ C \max_{\xi \in m} |e^{w(\xi - i\alpha_+)}|, & \theta \in [-\pi/2, 0), \end{cases} \tag{2.1}$$

where the constant C *depends only on the set* m.

We shall prove the lemma assuming, to be definite, that $\theta \in [0, \pi/2]$. It is sufficient to prove boundedness and obtain an estimate similar to (2.1) for the dual operator $T^+_{\theta, w}: L^2(l_{\theta, w}) \to L^2(m)$ defined by the relation

$$G \in L^2(l_{\theta, w}) \quad \Rightarrow \quad T^+_{\theta, w}: G \mapsto \int_{l_{\theta, w}} G(z) e^{z(\xi - i\alpha_-)}\, dz.$$

Putting $G_1(r) = G(w + re^{i\theta}) \in L^2(0, \infty)$, $\|G_1\|_{L^2(0, \infty)} = \|G\|_{L^2(l_{\theta, w})}$, we have

$$T^+_{\theta, w}: G_1 \mapsto e^{i\theta} e^{w(\xi - i\alpha_-)} \int_0^\infty G_1(r) \exp(re^{i\theta}(\xi - i\alpha_-))\, dr.$$

Therefore, to prove the lemma, it is sufficient to establish the boundedness and estimate the norm of the operator $T^+_\theta: L^2(0, \infty) \to L^2(m)$ defined by the relation

$$T^+_\theta : G_1 \mapsto \int_0^\infty G_1(r) \exp(re^{i\theta}(\xi - i\alpha_-))\, dr.$$

The set $m - i\alpha_-$ lies in the second quadrant. Hence, because $\theta \in [0, \pi/2]$, we have $\operatorname{Re}[e^{i\theta}(\xi - i\alpha_-)] \leqslant 0$ for all $\xi \in m$. Besides, the map

$$A: G_1 \mapsto g(s) = (2\pi)^{-1} \int_0^\infty G_1(r) e^{rs}\, dr$$

is an isomorphism between the space $L^2(0, \infty)$ and the Hardy space H^2_- in the left half-plane. Thus, it is sufficient to prove boundedness and estimate the norm of the operator $T^+_\theta A^{-1}: H^2_- \to L^2(m)$. The inverse map A^{-1} is given by the formula

$$A^{-1}: g \mapsto G_1(r) = \int_{-\infty}^\infty g(it) e^{-itr}\, dt$$

and we have for the function $g \in H^2_-$

$$T^+_\theta A^{-1}: g \mapsto \int_0^\infty \exp(re^{i\theta}(\xi - i\alpha_-)) \int_{-\infty}^\infty g(it) e^{-itr}\, dt\, dr$$

$$= i \int_{-i\infty}^{i\infty} \frac{g(t)\, dt}{e^{i\theta}(\xi - i\alpha_-) - t} = 2\pi g(e^{i\theta}(\xi - i\alpha_-)).$$

Thus, the operator $T^+_\theta A^{-1}$, up to the factor 2π, is the operator of restriction of functions from H^2_- to the set $m_\theta = \{s : s = e^{i\theta}(\xi - i\alpha),\ \xi \in m\}$, which for any $\theta \in [0, \pi/2]$ is located in the left half-plane. With each such set the following measure μ_θ is associated:

$$\mu_\theta(B) = \{\text{the linear Lebesgue measure of the set } B \cap m_\theta\}.$$

Obviously, the measure μ_θ is a Carleson measure [20, I.5] on the left half-plane, and its Carleson norm is bounded by a constant C_1 that does not depend on θ. By the Carleson embedding theorem [20, I.5], the operators $T_\theta^+ A^{-1}$ have norms not exceeding some constant C. Since $\|A^{-1}\| \leqslant 1$, we have

$$\|T_{\theta,w}\| = \|T_{\theta,w}^+\| \leqslant \max_{\xi\in m} |e^{w(\xi - i\alpha_-)}| \|T_{\theta,w}^+ A^{-1}\|$$

and thus Lemma 2.1 is proved.

Note that if m is a piecewise linear set, the statement of Lemma 2.1 can be obtained from an earlier result due to M. M. Dzhrbashyan [21, p. 444].

LEMMA 2.2. *Let $m \in \mathbb{C}$ be a piecewise Lyapunov set and let the operator $\mathscr{K} : f \mapsto (\mathscr{K} f)(z)$ assign to each function $f \in L^2(m)$ its Cauchy integral along m. Let $l \subset \mathbb{C}$ be a smooth curve of a finite length. Then the restriction operator $f \mapsto (\mathscr{K} f)|_l$ is a bounded operator from the space $L^2(m)$ into the space $L^2(l)$.*

The lemma can be proved by combining the Kellog theorem on the boundary behavior of the derivative of a conformal map [22] and the theorem on Carleson measures.

To prove completeness of the system $\mathfrak{H}_\alpha^+$, it is sufficient to prove that every function $H \in L^2(K)$ orthogonal to $\mathfrak{H}_\alpha^+$ is zero almost everywhere. Let $H \perp \mathfrak{H}_\alpha^+$. Setting $F(t) = H(t)\,dt/|dt|$, we have

$$\int_K e^{\alpha nt} \sin nt F(t)\,dt = 0, \qquad n \in \mathbb{Z}_+. \tag{2.2}$$

Denoting $\varphi(\lambda) = \int_K F(t)\exp(\lambda t)\,dt$, rewrite (2.2) as

$$[\varphi((\alpha+i)\lambda) - \varphi((\alpha-i)\lambda)]|_{\mathbb{Z}_+} = 0.$$

Hence it follows that the function

$$\psi(\lambda) = [\varphi((\alpha+i)\lambda) - \varphi((\alpha-i)\lambda)](\sin\pi\lambda)^{-1} \tag{2.3}$$

is holomorphic in the right half-plane $\mathbb{C}_+ = \{\lambda : \operatorname{Re}\lambda > 0\}$. Put

$$\begin{gathered} \mathfrak{R} = (\alpha+i)\mathfrak{R}_\alpha \cup (\alpha-i)\mathfrak{R}_\alpha, \quad K^\pm = (\alpha \pm i)K, \quad \theta = \arg(\alpha+i), \\ \beta = \max(\alpha, (\alpha^2+1)/2\alpha) = \pi^{-1} \max_{\lambda\in\mathfrak{R}} \operatorname{Re}\lambda. \end{gathered} \tag{2.4}$$

Everywhere below, unless the contrary is specified explicitly, we shall assume that

$$K \subset \mathfrak{R}_\alpha. \tag{2.5}$$

The entire function of exponential type (e.f.e.t.) $\Omega(\lambda) = \varphi((\alpha+i)\lambda) - \varphi((\alpha-i)\lambda)$ may obviously be represented as

$$\Omega(\lambda) = \frac{1}{\alpha+i}\int_{K^+} e^{\lambda t} F\left(\frac{t}{\alpha+i}\right) dt - \frac{1}{\alpha-i}\int_{K^-} e^{\lambda t} F\left(\frac{t}{\alpha-i}\right) dt. \tag{2.6}$$

It follows from (2.4) and (2.5) that the integration paths in (2.6) lie in the half-band $\{\lambda : |\operatorname{Im}\lambda| \leqslant \pi, \operatorname{Re}\lambda \leqslant \pi\beta\}$. Hence, in view of Lemma 2.1, the function $\psi(\lambda)\exp(-\beta\pi\lambda)$ belongs to the Hardy space H_+^{ε} in the right half-plane. By the Paley-Wiener theorem, there exists a function $\chi \in L^2(-\infty, \beta\pi)$ such that

$$\psi(\lambda) = \int_{-\infty}^{\beta\pi} e^{\lambda t}\chi(t)\,dt, \qquad \operatorname{Re}\lambda > 0.$$

Comparing this equality with relations (2.3) and (2.6), we get

$$\frac{1}{\alpha+i}\int_{K^+} e^{\lambda t}F\left(\frac{t}{\alpha+i}\right)dt - \frac{1}{\alpha-i}\int_{K^-} e^{\lambda t}F\left(\frac{t}{\alpha-i}\right)dt$$
$$= \frac{1}{2i}(e^{i\pi\lambda} - e^{-i\pi\lambda})\int_{-\infty}^{\beta\pi}\chi(t)e^{\lambda t}\,dt, \qquad \operatorname{Re}\lambda > 0. \quad (2.7)$$

Put

$$\mathfrak{L}_\beta^{\pm} = \{z : z = t \pm i\pi,\ t \leqslant \beta\pi\}, \qquad \mathfrak{L}_\beta = \mathfrak{L}_b^+ \cup \mathfrak{L}_\beta^-,\ K^0 = K^+ \cup K^-, \quad (2.8)$$

and define the functions $f(t)$ on K^0 and $x(t)$ on $\mathfrak{L}_\beta$ by the equalities

$$f(t) = \begin{cases} \dfrac{1}{\alpha+i}F\left(\dfrac{t}{\alpha+i}\right), & t \in K^+ \setminus K^-, \\ \dfrac{1}{\alpha-i}F\left(\dfrac{t}{\alpha-i}\right), & t \in K^- \setminus K^+, \\ \dfrac{1}{\alpha+i}F\left(\dfrac{t}{\alpha+i}\right) - \dfrac{1}{\alpha-i}F\left(\dfrac{t}{\alpha-i}\right), & t \in K^+ \cap K^-, \end{cases} \quad (2.9)$$

$$x(t) = (2i)^{-1}\chi(t \mp i\pi), \qquad t \in \mathfrak{L}_\beta^{\pm}.$$

These functions have the following concordance properties:

$$x(t_1) = x(t_2) \qquad t_1, t_2 \in \mathfrak{L}_\beta,\ \operatorname{Re} t_1 = \operatorname{Re} t_2, \quad (2.10)$$

$$(\alpha+i)f(t_1) = (\alpha-i)f(t_2),$$
$$t_1 \in K^+ \setminus K^-,\ t_2 \in K^- \setminus K^+,\ (\alpha-i)t_1 = (\alpha+i)t_2, \quad (2.11)$$

$$f(t) = -\frac{\alpha+i}{\alpha-i}f(te^{2i\theta}) + \frac{\alpha-i}{\alpha+i}f(te^{-2i\theta}), \qquad t \in K^+ \cap K^-. \quad (2.12)$$

If the orientations of K^+ and $\mathfrak{L}_\beta^-$ ($K^- \setminus K^+$ and $\mathfrak{L}_\beta^+$) are assumed to coincide with (to be opposite to) those induced by the orientations of the sets K and $(-\infty, \beta\pi]$, then relation (2.7) becomes

$$\int_{K^+\cup K^-} e^{\lambda t}f(t)\,dt + \int_{\mathfrak{L}_\beta} e^{\lambda t}x(t)\,dt = 0, \qquad \operatorname{Re}\lambda > 0. \quad (2.13)$$

Now put

$$E = K^0 \cup \mathfrak{L}_\beta, \qquad D = \mathbb{C} \setminus E,$$

and denote by D^- the connected component of the set D which contains the ray $(\pi\beta, +\infty)$. By applying the Laplace transform to equality (2.13), we find

$$\frac{1}{2i\pi}\int_{K^0}\frac{f(t)}{t-z}\,dt+\frac{1}{2i\pi}\int_{\mathfrak{L}_\beta}\frac{x(t)}{t-z}\,dt=0, \qquad z\in D^-. \tag{2.14}$$

By reversing the above arguments, we arrive at the following criterion of incompleteness for the system $\mathfrak{H}_\alpha^+$ in the space $L^2(K)$.

THEOREM 2.1. *Given a number $\alpha>0$ and a piecewise Lyapunov set $K\subset\mathfrak{R}_\alpha$ with connected complement, the following statements are equivalent:*

(i) *the system $\mathfrak{H}_\alpha^+$ is incomplete in $L^2(K)$;*
(ii) *there exist nonzero functions $f\in L^2(K)$, $x\in L^2(\mathfrak{L}_\beta)$, such that relation (2.14) and the conditions of concordance (2.10)–(2.12) are fulfilled.*

§3. Proof of Theorem 1.1

We shall start with a particular case.

LEMMA 3.1. *Let $K\subset\mathfrak{R}_\alpha$ be a simple Lyapunov curve joining the points 0 and π which does not contain any pair of complex conjugate points belonging to $I_1^+\setminus\{0\}$ and $I_1^-\setminus\{0\}$. Then the system $\mathfrak{H}_\alpha^+$ is complete in the space $L^2(K)$.*

PROOF. Consider first the "classical" case $K=[0,\pi]$. Here the set E consists of the segments $[0,\pi(\alpha+i)]$ and $[0,\pi(\alpha-i)]$ and the two rays $\{z:\operatorname{Im}z=\pm\pi,\ \operatorname{Re}z<\alpha\pi\}$. If, under the conditions of Theorem 2.1, the functions f and x have properties stated in (ii), then, by (2.14), they are the boundary values of the function

$$\Phi(z)=\frac{1}{2i\pi}\int_{K^0}\frac{f(t)\,dt}{t-z}+\frac{1}{2i\pi}\int_{\mathfrak{L}_\beta}\frac{x(t)\,dt}{t-z}, \qquad z\in D^+, \tag{3.1}$$

belonging to the Smirnov space $E^2(D^+)$ (see, e.g. [23]) in the domain $D^+=\operatorname{Int}(\mathbb{C}\setminus D^-)$.

Denote by $W(z)$ the conformal map of the region D^+ onto the plane cut along the positive real axis $\mathbb{R}_+$ such that $W(0)=0$, $W(\overline{z})=\overline{W(z)}$. Then the function $\varphi(w)=\Phi(W_-(w))$, where the function W_- is inverse to W, is analytic outside $\mathbb{R}_+$ and, by condition (2.10), may be extended to a function analytic outside the segment $[0,\rho]$, where $\rho=W(\pi(\alpha+i))$. It follows from (2.11) that on the segment $[0,\rho]$ we have $\varphi(u+i0)=e^{-2i\theta}\varphi(u-i0)$. Therefore, the function $\psi(w)=\varphi(w)(w/(w-\rho))^{-\theta/\pi}$ is analytic in the entire plane, except, maybe, at the points 0, ρ, ∞. Simple additional analysis shows that at these points the function ψ has removable singularities. Therefore, $\varphi\equiv0$, whence $\Phi\equiv0$ and $f\equiv0$. Thus, we have obtained another proof of the completeness of the system $\mathfrak{H}_\alpha^+$ in the space $L^2(0, \pi)$.

Now let us consider the system $\mathfrak{H}_\alpha^+$ in the space $L^2(K)$ with an arbitrary curve K satisfying the conditions of Lemma 3.1. Let the function $F \in L^2(K)$ be such that (2.2) is fulfilled, let the functions f and x satisfy relations (2.9), and let the piecewise analytic function $\Phi(z)$ be defined by (3.1).

The curve K divides the rhombus $\mathfrak{R}_\alpha$ into two parts $\mathfrak{R}_\alpha^+$ and $\mathfrak{R}_\alpha^-$ lying to the left and to the right of it, respectively. Assuming, to be definite, that $\mathfrak{R}_\alpha^+ \neq \varnothing$, we see that $F(z)$ is analytically continuable into $\mathfrak{R}_\alpha^+$ and the continuation belongs to the space $L^2(B)$,

$$B = [0, \pi(\alpha + i)/2\alpha] \cup [\pi(\alpha + i)/2\alpha, \pi].$$

Denote by ω an arbitrary maximal domain within $\mathfrak{R}_\alpha^+$. Its boundary $\partial\omega$ consists of a simple arc $l \subset K$ with ends on the boundary $\partial\mathfrak{R}_\alpha$ of the rhombus and the portion σ of the polygonal line B connecting these ends.

Since K contains no pairs of complex conjugate points on the half-open intervals $I_1^+ \setminus \{0\}$ and $I_1^- \setminus \{0\}$, the sets K^+ and K^- do not have common points other than the point 0. Therefore, every point of the arc $(\alpha + i)l$ may be reached by a path entirely lying in D^- and, by (2.14), $f(z)$ coincides on this arc with the boundary value of $\Phi(z)$ from the domain $D^+ = \mathbb{C} \setminus D^-$. In other words, the function $\Phi(z)$ analytically continues the function $f(z)$ from the arc $(\alpha + i)l$ into the domain $(\alpha + i)\omega$. But then, by virtue of (2.9), the function $F(z)$ is analytically continuable into ω, and

$$\int_l |F(z)|^2\, |dz| = (\alpha^2 + 1)^{1/2} \int_{(\alpha+i)l} |\Phi(z)|^2\, |dz|. \tag{3.2}$$

By applying the above arguments to each domain ω of the type described, we can reduce (2.2), by deforming the integration path, to the form

$$\int_B F(t)e^{\alpha nt} \sin nt\, dt = 0, \qquad n \in \mathbb{Z}_+,$$

where, according to Lemma 2.2 and (3.2), the function F belongs to the space $L^2(B)$. This means that the functional corresponding to F in $L^2(B)$ annihilates the system $\mathfrak{H}_\alpha^+$. Applying equalities (2.9) and (2.14) again, now to $K = B$, we conclude that F is analytically continued to the entire rhombus $\mathfrak{R}_\alpha$ and that, according to Lemma 2.2, $F \in L^2(0, \pi)$. Deforming the polygonal line B into the segment $[0, \pi]$, we find that equality (2.2) is fulfilled when $K = [0, \pi]$. As was shown earlier, it follows that $F \equiv 0$. The lemma is proved.

Below, in the proof of Theorem 1.1, we shall distinguish two cases:

(1) the set K contains no pairs of complex conjugate nonzero points on the intervals $I_1^\pm$ and $\pi \in K$;

(2) $\pi \notin K$.

Let us first consider the case of an arbitrary set K satisfying condition (1). If $0 \notin K$, then $D^- = \mathbb{C} \setminus E$ and the equalities $f \equiv 0$ and $x \equiv 0$ follow directly from (2.14). Therefore, we can assume $0 \in K$. Then there exists a simple curve $l \subset K$ with ends at the points 0 and π. The connectedness of the set $\mathbb{C} \setminus K$ means that this curve is unique and divides the rhombus $\mathfrak{R}_\alpha$ into the two parts $\mathfrak{R}_\alpha^{(1)}$ and $\mathfrak{R}_\alpha^{(2)}$ lying to the left and to the right of l, respectively, when moving from 0 to π. Set $K_j = (K \setminus l) \cap \mathfrak{R}_\alpha^{(j)}$, $K_j^\pm = (\alpha \pm i)K_j$, $j = 1, 2$; $l^\pm = (\alpha \pm i)l$. We have $K^\pm = K_1^\pm \cup l^\pm \cup K_2^\pm$. It follows from condition (1) that $K^+ \cap K^- = \{0\}$. In this case $K_1^-, K_2^+ \subset \operatorname{Int}(\operatorname{Clos} D^-)$, and from relation (2.14) we have $f|_{K_1^-} = 0$, $f|_{K_2^+} = 0$. Hence (2.9) implies $f|_{K_1^+} = 0$, $f|_{K_2^-} = 0$, so that the function H which is orthogonal to the system $\mathfrak{H}_\alpha^+$ in $L^2(K)$ is entirely supported by the curve l. Therefore, proceeding from Lemma 3.1, we have $H \equiv 0$.

In case (2), without loss of generality, the set K may be assumed to be connected and have all its end points on the boundary of the rhombus $\mathfrak{R}_\alpha$. We shall split the arguments into several steps.

(a) Let I_2^+ and I_2^- be the segments connecting the point π with the points $\pi(\alpha+i)/2\alpha$ and $\pi(\alpha-i)/2\alpha$ and let the set $\operatorname{supp} F \cap (I_2^+ \cup I_2^-)$ contain no pairs of complex conjugate points. If $\operatorname{supp} x \neq \varnothing$, then one can choose the points $c^\pm \in \mathfrak{L}_\beta^\pm \cap \operatorname{supp} x$ so that $\operatorname{Re} c^+ = \operatorname{Re} c^- = \max_{t \in \operatorname{supp} x} \operatorname{Re} t$. By the assumption made, at least one of them, say c^+, does not belong to the support of the function f. But then all points of the ray $\{t \in \mathfrak{L}_\beta^+,\ \operatorname{Re} t \leqslant \operatorname{Re} c^+\}$ sufficiently close to c^+ are accessible from both sides of the support by paths on which $\Phi(z) = 0$. By (2.14), at those points we have $x = 0$, in contradiction to the condition $c^+ \in \operatorname{supp} x$. Thus, $x(t) \equiv 0$.

Relation (2.10) implies $\varphi((\alpha+i)\lambda) \equiv \varphi((\alpha-i)\lambda)$. Therefore, the Taylor coefficients d_j of the function φ satisfy the relations $d_j(1-\exp(-2ij\theta)) = 0$ ($\theta = \arg(\alpha+i) \in (0, \pi/2)$). If θ/π is irrational, then $\alpha_j = 0$, $j = 1, 2, \dots$, and thus, $\varphi(\lambda) = \text{const}$. If θ/π is rational, $\theta/\pi = m/n$, $n > 1$, then $\alpha_j = 0$ for all $j \not\equiv 0 \pmod n$. But then the indicator of the function $\varphi(\lambda)$ is a $(2\pi/n)$-periodic function, and since the indicator diagram of $\varphi(\lambda)$ lies in the rhombus $\mathfrak{R}_\alpha$, then it is reduced to the point 0. Thus, for any value of θ, the function grows no faster than a function of minimal exponential type. Hence, the integral $\int_K F(t)/(t-z)\,dt$ may be uniquely continued analytically everywhere into the domain $\mathbb{C} \setminus \{0\}$ and since the set $\mathbb{C} \setminus K$ is connected, we have $F(t) \equiv 0$.

(b) Let now the set $(\operatorname{supp} F) \cap (I_2^+ \cup I_2^-)$ contain a nonempty part symmetric with respect to the real axis. Denote by $d^\pm$ the closest points of this part to the point π. Take the simple curve l connecting d^- and d^+, and let it be oriented from d^- to d^+. Since $\pi \notin K$, the curve l is not reduced to a point and divides the interior of the rhombus $\mathfrak{R}_\alpha$ into the parts $\mathfrak{R}_\alpha^-$ and

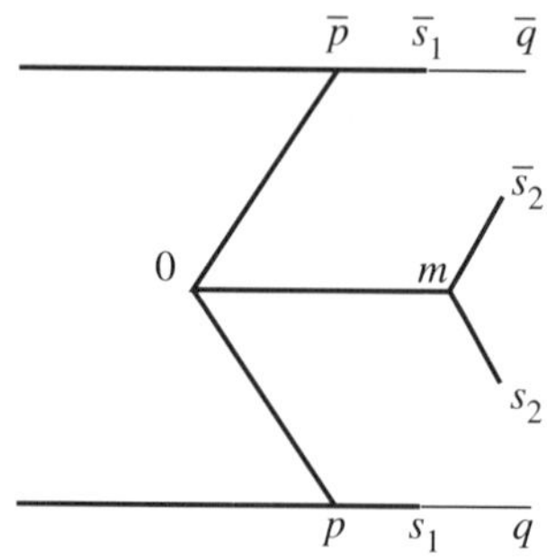

FIGURE 1

$\mathfrak{R}_\alpha^+$ lying to the right and to the left of it, respectively. In this case, $\mathfrak{R}_\alpha^- \neq \varnothing$.

As in the case (a), we conclude that $x(t) = 0$ at all points of the set $\mathfrak{L}_\beta$ located to the right of the points $c^\pm = (\alpha \pm i)d^\pm$ on the respective rays.

Now choose a simple point a of the set K lying in the closure of the set $\mathfrak{R}_\alpha^-$, but not belonging to the curve l. Since K does not divide the plane, there exist a simple closed curve p_a and a neighborhood V_a of the point a such that $K \cap p_a = \{a\}$ and the point a is approachable along p_a from both sides of the smooth curve $K \cap V_a$. If the curve p_a does not entirely lie in $\mathfrak{R}_\alpha^-$, then we shall replace it by the part containing a and bounded by the first points of its intersection with the boundary $\mathfrak{R}_\alpha^-$ as one moves from a in opposite directions. As a result, we shall obtain a curve p_a such that the curve $(\alpha + i)p_a$ with the point $(\alpha + i)a$ removed lies within D^- and the point $(\alpha + i)a$ itself is approachable along $(\alpha + i)p_a$ from both sides on the curve $(\alpha + i)(K \cap V_a)$. But then $f(t) \equiv 0$ for $t \in (\alpha + i)(K \cap V_a)$ and $F(t) \equiv 0$ for $t \in K \cap V_a$. Thus, the open set $\mathfrak{R}_\alpha^-$ does not contain points of the support of the function $F(t)$.

(c) According to (b), we shall assume below that the set K lies entirely in the closure of the set $\mathfrak{R}_\alpha^+$. If $\mathfrak{R}_\alpha^+ = \varnothing$, then K is reduced to the polygonal line b with vertices at the points d^-, $\pi(\alpha - i)/2\alpha$, 0, $\pi(\alpha + i)/2\alpha$, and d^+. The set K^0 in this case (Figure 1) consists of the segments

$$[p, s_1], \ [0, p], \ [0, m], \ [m, s_2], \ [m, \overline{s}_2], \ [0, \overline{p}], \ [\overline{p}, \overline{s}_1],$$

where

$$m = \pi(\alpha + \alpha^{-1})/2, \quad q = (\alpha - i)\pi, \quad p = -i\pi + \pi(\alpha - \alpha^{-1})/2,$$
$$\left.\begin{aligned} s_1 &= \pi[(1-\gamma) - i\gamma/\alpha](\alpha - i), \\ s_2 &= \pi[(1-\gamma) + i\gamma/a](\alpha - i), \end{aligned}\right. \quad \gamma \in [0, 1/2].$$

Obviously, the intervals $(\overline{p}, \overline{q}]$, $(0, m)$, $(p, q]$, $[m, s_2]$, $[m, \overline{s}_2]$ are contained in the domain $G = \operatorname{Int}(\operatorname{Clos} D^-)$, and according to (2.14),

$$x(t)|_{(s_1, q)\cup(\overline{s}_1, \overline{q})} \equiv 0, \qquad f(t)|_{(m, s_2)\cup(m, \overline{s}_2)} \equiv 0,$$
$$f(t)|_{[0, m]} \equiv 0, \qquad (f(t) + x(t))|_{(p, s_1)\cup(\overline{p}, \overline{s}_1)} \equiv 0.$$

From relations (2.10)–(2.12), it follows that

$$f(t) = e^{-2i\theta} f(te^{-2i\theta}), \qquad t \in (\overline{p}, \overline{s}_1) \cup (m, s_2).$$

Therefore, $f(t)|_{(p, s_1)\cup(\overline{p}, \overline{s}_1)} \equiv 0$ and (2.14) leads to $x(t)|_{(p, s_1)\cup(\overline{p}, \overline{s}_1)} \equiv 0$. Besides, from (2.12) and (2.14) we obtain

$$e^{2i\theta} f(te^{2i\theta}) - e^{-2i\theta} f(te^{-2i\theta}) = 0, \qquad t \in (0, m).$$

Rewrite relation (2.14) omitting the intervals corresponding to those parts of the set E where the integrand is surely zero:

$$\left(\int_{-\infty-i\pi}^{p} - \int_{\overline{p}}^{-\infty+i\pi}\right) \frac{x(t)}{t-z}\, dt + \int_{p\circ\overline{p}} \frac{f(t)}{t-z}\, dt = 0, \qquad z \in D^{-}.$$

In this case, relations (2.10) and (2.11) are true. By Theorem 2.1, the existence of nonzero functions $x(t)$ and $f(t)$ satisfying all these relations is equivalent to the incompleteness of $\mathfrak{H}_\beta^+$, where

$$\beta = \begin{cases} \cot 2\theta, & 0 < \theta < \pi/4, \\ \cot(\pi - 2\theta), & \pi/4 \leqslant \theta < \pi/2 \end{cases}$$

in the space $L^2(0, \pi)$. Since the system $\mathfrak{H}_\beta^+$ is complete in $L^2(0, \pi)$, the system $\mathfrak{H}_\alpha^+$ is also complete in $L^2(K)$.

(d) Now let $\mathfrak{R}_\alpha^+ \neq \varnothing$. Since $\mathbb{C} \setminus K$ is a connected set, the boundary of every connected component of the set $\mathfrak{R}_\alpha^+ \setminus K$ contains at least two points of the set $\partial\mathfrak{R}_\alpha$. We shall refer to a connected component U as *regular*, if the set $\partial U \cap K \cap \partial\mathfrak{R}_\alpha$ consists of exactly two points. Otherwise, the component will be called *irregular*. Let us prove that the set $\mathfrak{R}_\alpha^+ \setminus K$ has only a finite number of irregular connected components.[1]

Indeed, let $\{U_j\}_{j=1}^\infty$ be an infinite set of pairwise different irregular components. Choose the points $a_j \in \partial U_j \cap K \cap \partial\mathfrak{R}_\alpha^+$. It may be supposed that the sequence $\{a_j\}$ converges to a point $a \in \partial U_j \cap K \cap \partial\mathfrak{R}_\alpha^+$. Let the neighborhood V_a of the point a be so small that the nodes and end point of K do not fall within it, except maybe the point a itself, and the set $K \cap V_a$ is the union of a finite number of simple smooth arcs having a single common point a. An infinite number of various connected components in the set $(\mathfrak{R}_\alpha^+ \setminus K) \cap V_a$ can only occur if the polygonal line $\partial\mathfrak{R}_\alpha \cap V_a$ is tangent to one or two arcs originating at the point a. Every such component is a regular component of the set $\mathfrak{R}_\alpha^+ \setminus K$. This contradicts the choice of the point a.

(e) Denote by $n(K)$ the number equal to the sum of multiplicities of all nodes of K lying strictly within $\mathfrak{R}_\alpha$ and the number of irregular connected components of the set $\mathfrak{R}_\alpha^+ \setminus K$. It follows from the arguments of (d) that

[1] Unlike regular components, whose number can be infinite. An infinite number of regular components can appear, if one of the arcs of the set K is tangent to the sides of the rhombus $\mathfrak{R}_\alpha$ at a condensing sequence of points.

$n(K) < \infty$. We shall finish the proof of completeness of the system $\mathfrak{H}_\alpha^+$ under the condition $\pi \notin K$ by induction over $n(K)$.

If $n(K) = 0$, then all nodes of the set K lie on the boundary of the rhombus $\mathfrak{R}_\alpha$. Since all connected components of the set $\mathfrak{R}_\alpha^+ \setminus K$ are regular and all the ends of the sets K lie on $\partial\mathfrak{R}_\alpha$, it follows that the set $\partial\mathfrak{R}_\alpha^+$ does not contain the nodes of the set K either and that K is a simple line with ends at d^- and d^+. It follows from (2.14) that the function $F(t)$ is continuable from K to a function of the space $E^2(\mathfrak{R}_\alpha^+)$. By Lemma 2.2, its boundary values on the polygonal line b from (c) belong to the space $L^2(b)$ and are orthogonal in it to the system $\mathfrak{H}_\alpha^+$. According to the results of (c), these boundary values are zero, whence $F \equiv 0$.

(f) Let now $n(K) > 0$. Then, at least one of the following statements is true:

(i) the set l contains the nodes of the set K;
(ii) at least one of the points $d^\pm$ is not an end of the set K.

Let statement (i) be true. Denote by a the first node of K lying on l in the direction from d^- to d^+ along l and by l' the simple line which is the part of l between d^- and a. Let U be a connected component of the set $\mathfrak{R}_\alpha^+ \setminus K$, $a \in \partial U$, whose boundary contains some part of the arc $l' \setminus \{a\}$, and let $l'' = \partial U \setminus l'$. It follows from condition (2.14) that the function $F(t)$ is continuable from $l' \cap \partial U$ to a function from the space $E^2(U)$. Therefore, in (2.2) K may be replaced by the set $K' = (K \setminus (l' \cap \partial U)) \cup l''$. Since the multiplicity of the point a in K' is smaller than that in K by one and the multiplicities of the other nodes lying inside $\mathfrak{R}_\alpha^+$ do not increase if K is replaced by K', we have $n(K') < n(K)$.

Now let statement (ii) by true and, say, the point d^+ not be an end of the set K. One may suppose that there are no nodes of K on l. Let U be the connected component of the set $\mathfrak{R}_\alpha^+ \setminus K$, $d^+ \in \partial U$, whose boundary contains some part of $l \setminus \{d^+\}$. It follows from the fact that the point d^+ is not an end of K that the component U is irregular. Therefore, by replacing the line $l \cap \partial U$ in K by $(\partial U \setminus l) \cup \{d^-, d^+\}$, we obtain a set K' for which $n(K') < n(K)$.

Thus, in both cases (i) and (ii), the set K in the assumption of the theorem and the function F on it satisfying the orthogonality condition (2.2) may be replaced by the set K', the number $n(K')$ being smaller than $n(K)$, and by the function $\widetilde{F}$ for which (2.2) (with K replaced by K') is fulfilled. By the assumption of induction, $\widetilde{F} \equiv 0$, whence obviously $F \equiv 0$. Thus, Theorem 1.1 is proved.

§4. Proof of Theorem 1.2

In this section we shall study the boundary problems arising in Theorem 1.2 and give the proof of this theorem. The boundary problems for

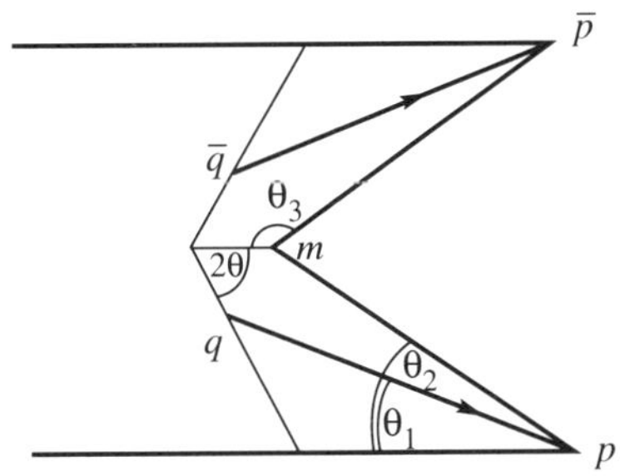

FIGURE 2

$\gamma = 1/2$ and $\gamma \in (0, 1/2)$ are essentially different. We shall restrict ourselves to the more complicated case of $\gamma \in (0, 1/2)$. Below, the notation of §§ 1 and 2 is used, unless other notation is specified.

Let us describe the structure of the set K^0 arising in our situation.

Put $p = (\alpha - i)\pi$, $q = \gamma\pi(1 - i/\alpha)(\alpha - i)$, $m = \gamma\pi(1 + i/\alpha)(\alpha - i)$. The set K^0 consists (Figure 2) of the segments $[q, p]$, $[m, p]$, $[m, \overline{p}]$ and $[\overline{q}, \overline{p}]$. As in the proof of the preceding theorem, it may be shown that the system $\mathfrak{H}_\alpha^+$ is incomplete in the space $L^2(K_\gamma)$ if and only if there exists a nonzero function $\Phi(z) \in E^2(D^+)$ satisfying the conditions:

$$\begin{cases} \Phi(t) = \Phi(\overline{t}), & t \in \mathfrak{L}_\beta, \\ \Phi^+(t) - \Phi^-(t) = e^{2i\theta}\Phi(te^{2i\theta}), & t \in [q, p], \\ \Phi^+(t) - \Phi^-(t) = e^{-2i\theta}\Phi(te^{-2i\theta}), & t \in [\overline{q}, \overline{p}]. \end{cases} \tag{4.1}$$

Here $\Phi^\pm(t)$ are the left- and the right-hand boundary values with respect to the orientation indicated in Figure 2 by the arrows.

Consider the conformal mapping $W: D^+ \to \mathbb{C} \setminus \mathbb{R}_+$ such that $W(z) = \overline{W(\overline{z})}$, $z \in D^+$. Put $M = 0 = W(m)$, $Q = W(q) = W(\overline{q})$. Each of the points p and $\overline{p}$ may be approached from D^+ by two nonequivalent paths. Therefore, each of them corresponds to the two points $P, P' \in \mathbb{R}$. We shall assume that the segments $[m, p]$ and $[m, \overline{p}]$ are mapped by W onto the segment $[M, P]$ and the rays $\mathfrak{L}_\beta^\pm$ onto the ray (P', ∞). Denote by W_- the mapping inverse to W. Then the mapping $\Gamma: \Phi(z) \mapsto \varphi(w) = \Phi(W_-(z))$ is an isomorphism of the spaces $E^2(D^-)$ and $\mathcal{F} \overset{\text{def}}{=} [W_-'(w)]^{1/2} E^2(\mathbb{C} \setminus \mathbb{R}_+)$. Recall that the norm in this latter space is defined by the relation

$$\|\varphi\|_{\mathcal{F}} = \|(W_-')^{1/2} \cdot \varphi\|_{E^2(\mathbb{C}\setminus\mathbb{R}_+)}.$$

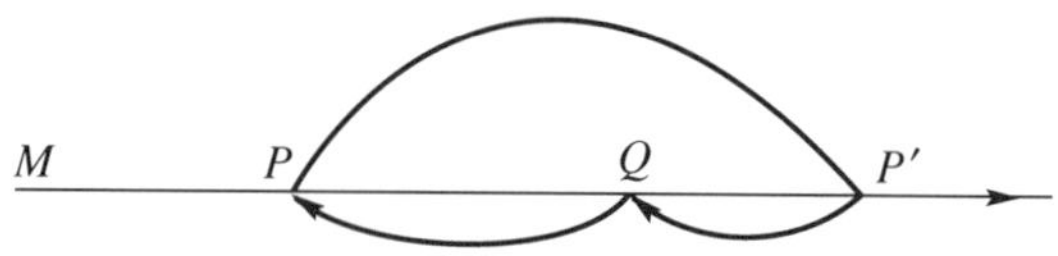

FIGURE 3

The mapping W induces (Figure 3) the following one-to-one mapping $\nu\colon [M, P'] \to [M, P']$ of the segment $[M, P']$ onto itself:

$$\nu\colon u \to \begin{cases} W^{-}[W_{-}(u-i0)], & u \in (Q, P'), \\ W[W_{-}(u-i0)e^{2i\theta}], & u \in (P, Q), \\ W^{+}[W_{-}(u+i0)e^{-2i\theta}], & u \in (M, P). \end{cases}$$

Here $W^{\pm}$ are the left- and the right-hand boundary values of the function W at the points of the set $[p, q] \cup [\overline{p}, \overline{q}]$ with respect to the orientation shown in Figure 2. From the definition of the map ν it is seen that

(i) the Carleman condition $\nu^3 = \mathrm{id}$ is satisfied;

(ii) $\nu(M) = Q$, $\nu(P') = P$, the points P and Q are discontinuity points of the first kind, $\nu(Q+0) = Q$, $\nu(Q-0) = M$, $\nu(P+0) = P$, $\nu(P-0) = P'$;

(iii) the segments $[M, P]$, $[P, Q]$, and $[Q, P']$ are mapped by ν onto the segments $[Q, P']$, $[M, P]$, and $[P, Q]$, respectively.

Denote by ν_{-} $(= \nu^2)$ the mapping inverse to ν. If the function $\Phi \in E^2(D^+)$ satisfies condition (4.1), then the function $\varphi = \Gamma\Phi$ satisfies the following relations:

$$\varphi^{+}(u) = \varphi^{-}(u), \qquad u > P', \tag{4.2}$$

$$\varphi^{-}(u) - \varphi^{-}(\nu(u)) = e^{2i\theta}\varphi^{+}(\nu^2(u)), \qquad u \in (Q, P'), \tag{4.3}$$

$$\varphi^{+}(u) - \varphi^{+}(\nu(u)) = e^{-2i\theta}\varphi^{-}(\nu^2(u)), \qquad u \in (Q, P'). \tag{4.4}$$

Here $\varphi^{\pm}$ are the left- and the right-hand boundary values of the function φ at the points of $\mathbb{R}_{+}$. From condition (4.2) it follows that the function $\varphi(w)$ is holomorphic in the domain $\mathbb{C} \setminus [M, P']$.

Therefore, the system $\mathfrak{H}_{\alpha}^{+}$ in $L^2(K_{\gamma})$ is not complete if and only if there exists a nonzero function $\varphi \in \mathscr{F}$ holomorphic in $\mathbb{C} \setminus [M, P']$ and satisfying the conditions (4.3)–(4.4).

The arguments to follow are rather cumbersome and we begin by presenting an outline.

(i) We shall prove the existence of any prescribed finite number of linearly independent functions $\varphi \in \mathscr{F}$ satisfying the conditions (4.3)–(4.4). Hence it will follow that the system $\mathfrak{H}_{\alpha}^{+}$ has an infinite defect in $L^2(K_{\gamma})$.

For some $p > 2$ these functions will belong to a narrower space:

$$\mathscr{F}_p \overset{\text{def}}{=} [W_{-}']^{-1/p} E^p(\mathbb{C} \setminus \mathbb{R}_{+}) \tag{4.5}$$

whose norm is $\|(W_{-}')^{1/p}\varphi\|_{E^p(\mathbb{C}\setminus\mathbb{R}_{+})} = \|\varphi\|_{\mathscr{F}_p}$.

(ii) We shall consider, instead of the system of equations (4.3)–(4.4), the system consisting of equation (4.3) and the equation

$$[\varphi^+(u) - \varphi^-(u)] - [\varphi^+(\nu(u)) - \varphi^-(\nu(u))]$$
$$= e^{-2i\theta}\varphi^-(\nu^2(u)) - e^{2i\theta}\varphi^+(\nu^2(u)), \qquad u \in [Q, P'], \quad (4.6)$$

obtained by subtracting (4.3) from (4.4).

(iii) Choosing arbitrary functions g_1 and g_2 satisfying certain restrictions on growth as the right-hand sides of the equations

$$\varphi^+(u) - \varphi^-(u) = g_1(u), \qquad u \in [Q, P'], \quad (4.7)$$
$$\varphi^+(u) - \varphi^-(u) = g_2(u), \qquad u \in [P, Q], \quad (4.8)$$

we shall reduce equations (4.6)–(4.8) to the Riemann boundary problem for the function $\varphi(w)$. The solution of this problem is determined by the "free parameters", i.e., the functions g_1 and g_2, and can be found explicitly.

(iv) When g_1 and g_2 are arbitrary, $\varphi(w)$ satisfies equation (4.6). It now remains to find the functions g_1 and g_2 so that (4.3) will hold. To do so, we shall substitute the boundary values of $\varphi(w)$ into (4.3). This will yield a relation between g_1 and g_2 which garantees that condition (4.3) is fulfilled. This relation is

$$T_L g_1 = T_R g_2, \quad (4.9)$$

where T_L and T_R are certain operators acting in their respective functional spaces.

(v) We shall prove that the operator T_L is Noetherian and the operator T_R is not finite-dimensional. This implies the existence of an arbitrary finite number of linearly independent pairs of functions g_1 and g_2 satisfying (4.9): this relation, if treated as an equation in g_1, has a solution for all right-hand sides belonging to a subspace of finite codimension, and the set of functions $T_R g_2$ is an infinite-dimensional linear manifold. Hence, according to the above, the incompleteness of the system $\mathfrak{H}_\alpha^+$ in the space $L^2(K_\gamma)$ follows.

Now we shall proceed to the proof in detail.

To indicate explicitly the spaces to which the "free parameters" should belong, let us describe the behavior of the derivative of the conformal mapping W_- in the neighborhoods of singular points. We have

$$|W_-'(w)| \asymp |w|^{-1}, \qquad w \to \infty, \quad (4.10)$$
$$|W_-'(w)| \asymp |w - P'|^{\theta_1/\pi - 1}, \qquad w \to P', \quad (4.11)$$
$$|W_-'(w)| \asymp |w - Q|, \qquad w \to Q, \quad (4.12)$$
$$|W_-'(w)| \asymp |w - P|^{\theta_2/\pi - 1}, \qquad w \to P, \quad (4.13)$$
$$|W_-'(w)| \asymp |w - M|^{\theta_3/\pi - 1}, \qquad w \to M, \quad (4.14)$$

where θ_1, θ_2 and θ_3 are the openings of the angles shown in Figure 2. Denote

$$\rho(w) = |w - P'|^{\theta_1/\pi - 1}|w - Q||w - P|^{\theta_2/\pi-1}|w - M|^{\theta_3/\pi-1}.$$

Fix the number $p > 2$ (its choice will be specified later) and seek the solution of the problem (4.4) in the space $\mathscr{F}_p$. From relations (4.5), (4.7), (4.8), (4.11)–(4.13), it follows that

$$g_1 \in \mathfrak{L}_1 \overset{\text{def}}{=} \left\{ g_1(u),\ u \in [Q, P'] : \int_Q^{P'} |g_1(u)|^p |u - Q||u - P'|^{\theta_1/\pi-1}\, du < \infty \right\},$$

$$g_2 \in \mathfrak{L}_2 \overset{\text{def}}{=} \left\{ g_2(u),\ u \in [P, Q] : \int_P^{Q} |g_2(u)|^p |u - P|^{\theta_2/\pi-1}|u - Q|\, du < \infty \right\}.$$

Choosing arbitrary functions g_1 and g_2 from these spaces, we unite the relations (4.6)–(4.8) into a Riemann problem for the function $\varphi \in \mathscr{F}_p$:

$$\begin{cases} \varphi^+(u) - \varphi^-(u) = g_1(u), & u \in [Q, P'], \\ \varphi^+(u) - \varphi^-(u) = g_2(u), & u \in [P, Q], \\ \varphi^+(u) - e^{-4i\theta}\varphi^-(u) & \\ \qquad = -e^{-2i\theta} g_1(\nu(u)) + e^{-2i\theta} g_2(\nu^2(u)), & u \in [M, P]. \end{cases} \tag{4.15}$$

Consider the auxiliary function $\varkappa(w) = (w-M)^{2\theta/\pi}(w-P)^{-2\theta/\pi}$ holomorphic in $\mathbb{C}\backslash[M, P]$. The single-valued branch of the function $\varkappa$ is determined by the condition $\varkappa(w) > 0$ for $w \in (P, +\infty)$. The substitution

$$\varphi(w) = \varkappa(w)\psi(w) \tag{4.16}$$

brings (4.15) to the following problem with a jump for the function ψ:

$$\psi^+(u) - \psi^-(u) = g_1(u)/\varkappa(u), \qquad u \in [Q, P'], \tag{4.17}$$

$$\psi^+(u) - \psi^-(u) = g_2(u)/\varkappa(u), \qquad u \in [P, Q], \tag{4.18}$$

$$\psi^+(u) - \psi^-(u) = -e^{-2i\theta}[g_1(\nu(u)) - g_2(\nu^2(u))/\varkappa^+(u)], \quad u \in [M, P], \tag{4.19}$$

whose solution we shall seek in the class of all functions ψ holomorphic in $\mathbb{C} \setminus [M, P']$, vanishing at infinity, and satisfying

$$\int_M^{P'} |\psi^\pm(u)|^p \rho_1(u)\, du < \infty, \qquad \rho_1(w) = \rho(w)|\varkappa(w)|^p.$$

The function $\psi(w)$ will be the sum of three summands, each of them giving a jump represented by one of equalities (4.17)–(4.19). Below we shall need the following proposition, which presents a particular case of Theorem 4.1 from [24, Chapter 1].

THEOREM A. *Let the number $p \in (1, \infty)$, the segment $[a, b] \subset \mathbb{R}$, the points $t_1, \dots, t_m \in [a, b]$ ($t_k \neq t_l$ for $k \neq l$) and the numbers $\beta_1, \dots, \beta_m$ such that $-1 < \beta_j < p - 1$ be given. If*

$$\widehat{\rho}(t) = \prod_{j=1}^{m} |t - t_j|^{\beta_j},$$

then the Hilbert operator is bounded in the space $L^p([a, b], \widehat{\rho})$ *with the norm*

$$\|g\| = \left(\int_a^b |g(t)|^p \rho(t)\, dt\right)^{1/p}.$$

Since $|\varkappa(u)| \geqslant c > 0$ for $u \in [Q, P']$, it follows from this theorem that the function

$$\psi_1(w) = \frac{1}{2i\pi}\int_Q^{P'} \frac{g_1(u)}{\varkappa(u)}\frac{du}{u-w}, \qquad w \in \mathbb{C}\setminus[Q, P'],$$

satisfies the condition

$$\int_Q^{P'} |\psi_1^{\pm}(w)|^p |w-P'|^{\theta_1/\pi-1}|w-Q|\, dw < \infty$$

and condition (4.17) in which $\psi = \psi_1$.

The function $\psi_2(w)$ giving the jump of $\psi(w)$ on the segment $[P, Q]$ is constructed similarly, taking into consideration the value of the power exponent of the factor $|w-P|$ in the weight $\rho_1(w)$. Set $\beta_P = \theta_2/\pi - 1 - 2p\theta/\pi$; $\beta_M = \theta_3/\pi - 1 + 2p\theta/\pi$. We have

$$\rho_1(w) = |w-P'|^{\theta_1/\pi-1}|w-Q||w-P|^{\beta_P}|w-M|^{\beta_M}.$$

To estimate the exponent β_P, note (see Figure 2) that $\theta \in (0, \pi/2)$ and $\theta_2 \in (0, 2\theta)$. Hence obviously $-1 < \beta_P + p < p-1$ and thus the relation

$$\psi_2(w) = \frac{w-P}{2i\pi}\int_P^Q \frac{g_2(u)}{(u-P)\varkappa(u)}\frac{du}{u-w}$$

defines the function $\psi_2(w)$ which solves the boundary problem (4.18) and has boundary values belonging to the space $L^p([P, Q], |w-Q||w-P|^{\beta_P})$.

Choose p so large that the exponent β_M satisfies the condition $-1 < \beta_M < p-1$. Then the function

$$\psi_3(w) = \frac{e^{-2i\theta}}{2i\pi}(w-P)\int_M^P \frac{-g_1(\nu(u)) + g_2(\nu^2(u))}{\varkappa^+(u)(u-P)(u-w)}\, du$$

solves the boundary problem (4.19) and its boundary values belong to the space $L^p([M, P], |w-P|^{\beta_P}|w-M|^{\beta_M})$.

The function $\psi(w) = \psi_1(w)+\psi_2(w)+\psi_3(w)$ solves the boundary problem (4.17)–(4.19). In this case the condition $\psi\varkappa \in \mathscr{F}_p$ is fulfilled, if

$$\text{(i)}\quad \psi(\infty) = 0, \qquad \text{(ii)}\quad \int_{P-\delta}^{P+\delta} |\psi^{\pm}(w)|^p|w-P|^{\beta_P}\, d\omega < \infty,$$

where $\delta > 0$ is a sufficiently small number. Note that

$$\int_{P-\delta}^{P+\delta} |\psi_j(w)|^p |w-P|^{\beta_P}\, dw < \infty \quad \text{with } j = 2, 3.$$

Conditions (i) and (ii) are fulfilled if the "free parameters" g_1 and g_2 lie, respectively, in the subspaces

$$\begin{aligned}\widetilde{\mathfrak{L}}_1 &= \left\{ g_1 \in \mathfrak{L}_1 : \int_Q^{P'} \frac{g_1(u)}{\varkappa(u)} \frac{du}{u-P} = \int_M^P \frac{g_1(\nu(u))}{\varkappa^+(u)(u-P)}\, du = 0 \right\}, \\ \widetilde{\mathfrak{L}}_2 &= \left\{ g_2 \in \mathfrak{L}_2 : \int_P^Q \frac{g_2(u)}{\varkappa(u)} \frac{du}{u-P} = \int_M^P \frac{g_2(\nu^2(u))}{\varkappa^+(u)(u-P)}\, du = 0 \right\}.\end{aligned} \tag{4.20}$$

Each of the integrals in (4.20) defines continuous linear functionals on $\mathfrak{L}_1$ and $\mathfrak{L}_2$. Therefore, $\widetilde{\mathfrak{L}}_1$ and $\widetilde{\mathfrak{L}}_2$ are closed subspaces of codimension 2 in $\mathfrak{L}_1$ and $\mathfrak{L}_2$, respectively. Subsequently, we shall assume that $g_1 \in \widetilde{\mathfrak{L}}_1$ and $g_2 \in \widetilde{\mathfrak{L}}_2$. In this case, clearly, the solution of the problem (4.15) determined by equality (4.16) belongs to the space $\mathscr{F}_p$.

Let us express the boundary values of the function $\varphi(w)$ which are presented in equation (4.13) via the functions g_1 and g_2. For $u \in [Q, P']$,

$$\begin{aligned}\varphi^-(u) = &-\frac{g_1(u)}{2} + \frac{\varkappa(u)}{2i\pi} \int_Q^P \frac{g_1(\sigma)}{\varkappa(\sigma)} \frac{d\sigma}{(\sigma-u)} \\ &+ \frac{\varkappa(u)}{2i\pi}(u-P) \int_P^Q \frac{g_2(\sigma)}{(\sigma-P)\varkappa(\sigma)} \frac{d\sigma}{\sigma-u} \\ &- \frac{\varkappa(u)}{2i\pi} e^{-2i\theta}(u-P) \int_M^P \frac{g_1(\nu(\sigma))\, d\sigma}{\varkappa^+(\sigma)(\sigma-P)(\sigma-u)} \\ &+ \frac{\varkappa(u)}{2i\pi} e^{-2i\theta} \int_M^P \frac{g_2(\nu^2(\sigma))\, d\sigma}{\varkappa^+(\sigma)(\sigma-P)(\sigma-u)}.\end{aligned}$$

Similar representations may be obtained for $\varphi^-(\nu(u))$ and $\varphi^+(\nu^2(u))$. By substituting the expressions found into (4.3) and rearranging the terms so that those containing the functions g_1 and g_2 appear in the left- and the right-hand sides, respectively, we obtain

$$\begin{aligned}&\left[\frac{\varkappa(u)}{2i\pi} \int_Q^{P'} \frac{g_1(\sigma)\, d\sigma}{\varkappa(\sigma)(\sigma-u)} \right. \\ &\qquad \left. + \frac{\varkappa^+(\nu^2(u))}{2i\pi}(\nu^2(u)-P) \int_M^P \frac{g_1(\nu(\sigma))\, d\sigma}{\varkappa^+(\sigma)(\sigma-P)(\sigma-\nu^2(u))}\right] \\ &+ \left[-\frac{\varkappa(\nu(u))}{2i\pi} \int_Q^{P'} \frac{g_1(\sigma)\, d\sigma}{\varkappa(\sigma)(\sigma-\nu(u))} \right. \\ &\qquad \left. + \frac{e^{-2i\theta}}{2i\pi} \varkappa(\nu(u))(\nu(u)-P) \int_M^P \frac{g_1(\nu(\sigma))\, d\sigma}{\varkappa^+(\sigma)(\sigma-P)(\sigma-\nu(u))}\right] \\ &- \left[\frac{e^{-2i\theta}}{2i\pi} \varkappa(u)(u-P) \int_M^P \frac{g_1(\nu(\sigma))\, d\sigma}{\varkappa^+(\sigma)(\sigma-P)(\sigma-u)} \right. \\ &\qquad \left. + \frac{e^{2i\theta}}{2i\pi} \varkappa^+(\nu^2(u)) \int_Q^{P'} \frac{g_1(\sigma)\, d\sigma}{\varkappa(\sigma)(\sigma-\nu^2(u))}\right]\end{aligned}$$

$$
\begin{aligned}
&= \left[\frac{\varkappa(\nu(u))}{2i\pi}(\nu(u)-P) \fint_P^Q \frac{g_2(\sigma)\,ds}{(\sigma-P)\varkappa(\sigma)(\sigma-\nu(u))} \right. \\
&\qquad \left. + \frac{\varkappa^+(\nu^2(u))}{2i\pi}(\nu^2(u)-P) \fint_M^P \frac{g_2(\nu^2(\sigma))\,d\sigma}{(\sigma-P)\varkappa^+(\sigma)(\sigma-\nu^2(u))} \right] \\
&\quad + \left[-\frac{\varkappa(u)}{2i\pi}(u-P) \int_P^Q \frac{g_2(\sigma)\,d\sigma}{(\sigma-P)\varkappa(\sigma)(\sigma-u)} \right. \\
&\qquad + \frac{e^{-2i\theta}}{2i\pi}\varkappa(\nu(u))(\nu(u)-P) \int_M^P \frac{g_2(\nu^2(\sigma))\,d\sigma}{\varkappa^+(\sigma)(\sigma-P)(\sigma-\nu(u))} \\
&\qquad \left. + \frac{e^{2i\theta}}{2i\pi}\varkappa^+(\nu^2(u))(\nu^2(u)-P) \int_Q^P \frac{g_2(\sigma)\,d\sigma}{(\sigma-Q)\varkappa(\sigma)(\sigma-u)} \right] \\
&\quad - \left[\frac{e^{-2i\theta}}{2i\pi}\varkappa(u)(u-P) \int_M^P \frac{g_2(\nu^2(\sigma))\,d\sigma}{\varkappa^+(\sigma)(\sigma-P)(\sigma-u)} \right], \qquad u \in [Q, P']. \qquad (4.21)
\end{aligned}
$$

This relation is an expanded version of relation (4.9). Accordingly, denote the left- and right-hand sides of (4.21) by $T_L g_1$ and $T_R g_2$, respectively. Also, denote the expressions in the first, second and third square brackets in the left- (right-) hand side of (4.21) by $S_L g_1$ ($S_R g_2$), $N_L g_1$ ($N_R g_2$) and $K_L g_1$ ($K_R g_2$), respectively. The operators with the subscripts " L " and " R " act from the spaces $\widetilde{\mathfrak{L}}_1$ and $\widetilde{\mathfrak{L}}_2$ respectively into the space $\widetilde{\mathfrak{L}}_1$. According to the outline described above, it only remains to prove that the operator T_L is Noetherian and the operator T_R is not finite-dimensional.

The proof will be based on the stability of the Noether property under compact perturbations.

Note that if $\rho_2(u) = |u-P'|^{\theta_1/\pi - 1}|u-Q|$, then the map $g(u) \mapsto \tilde{g}(u) = g(u)[\rho_2(u)]^{1/p}$ is an isomorphism of the spaces $\mathfrak{L}_1$ and $L^p(Q, P')$. Every operator $A\colon \mathfrak{L}_1 \to \mathfrak{L}_1$ is similar to the operator

$$
\widetilde{A} = \rho_2^{1/p} A \rho_2^{-1/p}\colon L^p(Q, P') \to L^p(Q, P).
$$

Thus, it is possible to use the compactness criterion for integral operators in the space $L^p(Q, P)$ without weight.

Let us prove that the operator $K_L\colon \widetilde{\mathfrak{L}}_1 \to \mathfrak{L}_1$ is compact. To do so, let us represent the operator K_L as $K_L = K_1 + K_2$, where

$$
\begin{aligned}
K_1\colon g &\mapsto \frac{e^{-2i\theta}}{2i\pi}\varkappa(u)(u-P) \int_M^P \frac{g(\nu(\sigma))\,d\sigma}{\varkappa^+(\sigma)(\sigma-P)(\sigma-u)}, \\
K_2\colon g &\mapsto \frac{e^{2i\theta}}{2i\pi}\varkappa^+(\nu^2(u)) \int_Q^{P'} \frac{g(\sigma)\,d\sigma}{\varkappa(\sigma)(\sigma-\nu^2(u))},
\end{aligned}
$$

and verify the compactness of each term separately.

By the substitution $s = \nu(\sigma)$, the operator K_1 is reduced to the form

$$K_1 \colon g \mapsto \frac{e^{-2i\theta}}{2i\pi} \varkappa(u)(u-P) \int_Q^{P'} \frac{g(s)\nu'_-(s)\,ds}{\varkappa^+(\nu_-(s))(\nu_-(s)-P)(\nu_-(s)-u)}.$$

Simple estimates (given below in Lemma 4.1) show that the kernel of the integral operator $\widetilde{K}_1 = \rho_2^{1/p} K_1 \rho_2^{-1/p} \colon L^p(Q, P') \to L^p(Q, P')$ may be represented as

$$\tilde{k}_1(u, s) = |u - P'|^{\beta_1} |u - Q|^{\beta_2} |s - P'|^{\beta_3} |s - Q|^{\beta_4} r(u, s), \tag{4.22}$$

where $\beta_1 = (\theta_1/\pi - 1)/p$, $\beta_2 = 1/p$, $\beta_3 = -(\theta_1/\pi - 1)/p + 2\theta_1\theta/(\theta_2\pi) - 1$, $\beta_4 = -1/p - 4\theta/\theta_3 + 2\pi/\theta_3 - 1$, $\sup_{u,s} |r(u, s)| < \infty$.

To prove the compactness of the operator $\widetilde{K}_1$ and of the other operators appearing below, we shall need the notion of regularity of an integral operator and one-dimensional version of the compactness theorem using this notion. We shall formulate them, fixing the measurable set $\Omega \subset \mathbb{R}$ of finite Lebesgue measure and the integral operator

$$K \colon f(\cdot) \mapsto \int_\Omega f(s) k(\cdot, s)\,ds$$

mapping the set of functions defined on Ω into itself. Denote by $L^p(\Omega)$ the spaces constructed by using the Lebesgue measure. A linear operator A is called *regular*, if it can be represented as $A = A_1 - A_2$, where A_1 and A_2 are positive linear operators.

THEOREM B ([25, Theorem 4.2]). *The linear integral operator K with a kernel k is a regular operator from $L^p(\Omega)$ into $L^q(\Omega)$, $1 \leqslant p$, $q \leqslant \infty$, if and only if the operator $|K| \colon f(\cdot) \mapsto \int_\Omega f(s)|k(\cdot, s)|\,ds$ is a bounded operator from $L^p(\Omega)$ into $L^q(\Omega)$.*

THEOREM C ([25, Theorem 5.4]). *Let K be a regular integral operator from $L^p(\Omega)$ into $L^q(\Omega)$. For the operator $K \colon L^{p_1}(\Omega) \to L^{q_1}(\Omega)$ to be compact it is sufficient that at least one of the following conditions holds: (i) $p_1 > p > 1$, $q_1 \leqslant q$; (ii) $p_1 \geqslant p > 1$, $q_1 < q < \infty$.*

Let us return to proving the compactness of the operator K_1. According to Theorems B and C, it is sufficient to prove the regularity of K_1 as a map from the space $L^p(Q, P')$ into $L^q(Q, P')$ for some $q > p$.

To estimate the exponents β_3 and β_4, we shall again use the inequalities $0 < \theta < \pi/2$, $0 < \theta_2 < 2\theta$, $\pi - 2\theta < \theta_3 < \pi - \theta$ and $\beta_M < p - 1$. Denoting $p' = p(p-1)^{-1}$, we obtain

$$\begin{aligned} \beta_3 &= \left(\frac{1}{p} - 1\right) + \frac{\theta_1}{\pi}\left(\frac{2\theta}{\theta_2} - \frac{1}{p}\right) > -\frac{1}{p'}, \\ \beta_4 &= 2\frac{\pi - 2\theta}{\theta_3} - \frac{1}{p} - 1 > \frac{1}{p} - 1 = -\frac{1}{p'}. \end{aligned} \tag{4.23}$$

Now choose the number $q > p$ so that $q\beta_1 > -1$. Then, for a function $f \in L^p(Q, P')$, we have

$$\int_Q^{P'} \left(\int_Q^{P'} |f(s)||\tilde{k}_1(u, s)|\, ds \right)^q du$$
$$\leqslant C \int_Q^{P'} \left(|u - P'|^{\beta_1} \int_Q^{P'} |f(s)||s - P'|^{\beta_3}|s - Q|^{\beta_4}\, ds \right)^q du$$
$$\leqslant C_1 \|f\|_{L^p}^q \int_Q^{P'} |u - P'|^{\beta_1 q}\, du \leqslant C_2 \|f\|_{L^p}^q.$$

The boundedness of the operator $|\widetilde{K}_1|$ from $L^p(Q, P')$ into $L^q(Q, P')$ and thus the compactness of the operator $\widetilde{K}_1 : L^p(Q, P') \to L^p(Q, P')$ are proved. Hence the compactness of the operator $K_1 : \widetilde{\mathfrak{L}}_1 \to \mathfrak{L}_1$ follows.

The compactness of the operator $K_2 : \widetilde{\mathfrak{L}}_1 \to \mathfrak{L}_1$ follows from the fact that its kernel on the space $\widetilde{\mathfrak{L}}_1$ coincides, according to (4.20), with the function $(2i\pi)^{-1} e^{2i\theta} \varkappa^+(\nu^2(u))[(\sigma - \nu^2(u))^{-1} - (\sigma - P)^{-1}]/\varkappa(\sigma)$ which is continuous with respect to all its variables.

Thus, the Noether property of the operator T_L is equivalent to that of the operator $T_L' = S_L + N_L$. Note that this operator is of a local type[2] and use the local principle [26], which means that T_L' is Noetherian if and only if T_L' is locally Noetherian at every point of the segment $[Q, P']$. At every inner point of the segment $[Q, P']$, the operator T_L' is clearly locally equivalent to the singular integration operator

$$S_0 : g \mapsto (i\pi)^{-1} \int_Q^{P'} \frac{g(s)}{s - u}\, ds$$

and therefore is locally Noetherian. Hence it is sufficient to prove that T_L' is locally Noetherian at the points Q and P'.

The "singular part" S_L of the operator T_L' is

$$(S_L g)(u) = \frac{\varkappa(u)}{2i\pi} \int_Q^{P'} \frac{g(\sigma)\, d\sigma}{\varkappa(\sigma)(\sigma - u)}$$
$$+ \frac{\varkappa^+(\nu\_(u))(\nu\_(u) - P)}{2i\pi} \int_M^P \frac{g(\nu(\sigma))\, d\sigma}{\varkappa^+(\sigma)(\sigma - P)(\sigma - \nu\_(u))}$$
$$= (S_1 g)(u) + (S_2 g)(u).$$

The operator S_1 may be represented as

$$(S_1 g)(u) = \frac{1}{2i\pi} \int_Q^{P'} \frac{g(\sigma)\, d\sigma}{\sigma - u} + \frac{1}{2i\pi} \int_Q^{P'} \frac{\varkappa(u) - \varkappa(\sigma)}{u - \sigma} \frac{g(\sigma)}{\varkappa(\sigma)}\, d\sigma$$
$$= (1/2)(S_0 g)(u) + (K_3 g)(u).$$

---

[2]Definitions of local type operators, the local Noether property, local equivalence and other notions associated with this topic are presented in [24] and [26].

The operator K_3 is compact in the space $\mathfrak{L}_1$, since its kernel is bounded.

Consider the operator S_2. We have

$$
\begin{aligned}
(S_2 g)(u) &= \frac{1}{2i\pi}\int_Q^{P'} \frac{\nu'_-(s)}{\nu_-(s)-\nu_-(u)} g(s)\,ds \\
&+ \frac{1}{2i\pi}\int_Q^{P'} \frac{(\varkappa^+(\nu_-(u))(\nu_-(u)-P) - \varkappa^+(\nu_-(s))(\nu_-(s)-P))g(s)\nu'_-(s)\,ds}{(\nu_-(s)-\nu_-(u))\varkappa^+(\nu_-(s))(\nu_-(s)-P)} \\
&= (Mg)(u) + (N_1 g)(u).
\end{aligned}
$$

To investigate this operator near the ends of the segment, let us clarify the structure of the map $\nu_-(s)$.

LEMMA 4.1. *For some $\varepsilon > 0$ we have*

$$\nu_-(s) - M = (s-Q)^{2\pi/\theta_3} f_1(s-Q), \qquad s \in (Q, Q+\varepsilon), \tag{4.24}$$

$$P - \nu_-(s) = (P'-s)^{\theta_1/\theta_2} f_2(P'-s) f_3(P'-s)^{\theta_1/\pi} f_4(P'-s), \qquad s \in (P'-\varepsilon, P'), \tag{4.25}$$

where $f_j(z)$ are functions analytic for $|z| < \varepsilon$, and $f_j(0) \neq 0$, $j = 1, 2, 3, 4$.

PROOF. By definition $\nu_-(s) = W[W_-(s-i0)e^{2i\theta}]$, $s \in [Q, P']$. The representation (4.25) can therefore simply be obtained by using the expansions

$$W_-(s-i0) - p = (P'-s)^{\theta_1/\pi} \sum_0^\infty c_k (P'-s)^k, \qquad c_0 \neq 0,\ s \in (P'-\varepsilon, P'),$$

$$P - W(z) = (z-\overline{p})^{\pi/\theta_2} \sum_0^\infty b_l (z-\overline{p})^l,$$

$$b_0 \neq 0,\ z \in (m, \overline{p}) \cap \{z : |z-\overline{p}| < \varepsilon\}.$$

The representation (4.24) is obtained similarly.

Choose the point $Q_0 \in (Q, P')$ and consider the operators

$$M_Q : g \mapsto \frac{1}{2i\pi}\int_Q^{Q_0} \frac{g(s)\nu'_-(s)}{\nu_-(s)-\nu_-(u)}\,ds,$$

$$M_{P'} : g \mapsto \frac{1}{2i\pi}\int_{Q_0}^{P'} \frac{g(s)\nu'_-(s)}{\nu_-(s)-\nu_-(u)}\,ds,$$

acting in the spaces

$$\mathfrak{L}_Q \stackrel{\text{def}}{=} \left\{ g(u),\ u \in [Q, Q_0] : \int_Q^{Q_0} |g(u)|^p |u-Q|\,du < \infty \right\},$$

$$\mathfrak{L}_{P'} \stackrel{\text{def}}{=} \left\{ g(u),\ u \in (Q_0, P') : \int_{Q_0}^{P'} |g(u)|^p |u-P'|^{\theta_1/\pi - 1}\,du < \infty \right\}.$$

These operators are locally equivalent to the operator M at the points Q and P', respectively. Let us single out the main parts of their kernels.

LEMMA 4.2. *The operators* $K_Q: \mathfrak{L}_Q \to \mathfrak{L}_Q$ *and* $K_{P'}: \mathfrak{L}_{P'} \to \mathfrak{L}_{P'}$ *defined by the relations*

$$K_Q: g \mapsto \frac{1}{2i\pi}\int_Q^{Q_0} g(s)\left[\frac{\nu_-'(s)}{\nu_-(s)-\nu_-(u)} - \frac{(2\pi/\theta_3)(s-Q)^{2\pi/\theta_3-1}}{(s-Q)^{2\pi/\theta_3}-(u-Q)^{2\pi/\theta_3}}\right]ds,$$

$$K_{P'}: g \mapsto \frac{1}{2i\pi}\int_{Q_0}^{P'} g(s)\left[\frac{\nu_-'(s)}{\nu_-(s)-\nu_-(u)} - \frac{(\theta_1/\theta_2)(P'-s)^{\theta_1/\theta_2-1}}{(P'-s)^{\theta_1/\theta_2}-(P'-u)^{\theta_1/\theta_2}}\right]ds$$

are compact.

We shall present the proof for the operator K_Q; the compactness of $K_{P'}$ is provided similarly.

(a) Without loss of generality, the representation (4.24) may be assumed true for all $s \in [Q, Q_0]$.

(b) The change of the variables $\sigma = (s-Q)^{2\pi/\theta_3}$, $\sigma = (u-Q)^{2\pi/\theta_3}$, $\mu(\sigma) = \nu_-(s) - M$, $\tilde{g}(\sigma) = g(s)$ reduces the operator K_Q to

$$\widetilde{K}_Q: \tilde{g} \mapsto \frac{1}{2i\pi}\int_0^{Q_1} \tilde{g}(\sigma)\tilde{k}_Q(v, \sigma)\,d\sigma, \qquad \tilde{k}_Q(v, \sigma) = \frac{\mu'(\sigma)}{\mu(\sigma)-\mu(v)} - \frac{1}{\sigma - v},$$

acting in the space

$$\widetilde{\mathfrak{L}}_Q = \left\{\tilde{g}(v),\ v \in (0, Q_1): \int_0^{Q_1} |\tilde{g}(v)|^p v^{\theta_3/\pi-1}\,dv < \infty\right\}.$$

(c) Let us estimate the kernel $\tilde{k}_Q(v, \sigma)$. We have

$$\tilde{k}_Q(v, \sigma) = -\frac{(v-\sigma)\mu'(\sigma) - [\mu(v)-\mu(\sigma)]}{(\mu(\sigma)-\mu(v))(\sigma - v)}.$$

Without loss of generality, we may assume $\mu'(\sigma) \neq 0$, $\sigma \in [0, Q_1]$ and therefore $[\mu(\sigma)-\mu(v)](\sigma-v) \asymp (\sigma-v)^2$, $\sigma, v \in [0, Q_1]$. The representation (4.24) gives $\mu(\sigma) = \sigma f_1(\sigma^\alpha)$, where $\alpha = \theta_3/(2\pi) < 1$. Therefore

$$\mu'(\tau) - \mu'(\sigma) = [f_1(\tau^\alpha) - f_1(\sigma^\alpha)] + \alpha(\tau^\alpha - \sigma^\alpha)f_1'(\tau^\alpha) + \alpha\sigma^\alpha[f_1'(\tau^\alpha) - f_1'(\sigma^\alpha)].$$

Using the inequalities

$$|f_1'(\tau^\alpha) - f_1'(\sigma^\alpha)| \leqslant \mathrm{const}\,|\tau^\alpha - \sigma^\alpha|, \qquad |f_1(\tau^\alpha) - f_1(\sigma^\alpha)| \leqslant \mathrm{const}\,|\tau^\alpha - \sigma^\alpha|,$$
$$|\tau^\alpha - \sigma^\alpha| \leqslant |v^\alpha - \sigma^\alpha|, \qquad \tau \in (\sigma, v),$$

we obtain

$$|\tilde{k}_Q(v, \sigma)| \leqslant \mathrm{const}\,|v^\alpha - \sigma^\alpha||v - \sigma|^{-1}.$$

To be definite, let $v > \sigma$. We have $|v^\alpha - \sigma^\alpha||v-\sigma|^{-1} = v^{\alpha-1}\varphi_\alpha(t)$, where $t = \sigma/v \in (0, 1)$, $\varphi_\alpha(t) = (1-t^\alpha)/(1-t)$. Clearly, $\max_{t\in[0,1]}\varphi_\alpha(t) < \infty$. A similar estimate in the case of $\sigma > v$ yields

$$|\tilde{k}_Q(v, \sigma)| \leqslant \mathrm{const}\,[\max(v, \sigma)]^{\alpha-1}. \tag{4.26}$$

(d) To prove the compactness of the operator $\widetilde{K}_Q$, as in proving that of the operator K_1, we shall pass to the space $L^p(0, Q_1)$ without weight, in which we shall consider the operators

$$\widehat{K}_{Q,j}\colon \hat{g} \mapsto \int_0^{Q_1} \hat{g}(\sigma)\left(\frac{v}{\sigma}\right)^{(\theta_3/\pi-1)/p} |\hat{k}_{Q,j}(v,\sigma)|\,d\sigma, \qquad j=1,2,$$

where

$$\hat{k}_{Q,1}(v,\sigma)=\begin{cases} v^{\alpha-1}, & \sigma<v,\\ 0, & \sigma\geqslant v,\end{cases} \qquad \hat{k}_{Q,2}(v,\sigma)=\begin{cases} 0, & \sigma\leqslant v,\\ \sigma^{\alpha-1}, & \sigma>v.\end{cases}$$

Consider, to be definite, the operator $\widehat{K}_{Q,1}$. We have

$$\begin{aligned} |(\widehat{K}_{Q,1}\hat{g})(v)| &\leqslant v^{(\theta_3/\pi-1)/p+(\theta_3/(2\pi)-1)} \int_0^v |\hat{g}(\sigma)|\sigma^{-(\theta_3/\pi-1)/p}\,d\sigma \\ &\leqslant \|\hat{g}\|_{L^p(0,Q_1)} v^{\theta_3/(2\pi)-1/p}. \end{aligned}$$

This clearly implies the boundedness of the operator $\widehat{K}_{Q,1}$ from the space $L^p(0, Q_1)$ into the space $L^q(0, Q_1)$ for numbers $q>p$ satisfying the condition $q(\theta_3/(2\pi)-1/p)>-1$; this implies compactness.

COROLLARY. *The operator M at the points Q and P' is locally equivalent to the operators*

$$\widetilde{M}_Q\colon g\mapsto \frac{1}{2i\pi}\int_Q^{P'} \frac{2\pi}{\theta_3}\frac{(s-Q)^{2\pi/\theta_3-1}}{(s-Q)^{2\pi/\theta_3}-(u-Q)^{2\pi/\theta_3}} g(s)\,ds,$$

$$\widetilde{M}_{P'}\colon g\mapsto \frac{1}{2i\pi}\int_Q^{P'} \frac{\theta_1}{\theta_2}\frac{(P'-s)^{\theta_1/\theta_2-1}}{(P'-s)^{\theta_1/\theta_2}-(P'-u)^{\theta_1/\theta_2}} g(s)\,ds,$$

respectively.

Arguments similar to those used in Lemma 4.2 show that the operator N_1 is locally equivalent at the points Q and P' to the operators

$$\widetilde{N}_{Q,1}\colon g\mapsto -\frac{1}{2i\pi}\int_Q^{P'} \frac{2\pi}{\theta_3}\frac{(s-Q)^{4\theta/\theta_3}-(u-Q)^{4\theta/\theta_3}}{(s-Q)^{2\pi/\theta_3}-(u-Q)^{2\pi/\theta_3}}(s-Q)^{\frac{2\pi-4\theta}{\theta_3-1}} g(s)\,ds,$$

$$\widetilde{N}_{P',1}\colon g\mapsto$$

$$-\frac{1}{2i\pi}\int_Q^{P'} \frac{\theta_1}{\theta_2}\frac{(P'-s)^{(1-2\theta/\pi)\theta_1/\theta_2}-(P'-u)^{(1-2\theta/\pi)\theta_1/\theta_2}}{(P'-s)^{\theta_1/\theta_2}-(P'-u)^{\theta_1/\theta_2}}(P'-u)^{\frac{2\theta\theta_1}{\pi\theta_2}-1} g(s)\,ds,$$

respectively.

Let us study the part N_L of the operator T_L. The change of variable $\sigma=\nu_-(s)$ in the second summand of the expression defining N_L (cf. (4.21)) gives

$$\begin{aligned} N_L\colon g \mapsto & -\frac{\varkappa(\nu(u))}{2i\pi}\int_Q^{P'}\frac{g(s)\,ds}{\varkappa(s)(s-\nu(u))}+\frac{1}{2i\pi}e^{-2i\theta}\varkappa(\nu(u))(\nu(u)-P) \\ & \times\left(\int_{Q_0}^{P'}+\int_Q^{Q_0}\right)\frac{g(s)}{\varkappa^+(\nu_-(s))(\nu_-(s)-P)}\,\frac{\nu_-'(s)\,ds}{\nu_-(s)-\nu_-(u)} \\ = & \, N_{Q,2}g+N_{P',2}g+K_4g. \end{aligned}$$

On the space $\widetilde{\mathfrak{L}}_1$, the operator $N_{Q,2}$ coincides (see (4.20)) with an operator having the kernel $-\varkappa(\nu(u))(\nu(u)-P)[2i\pi\varkappa(s)(s-P)(s-\nu(u))]^{-1}$, which, as is easy to see, is bounded when the points u and s lie in the neighborhood of the point P'. Hence the operator $N_{Q,2}$ is locally equivalent to a compact one at all points of the half-open interval $(Q, P']$. We have

$$\begin{aligned} N_{Q,2}g = & -\frac{1}{2i\pi}\int_Q^{P'}\frac{g(s)}{s+u-2Q}\,ds-\frac{1}{2i\pi}\int_Q^{P'}\frac{\varkappa(\nu(u))-\varkappa(s)}{s-\nu(u)}\,\frac{g(s)}{\varkappa(s)}\,ds \\ & -\frac{1}{2i\pi}\int_Q^{P'}\frac{u+\nu(u)-2Q}{(s+u-2Q)(s-\nu(u))}g(s)\,ds \\ = & \, \widetilde{N}_{Q,2}g+K_5g+K_6g. \end{aligned}$$

The operator K_5 is obviously locally equivalent at the point Q to a compact one. It follows from the definition of $\nu(t)$ that $\nu(u)-Q=Q-u+O(|Q-u|^2)$ for $u\in(Q, Q+\varepsilon)$. Hence the operator K_6 has a bounded kernel and is compact.

Using Lemma 4.1, it is easy to prove that the kernel $k_4(u, s)$ of K_4 admits the bound $|k_4(u, s)|\leqslant C|s-Q|^\gamma$, $\gamma=2\pi(1-2\theta/\pi)/\theta_3-1$. Now Theorems B and C combined with $\beta_M<\beta-1$ yield the compactness of K_4.

The principal part of the operator $N_{P',2}$ is singled out by arguments similar to those used to prove Lemma 4.2; they show that the operator N_L at the points Q and P' is locally equivalent to the operators

$$\begin{aligned} \widetilde{N}_{Q,2}\colon g &\mapsto -\frac{1}{2i\pi}\int_Q^{P'}\frac{g(s)}{s+u-2Q}\,ds, \\ \widetilde{N}_{P',2}\colon g &\mapsto \frac{1}{2i\pi}\int_Q^{P'}\frac{(P'-u)^{(1-2\theta/\pi)\theta_1/\theta_2}(P'-s)^{2\theta\theta_1/(\pi\theta_2)-1}}{a(P'-u)^{\theta_1/\theta_2}+b(P'-s)^{\theta_1/\theta_2}}g(s)\,ds, \end{aligned}$$

respectively, where $a, b>0$ are constant.

Thus, at the points Q and P', the operator T_L is locally equivalent to the operators

$$\begin{aligned} T_Q &= (1/2)S_0+\widetilde{M}_Q+\widetilde{N}_{Q,1}+\widetilde{N}_{Q,2}, \\ T_{P'} &= (1/2)S_0+\widetilde{M}_{P'}+\widetilde{N}_{P',1}+\widetilde{N}_{P',2}, \end{aligned}$$

respectively.

To prove that T_Q is locally Noetherian at the point Q, let us substitute $\sigma = s - Q$, $v = u - Q$ and note that the operator obtained from T_Q is locally equivalent, at zero, to the operator

$$\widetilde{T}_Q\colon g \mapsto \frac{1}{2i\pi}\int_0^\infty g(\sigma)\left[\frac{1}{\sigma - v} + 2\frac{\pi}{\theta_3}\frac{\sigma^{2\pi/\theta_3 - 1}}{\sigma^{2\pi/\theta_3} - v^{2\pi/\theta_3}} - 2\frac{\pi}{\theta_3}\frac{\sigma^{4\theta/\theta_3} - v^{4\theta/\theta_3}}{\sigma^{2\pi/\theta_3} - v^{2\pi/\theta_3}}\sigma^{(2\pi - 4\theta)/\theta_3 - 1} - \frac{1}{\sigma + v}\right] d\sigma$$

acting in the space $L^p(\mathbb{R}_+, \sigma)$. Let us prove, following the outline presented in [27, Chapter 13], that the operator $\widetilde{T}_Q$ is invertible. To do so, it is sufficient to prove that the operator $\widetilde{T}_Q$ belongs to the algebra $\mathfrak{A}_{1,p}$ generated by the operator of singular integration in the space $L^p(\mathbb{R}_+, \sigma)$ and that this operator does not belong to any maximal ideal of this algebra. The substitution $\sigma = e^t$, $v = e^x$, $G(x) = g(e^x)e^{2x/p}$ transforms the operator $\widetilde{T}_Q$ into the convolution operator $\widehat{T}_Q$ acting in the space $L^p(\mathbb{R})$:

$$\begin{aligned}\widehat{T}_Q\colon G \mapsto \frac{1}{2i\pi}\int_{-\infty}^\infty G(t)\Bigg[&\frac{e^{2p^{-1}(x-t)}}{1 - e^{x-t}} + \frac{2\pi}{\theta_3}\frac{e^{2p^{-1}(x-t)}}{1 - e^{2\pi\theta_3^{-1}(x-t)}} \\ &- \frac{2\pi}{\theta_3}\frac{1 - e^{4\theta\theta_3^{-1}(x-t)}}{1 - e^{2\pi\theta_3^{-1}(x-t)}}e^{2p^{-1}(x-t)} + \frac{e^{2p^{-1}(x-t)}}{1 + e^{x-t}}\Bigg] dt \\ = \frac{1}{2i\pi}\int_{-\infty}^\infty G(t)[k_1(x-t) + k_2(x-t) &+ k_3(x-t) + k_4(x-t)]\, dt.\end{aligned}$$

The symbol of this operator is

$$\begin{aligned}\hat{k}_Q(\xi) &= \frac{1}{2i\pi}\sum_{j=1}^4 \int_{-\infty}^\infty e^{-i\xi t}k_j(t)\, dt \\ &= (1/2)\coth(\pi\xi + 2i\pi/p) + (1/2)(\sinh(\pi\xi + 2i\pi/p))^{-1} \\ &\quad + (1/2)\coth(\theta_3\xi/2 + i(2\theta + \theta_3/p)), \qquad \xi \in \mathbb{R}.\end{aligned}$$

Let us prove that the operator $\widetilde{T}_Q$ belongs to the algebra $\mathfrak{A}_{1,p}$. Introduce, as in [24], the following notation. Let $\delta \in (0, 1/2]$ and $l(\delta)$ be the arc of circle which connects the points $\pm s$, lies in the lower half-plane and is such that the segment $[-1, 1]$ is viewed from the points $z \in l(d)\setminus\{-1, 1\}$ under the angle $2\pi\delta$. When $\delta \in [1/2, 1)$, put $l(\delta) = -l(1-\delta)$.

Also put $\delta_Q = 2/p$ and $z = \widehat{S}_Q(\xi) = \coth(\pi\xi + i\pi\delta_Q)$. The function $\widehat{S}_Q$ is a one-to-one map of $\mathbb{R}$ onto $l(\delta_Q)$. Therefore, the function $\widehat{\varkappa}_Q\colon l(\delta_Q) \to \mathbb{C}$ may be defined by the relation $\widehat{\varkappa}_Q(\widehat{S}_Q(\xi)) = \hat{k}_Q(\xi)$, which will also be called the symbol. Define the norm $\|\cdot\|_Q$ in the space of functions on $l(\delta_Q)$ by the relation

$$\|f\|_Q = \max\left\{\max_z |f(z)|, \max_z |(1 - z^2)\log[(1-z)(1+z)^{-1}]f'(z)|\right\},$$

the inner maximums being taken over $l(\delta_Q)$. According to Theorems 13.5 and 13.6 and Corollaries 13.3 and 13.4 of [27], the spectrum of the algebra $\mathfrak{A}_{1,p}$ coincides with the arc $l(\delta_Q)$. To prove that $\widetilde{T}_Q$ belongs to this algebra, it is sufficient to show that the symbol $\widehat{\varkappa}_Q(z)$ may be approximated by rational functions of z with respect to the norm $\|\cdot\|_Q$. Now the invertibility of $\widetilde{T}_Q$ will be equivalent to the relation $\widehat{\varkappa}_Q(z) \neq 0$, $z \in l(\delta_Q)$, or, which is the same, $\hat{k}_Q(\xi) \neq 0$, $\xi \in \mathbb{R}$.

The expression for the function $\widehat{\varkappa}_Q(z)$ can be found explicitly

$$\widehat{\varkappa}_Q(z) = \frac{1}{2}\left\{ z - \sqrt{z^2 - 1} \right.$$
$$\left. + \left[\left(\frac{z+1}{z-1}\right)^{\theta_3/(2\pi)} + e^{-4i\theta}\right]\left[\left(\frac{z+1}{z-1}\right)^{\theta_3/(2\pi)} - e^{-4i\theta}\right]^{-1}\right\}.$$

Here we choose those branches of the power functions which are defined by the condition $\arg(z+1) = \arg(z-1) = 0$ with $z \in (1, \infty)$.

Assume that $0 \notin l(\delta_Q)$. For every number $\mu > 0$, we shall consider the space $C_{1,\mu}(\lambda(\delta_Q))$ consisting of functions f, continuous on the arc $l(\delta_Q)$, whose derivatives on the arc $l(\delta_Q)$ are continuous at the inner points, with the norm

$$\|f\|_{1,\mu} = \max\left(\max_z |f(z)|, \max_z(|1-z^2|^{\mu} - |f'(z)|)\right),$$

the inner maximums being taken over $l(\delta_Q)$. Obviously, the system of the functions $\{z^n\}_{-\infty}^{\infty}$ is complete in $C_{1,\mu}$. Hence it is possible to approximate $\widehat{\varkappa}_Q(z)$ with respect to the norm $\|\cdot\|_Q$ by rational functions and the inclusion $\widetilde{T}_Q \in \mathfrak{A}_{1,p}$ is fulfilled.

Note that the expression for $\widehat{\varkappa}_Q(z)$ does not depend on p. Therefore, if $\widehat{\varkappa}_Q(\pm 1) \neq 0$, then it is always possible to choose p large enough for the arc $l(\delta_Q)$ not to run through the origin and the zeros of the function $\widehat{\varkappa}_Q(z)$; thus the operator is invertible into $\mathfrak{A}_{1,p}$. For these values of p, the operator T_L is locally Noetherian at the point Q. The relation $\widehat{\varkappa}_Q(\pm 1) \neq 0$ can be verified directly.

The arguments for the point P' are similar. Here we shall only give the expressions for the exponent $\delta_{P'}$ and the symbol $\widehat{\varkappa}_{P'}(z)$. In this case, $\delta_{P'} = \theta_1(\pi p)^{-1}$, and

$$\widehat{\varkappa}_{P'}(z) = \frac{1}{2}\left\{ z + \left[\left(\frac{z+1}{z-1}\right)^{\theta_2/\theta_1} + e^{4i\theta}\right]\left[\left(\frac{z+1}{z-1}\right)^{\theta_2/\theta_1} - e^{4i\theta}\right]^{-1} \right.$$
$$\left. + \left[a\left(\frac{z+1}{z-1}\right)^{\theta_1/\theta_2 - 2i\theta} - b\left(\frac{z+1}{z-1}\right)^{-\theta_1/\theta_2 + 2i\theta}\right]^{-1}\right\}.$$

The local Noether property of the operator T_L at Q and P' is proved. Therefore, T_L is a Noetherian operator and it only remains, for the proof of

Theorem 1.2, to verify that the operator T_R defined by relation (4.21) is not finite-dimensional. But this is obvious: T_R is a local operator and at every inner point of the segment $[P, Q]$ it is locally equivalent to the singular integration operator and thus cannot be even compact, to say nothing of being finite-dimensional. Theorem 1.2 is proved.

Every nontrivial function $F \in L^2(K_\gamma)$ satisfying condition (2.2) for $K = K_\gamma$ can be used to prove the incompleteness of the system $\mathfrak{H}_\alpha^+$ in the spaces $L^2(K)$ on more general sets K. Indeed, let $\pi \in K$ and for some $\gamma \in (0, 1/2]$ the inclusions $\gamma\pi(\alpha \pm i)/2\alpha \in K$ be fulfilled. Then, there exist simple curves l^+ and l^- joining the point π to $\gamma\pi(\alpha + i)/2\alpha$ and $\gamma\pi(\alpha - i)/2\alpha$ in the set K. Assume moreover that the union $l^+ \cup l^-$ lies in the quadrangle T_γ with the vertices $0, \pi$ and $\gamma\pi(\alpha \pm i)/2\alpha$. From identity (2.14) one can conclude that the function F can be continued to a function F_* of the space $E^2(T_\gamma)$. In this case, $F_* \in L^2(l^+ \cup l^-)$, and by deforming the set K_γ in $l^+ \cup l^-$ and using equality (2.2), we see that the function $\overline{F(t)}\,\overline{dt}/dt$ continued by zero to the set $K \setminus (l^+ \cup l^-)$ cancels the system $\mathfrak{H}_\alpha^+$ in the space $L^2(K)$.

References

1. M. V. Keldysh, *On the eigenvalues and eigenfunctions of some classes of nonselfadjoint equations*, Dokl. Akad. Nauk SSSR **77** (1951), no. 1, 11–14. (Russian)

2. ———, *On the completeness of the eigenfunctions of some classes of nonselfadjoint linear operators*, Uspekhi Mat. Nauk **27** (1972), no. 4, 15–47; English transl. in Russian Math. Surveys **27** (1972).

3. M. G. Zhavadov, *On the completeness of a certain part of the eigenfunctions of a nonselfadjoint differential operator*, Dokl. Akad. Nauk SSSR **159** (1964), no. 4, 723–725; English transl. in Soviet Math. Dokl. **5** (1964).

4. M. G. Gasymov, *On the theory of polynomial operator pencils*, Dokl. Akad. Nauk SSSR **199** (1971), no. 4, 747–750; English transl. in Soviet Math. Dokl. **12** (1971).

5. ———, *On the multiple completeness of a part of the eigenvectors and adjacent vectors of polynomial operator pencils*, Izv. Akad. Nauk Armyan. SSR Ser. Mat. **6** (1971), no. 2–3, 131–147.

6. M. G. Gasymov and M. G. Zhavadov, *Multiple completeness of a part of the eigenfunctions and adjacent functions of differential operator pencils*, Dokl. Akad. Nauk SSSR **203** (1972), no. 6, 1235–1237; English transl. in Soviet Math. Dokl. **13** (1972).

7. G. V. Radzievskiĭ, *On a method to prove completeness of root vectors of operator functions*, Dokl. Akad. Nauk SSSR **214** (1974), no. 2, 291–294; English transl. in Soviet Math. Dokl. **15** (1974).

8. B. Ya. Levin, *Entire functions*, Izdat. Moskov. Univ., Moscow, 1971. (Russian)

9. R. Feinerman and D. J. Newman, *Completeness of* $\{A \sin nx + B \cos nx\}$ *on* $[0, \pi]$, Michigan Math. J. **15** (1968), no. 3, 305–312.

10. A. A. Shkalikov, *On some system of functions*, Mat. Zametki **18** (1975), no. 6, 855–860; English transl. in Math. Notes **18** (1975).

11. Yu. A. Kazmin, *On the closure of linear hulls of two systems of functions*, Dokl. Akad. Nauk SSSR **236** (1977), no. 3, 535–537; English transl. in Soviet Math. Dokl. **18** (1977).

12. ———, *Closures of the linear hull of one system of functions*, Sibirsk. Mat. Zh. **18** (1977), no. 4, 799-805; English transl. in Siberian Math. J. **18** (1977).

13. ———, *On linear hulls of two sequences of functions*, Topics of Mathematical Analysis, Izdat. Rostov. Univ., Rostov-on-Don, 1978, pp. 72–83. (Russian)

14. A. N. Barmenkov and Yu. A. Kazmin, *Completeness of some special systems of functions*, Theory of Mappings, its Generalizations and Applications, "Naukova Dumka", Kiev, 1982, pp. 29–43. (Russian)

15. A. G. Tumarkin, *On completeness of some systems of functions*, Funktsional. Anal. i Prilozhen. **14** (1980), no. 2, 81–82; English transl. in Functional Anal. Appl. **14** (1980).

16. ______, *On completeness of real and imaginary parts of powers of complex-valued functions*, Sibirsk. Mat. Zh. **23** (1982), no. 6, 160–169; English transl. in Siberian Math. J. **23** (1982).

17. G. S. Litvinchuk, *Boundary problems and singular integral equations with a shift*, "Nauka", Moscow, 1977. (Russian)

18. Yu. I. Karlovich and V. G. Kravchenko, *Algebra of singular integral operators with piecewise continuous coefficients and a piecewise-smooth shift on a composite contour*, Izv. Akad. Nauk SSSR Ser. Mat. **47** (1983), no. 5, 1030–1077; English transl. in Math. USSR-Izv. **22** (1984).

19. Yu. I. Lyubarskiĭ and V. A. Tkachenko, *On the system* $\{e^{anz}\sin nz\}_1^\infty$, Funktsional. Anal. i Prilozhen. **18** (1984), no. 2, 69–70; English transl. in Functional Anal. Appl. **18** (1984).

20. J. B. Garnett, *Bounded analytic functions*, Academic Press, New York, 1981.

21. M. M. Dzhrbashyan, *Integral transforms and representations of functions in a complex domain*, "Nauka", Moscow, 1966. (Russian)

22. G. M. Goluzin, *Geometric theory of functions of a complex variable*, "Nauka", Moscow, 1966; English transl., Amer. Math. Soc., Providence, R.I., 1969.

23. I. I. Privalov, *Boundary properties of analytic functions*, GITTL, Moscow, 1950; German transl., VEB Deutscher Verlag Wiss., Berlin, 1956.

24. I. Ts. Gokhberg and N. Ya. Krupnik, *Introduction to the theory of one-dimensional singular integral operators*, "Shtiintsa", Kishinev, 1973; German transl., Birkhäuser, Basel, 1978.

25. M. A. Krasnosel′skiĭ, T. P Zabreĭko, E. I. Pustyl′nik, and P. E. Sobolevskiĭ, *Integral operators in spaces of summable functions*, "Nauka", Moscow, 1966; English transl., Noordhoff, Groningen, 1975.

26. I. B. Simonenko, *A new general method to study linear operator equations of the singular integral type*. I, Izv. Akad. Nauk SSSR Ser. Mat. **29** (1965), no. 3, 567–586.

27. N. Ya. Krupnik, *Banach algebras with a symbol and singular integral operators*, "Shtiintsa", Kishinev, 1984; English transl., Birkhäuser, Basel, 1987.

28. Yu. I. Lyubarskiĭ, *Properties of systems of linear combinations of powers*, Algebra i Analiz **1** (1989), no. 6, 1–70; English transl., Leningrad Math. J. **1** (1990), no. 6, 1297–1370.

INSTITUTE FOR LOW TEMPERATURE PHYSICS & ENGINEERING, 47 LENIN AVENUE, KHARKOV 310164, UKRAINE

Translated by THE AUTHORS

ADVANCES IN SOVIET MATHEMATICS
Volume 11, 1992

Frames in the Bargmann Space of Entire Functions

YU. I. LYUBARSKIĬ

§1. Introduction

a. This paper is inspired by the paper of I. Daubechies and A. Grossmann [1] who studied the Bargmann space, i.e., the Hilbert space $\mathscr{B}$ of entire functions with the scalar product

$$\langle f, g\rangle = \frac{1}{2\pi}\iint_{\mathbb{C}} f(z)\overline{g(z)}\exp(-|z|^2/2)\,dm_z.$$

Here and in what follows dm is the plane Lebesgue measure. The space $\mathscr{B}$ appears in many questions (a short survey is given in [1]). The possibility of representing functions from $\mathscr{B}$ by means of exponential series with exponents from the set of lattice points $\{z_{mn} = am + ibn;\ m, n \in \mathbb{Z}\}$, with given $a, b > 0$, is considered in [1]. For the case when this set forms a frame in $\mathscr{B}$, i.e.,[1]

$$\|f\|^2 \asymp \sum |f(z_{mn})|^2 \exp(-|z_{mn}|^2/2) \tag{1}$$

a method is given in [1] for the expansion of an arbitrary function $f \in \mathscr{B}$ in an exponential series in terms of the system

$$\mathscr{E}_{ab} = \{\exp((am + ibn)z/2);\ m, n \in \mathbb{Z}\}.$$

Bounds (1) are obtained in [1] for $ab = 2\pi/N$, $N = 2, 3, \dots$, and are announced for $ab < 0.996 \times 2\pi$. For $ab \geq 2\pi$ estimate (1) is false. The cause for that in the case $ab > 2\pi$ is the incompleteness of $\mathscr{E}_{ab}$ in $\mathscr{B}$ with infinite defect. When $ab = 2\pi$, the system $\mathscr{E}_{ab}\setminus\{1\}$ is complete and minimal in $\mathscr{B}$, but it does not form an unconditional basis. So estimate (1) is invalid again.

b. The goal of this paper is to describe a wide class of frames in $\mathscr{B}$ as well as to describe a class of sets $Z = \{z_\nu\} \subset \mathbb{C}$ such that the system

$$\mathscr{E}(Z) = \{\exp(z_\nu z/2);\ z_\nu \in Z\} \tag{2}$$

1991 *Mathematics Subject Classification*. Primary 30D20.

[1] Here and in the sequel $\asymp$ means that the ratio of the two parts is bounded away from zero and infinity by two positive quantities.

is complete and minimal in $\mathscr{B}$ and a summability method can be given to expand functions of $\mathscr{B}$ over this system. A characterization of such sets is given in terms of the behavior of the entire function for which Z is the zero set. The notions concerning growth and zero distribution of entire functions used below are presented for example in [2].

DEFINITION 1. An entire function F belongs to the class $\mathscr{F}$ if there exist $\eta > 0$, $\delta \in (0, \eta/2)$ such that

(1) F has completely regular growth with respect to order 2 and its indicator $h(\theta)$ satisfies the relation $h(\theta) > \eta + 1/4$, $\theta \in (0, 2\pi)$;

(2) the zeros $Z = \{z_\nu\}$ of F are simple and separated:
$$\inf\{|z_\nu - z_\mu|,\ \mu \neq \nu\} = d_0 > 0;$$

(3) if we let
$$P(z) = \begin{cases} |F(z)| \exp(-|z|^2 h(\arg z)), & \operatorname{dist}(z, Z) > d_0/2, \\ |z - z_\nu|^{-1} |F(z)| \exp(-|z|^2 h(\arg z)), & |z - z_\nu| < d_0/2, \end{cases}$$
then the following inequality is valid
$$P(z + \zeta) < \text{Const}\, P(z) \exp(\delta |\zeta|^2), \qquad \zeta, z \in \mathbb{C}. \tag{3}$$

EXAMPLE. Let the lattice $Z = \mathbb{Z} \times \lambda e^{i\alpha}\mathbb{Z}$ be fixed with main periods 1, $\lambda e^{i\alpha}$ ($\lambda > 0$, $\alpha \in (0, \pi)$), and let $\zeta(z)$, $\sigma(z)$ be the Weierstrass functions corresponding to this lattice (see for example [6]). They satisfy the following relations
$$\zeta(z+1) = \zeta(z) + \eta_1, \qquad \eta_1 = 2\zeta(1/2),$$
$$\zeta(z + \lambda e^{i\alpha}) = \zeta(z) + \eta_2, \qquad \eta_2 = 2\zeta(\lambda e^{i\alpha}/2),$$
$$\eta_1 \lambda e^{i\alpha} - \eta_2 = 2i\pi,$$
$$\sigma(z+1) = -\sigma(z) \exp(\eta_1(z + 1/2)),$$
$$\sigma(z + \lambda e^{i\alpha}) = -\sigma(z) \exp(\eta_2(z + \lambda e^{i\alpha}/2)).$$
Consider the following homogeneous function of degree 2
$$V(z) = \frac{\pi |z|^2}{2\lambda \sin\alpha} + \operatorname{Re}(\alpha z^2) = r^2 h(\theta), \qquad z = re^{i\theta},$$
where $\alpha = \eta_1/2 - \pi/(2\lambda \sin\alpha)$. By a direct verification one can see that $|\sigma(z)| \exp(-V(z))$ is a doubly-periodic function with main periods 1, $\lambda e^{i\alpha}$. Hence
$$|\sigma(z) \exp(-\alpha z^2)| \asymp \exp\left(\frac{\pi |z|^2}{2\lambda \sin\alpha}\right), \qquad \operatorname{dist}(z, Z) > \varepsilon,$$
and the relation $P(z) \asymp 1$ is fulfilled for the function P corresponding to $F(z) = \sigma(z) \exp(-\alpha z^2)$; this is stronger than (3). So for $\lambda \sin\alpha < 2\pi$, the function $\sigma(z) \exp(-\alpha z^2)$ belongs to $\mathscr{F}$. It is noteworthy that $\lambda \sin\alpha$ is exactly the area of a cell of the lattice Z. The class $\mathscr{F}$ also contains functions

of much more general form. Roughly speaking, condition (3) means that the function F does not oscillate very rapidly outside the d_0-neighborhood of Z after growth compensation, i.e., after multiplication by $\exp(-|z|^2 h(\arg z))$.

THEOREM 1. *Let an entire function F belong to the class $\mathscr{F}$ and $Z = \{z_\nu\}$ be its zero set. Then Z is a frame in $\mathscr{B}$.*

c. DEFINITION 2. Let $\gamma \in \mathbb{R}$ be given. An entire function S belongs to the class $\mathscr{S}_\gamma$ iff

(1) all zeros $\{z_\nu\}$ of S are simple and separated;
(2) for any $\varepsilon > 0$ the following relation holds:

$$|S(z)| \asymp (1+|z|)^{-\gamma} \exp(|z|^2/4), \qquad \operatorname{dist}(z, \{z_\nu\}) > \varepsilon. \tag{4}$$

In the space $\mathscr{B}$ this class plays the same role as the class of sine-type functions introduced by B. Ya. Levin [3] for the space of all bounded on $\mathbb{R}$ entire functions of exponential type not exceeding π. Class $\mathscr{S}_0$ contains, for example, the function constructed in the previous example by means of the Weierstrass σ-function corresponding to a lattice of cell area 2π. Other examples of functions from $\mathscr{S}_\gamma$ (and more general classes) were constructed by M. L. Sodin and the author in [4], where an estimate of the derivative of any function $S \in \mathscr{S}_\gamma$ at its zeros was also given

$$|S'(z_\nu)| \asymp (1+|z_\nu|)^{-\gamma} \exp(|z_\nu|^2/4).$$

THEOREM 2. *Let $\gamma \in (0, 1]$, $S \in \mathscr{S}_\gamma$ be given and $Z = \{\zeta_\nu\}$ be the zero set of S. Then the system $\mathscr{E}(Z)$ (defined by (2)) is complete and minimal in $\mathscr{B}$.*

Under the conditions of this theorem a unique series over the system $\mathscr{E}(Z)$ may be associated with every function $f \in \mathscr{B}$

$$f \sim \sum c_\nu(f) \exp(z_\nu z/2).$$

For the case $\gamma \in (0, 1)$ we give a summability method for this series. Let us fix a function $\alpha \in C^2(0, \infty)$ such that $\alpha(t) = 1$ when $t \in [0, 1]$, $\alpha(t) \in [0, 1]$ when $t \in [1, 2]$, and $\alpha(t) = 0$ when $t > 2$.

THEOREM 3. *Let $S \in \mathscr{S}_\gamma$, $\gamma \in (0, 1)$, and $Z = \{z_\nu\}$ be its zero set. Then for any $f \in \mathscr{B}$*

$$\lim_{R\to\infty} \left\| f(z) - \sum_\nu \alpha(|z_\nu|/R) c_\nu(f) e^{z_\nu z/2} \right\| = 0.$$

Unfortunately, this theorem cannot be applied directly to a lattice of cell area 2π. The σ-function corresponding to such a lattice belongs to $\mathscr{S}_0$ and after division by z it goes to $\mathscr{S}_1$. Nevertheless it can be easily seen that if $\delta \in (0, 1/2)$, $a, b > 0$ and $ab = 2\pi$, then

$$Z = \{an + ibm;\ n \in \mathbb{Z},\ m \in \mathbb{Z} \setminus \{0\}\} \cup \{a(n + \delta \operatorname{sign}(n));\ n \in \mathbb{Z}\}$$

is the zero set of a function from $\mathscr{S}_{2\delta}$ (cf. [4]).

By applying the Bargmann transform (see for example [1]) one can reformulate Theorems 2 and 3 as theorems concerning systems of exponentials in the space $L^2(\mathbb{R})$.

d. The author is thankful to M. L. Sodin for useful discussions.

e. After this paper was prepared for publication, the author was informed about the article [7] which gives a criterion for a set $Z \subset \mathbb{C}$ to be a frame in terms of the lower density of this set.

§2. Proof of Theorem 1

a. This theorem is an analog of M. Cartwright's classical theorem which estimates an entire function of exponential type less than π via its values at integer points. In what follows we use the idea of the proof of M. Cartwright's theorem given in [2, Chapter IV]. This proof is based on the Boas-Bernstein interpolating formula in the appropriate spaces of entire functions.

b. Let the conditions of Theorem 1 be fulfilled. First we shall prove the inequality

$$\sum_{\nu} |f(z_\nu)|^2 \exp(-|z_\nu|^2/2) < \text{Const}\,\|f\|^2. \tag{5}$$

To this end we choose $C, c > 0$, $d \in (0, d_0/2)$ and functions $a_\nu(z)$ analytic for $|z - z_\nu| < d$ so that

$$c\exp(-|z_\nu|^2/4) < |a_\nu(z)| < C\exp(-|z_\nu|^2/4), \qquad |z - z_\nu| < d. \tag{6}$$

(Such functions can easily be constructed from the corresponding σ-functions.) The disks $\{z;\ |z - z_\nu| < d\}$ are pairwise disjoint. Thus estimate (6) and the mean value theorem yield

$$\begin{aligned}\sum_{\nu} &|f(z_\nu)|^2 \exp(-|z_\nu|^2/2) \\ &< C^2 \sum_{\nu} \left|\frac{f(z_\nu)}{a(z_\nu)}\right|^2 < \frac{C^2}{2\pi d^2} \sum_{\nu} \iint_{|z-z_\nu|<d} \left|\frac{f(z_\nu)}{a(z_\nu)}\right|^2 dm_z \\ &< \frac{C^2}{2\pi c^2 d^2} \sum_{\nu} \iint_{|z-z_\nu|<d} |f(z)|^2 \exp(-|z|^2/2)\, dm_z < \text{Const}\,\|f\|^2,\end{aligned}$$

which is the desired inequality.

c. To prove the opposite inequality

$$\|f\|^2 < \text{Const} \sum |f(z_\nu)|^2 \exp(-|z_\nu|^2/2)$$

we shall assume for simplicity that the indicator of the function F is constant, $h(\theta) = 1/4 + \eta$.

First, it follows from [2, Chapter II] that

$$|F(z)| = \exp[(\eta + 1/4)|z|^2 + o(|z|^2)], \qquad |z - z_\nu| \geq d_0. \tag{7}$$

By applying (7) and the minimum principle to the function $F(z)(z - z_\nu)^{-1}$, which is analytic and nonvanishing in the disk $|z - z_\nu| < d_0$, we obtain

$$|F'(z_\nu)| = \exp[(\eta + 1/4)|z_\nu|^2 + o(|z_\nu|^2)]. \tag{8}$$

d. Consider some function $S \in \mathscr{S}_0$ such that $S(0) = 1$. Let

$$s_0(z) = S(2\sqrt{\varepsilon} z)\exp(-2\varepsilon z^2),$$
$$s_\theta(z) = s_0(z\exp(-i\theta)), \qquad \theta \in [0, 2\pi].$$

The value of $\varepsilon > 0$ will be chosen below. It is clear that

$$s_\theta(z) \leq \text{Const}\exp[-\varepsilon\operatorname{Re}(ze^{-i\theta})^2 + 3\varepsilon\operatorname{Im}(ze^{-i\theta})^2], \qquad s(0) = 1. \tag{9}$$

Also put $F_\theta(z) = F(z)\exp(-\eta(ze^{-i\theta})^2)$, $\theta \in [0, 2\pi]$. The desired interpolation formula is given by

LEMMA 1. *Let $\varepsilon < \eta/2$. Then for any $f \in \mathscr{B}$, $\theta \in [0, 2\pi]$*

$$f(z) = F_\theta(z)\sum_\nu f(z_\nu)\frac{s_\theta(z - z_\nu)}{F'_\theta(z_\nu)(z - z_\nu)}. \tag{10}$$

The series in the right-hand side converges uniformly on every compact set in $\mathbb{C}$.

PROOF. Let $\theta \in [0, 2\pi]$ be fixed. Consider the auxiliary function

$$A(z, \zeta) = \sum_\nu f(z_\nu)\frac{s_\theta(z - z_\nu)}{F'_\theta(z_\nu)(\zeta - z_\nu)}, \qquad \zeta, z \in \mathbb{C}. \tag{11}$$

The inequality (5) yields the representation

$$f(z_\nu) = c_\nu\exp(|z_\nu|^2/4), \qquad \{c_\nu\} \in l^2.$$

To prove the convergence of the series (11), let

$$z = (x' + iy')\exp(i\theta), \qquad x'_\nu, x', y'_\nu, y' \in \mathbb{R}.$$
$$z_\nu = (x'_\nu + iy'_\nu)\exp(i\theta),$$

Then

$$|F'_\theta(z_\nu)| = |F'(z_\nu)|\exp[-\operatorname{Re}(\eta(z_\nu e^{-i\theta})^2)]$$
$$= \exp[|z_\nu|^2/4 + 2\eta y'^2_\nu + o(|z_\nu|^2)]. \tag{12}$$

Combining (8), (9) and (12) we arrive at

$$\left|f(z_\nu)\frac{s_\theta(z - z_\nu)}{F'_\theta(z_\nu)(\zeta - z_\nu)}\right|$$
$$\leq \frac{|c_\nu|}{|\zeta - z_\nu|}\exp(-2\eta y'^2_\nu - \varepsilon(x_\nu - x'_\nu)^2 + 3\varepsilon(y_\nu - y'_\nu)^2 + o(|z_\nu|^2)).$$

This yields uniformly compact convergence of the series (11). By virtue of the last inequality and the bounded convergence theorem, we have

$$A(z, \zeta) \to 0, \qquad \zeta, z \to \infty, \ \operatorname{dist}(z, Z) > d_0 > 0. \tag{13}$$

The following entire function of ζ

$$\psi(z, \zeta) = f(\zeta) s_\theta(\zeta - z) - F_\theta(\zeta) A(z, \zeta)$$

vanishes when $\zeta \in Z$. So the function

$$\Psi(z, \zeta) = \psi(z, \zeta)/F_\theta(\zeta) = f(\zeta) s_\theta(\zeta - z)/F_\theta(\zeta) - A(z, \zeta)$$

is also entire. The inclusion $f \in \mathscr{B}$ implies $|f(\zeta)| < \operatorname{Const}\exp(|\zeta|^2/4)$. Together with (7), (9) and (13), this implies that $\Psi(z, \zeta)$ vanishes as ζ tends to infinity outside the d_0-neighborhood of Z. By the maximum principle, $\Psi(z, \zeta) \to \infty$ as $\zeta \to \infty$ everywhere in $\mathbb{C}$, so $\Psi \equiv 0$ and $\psi \equiv 0$. Equality (10) now can be obtained by the substitution $\zeta = z$.

e. To estimate f at a point $z \in \mathbb{C}$ we use representation (10) in which $\theta = \arg z$. For $z = r\exp(i\theta)$ this gives

$$\begin{aligned} &|f(z)|e^{-|z|^2/4} \\ &\quad \le |F(z)|e^{-|z|^2(\eta + 1/4)} \left(\sum_{|z - z_\nu| \le d_0} + \sum_{|z - z_\nu| > d_0} \right) \left| f(z_\nu) \frac{s_\theta(z - z_\nu)}{F'_\theta(z_\nu)(z - z_\nu)} \right| \\ &\quad = \Sigma_1(z) + \Sigma_2(z). \end{aligned}$$

We must establish the estimates

$$\iint_{\mathbb{C}} |\Sigma_j(z)|^2 \, dm_z \le \operatorname{Const} \|\{f(z_\nu)e^{-|z_\nu|^2/4}\}\|^2_{l^2}, \qquad j = 1, 2. \tag{14}$$

The first sum Σ_1 contains at most one summand. Since for $|z - z_\nu| < d_0$ we have $|\exp(-\eta(z_\nu e^{-i\theta})^2)| \asymp \exp(-\eta|z_\nu|^2)$ it follows that

$$\begin{aligned} |\Sigma_1(z)| &= \frac{|F(z)|\exp[-|z|^2(\eta + 1/4)]}{|z - z_\nu|} \frac{|c_\nu|}{|F'(z_\nu)|\exp[-(\eta(z_\nu e^{-i\theta})^2 + |z_\nu|^2/4)]} \\ &\asymp \frac{P(z)}{P(z_\nu)}|c_\nu|, \qquad |z - z_\nu| < d_0, \end{aligned}$$

and by virtue of (3) $|\Sigma_1(z)| \asymp |c_\nu|$, $|z - z_\nu| < d_0$. Thus,

$$\iint_{\mathbb{C}} |\Sigma_1(z)|^2 \, dm_z \asymp \sum_\nu |c_\nu|^2 \iint_{|z - z_\nu| < d_0} dm_z < \operatorname{Const} \sum_\nu |c_\nu|^2.$$

We have obtained (14) for $j = 1$.

To estimate Σ_2, let $z_\nu = (x'_\nu + iy'_\nu)\exp(i\theta)$ for every ν. Then for $|z - z_\nu| > d_0$ we have

$$\begin{aligned} &\left| f(z_\nu) \frac{s_\theta(z - z_\nu)}{F'(z_\nu)\exp[-\eta(z_\nu e^{-i\theta})^2](z - z_\nu)} \right| \\ &\qquad < \frac{\operatorname{Const}|c_\nu|}{F'(z_\nu)\exp[-(\eta + 1/4)|z_\nu|^2]} |s_\theta(z - z_\nu)\exp[-\eta(|z_\nu|^2 - (z_\nu e^{-i\theta})^2)]|. \end{aligned}$$

Taking into consideration the relations $\arg z = \theta$ and $3\varepsilon - 2\eta < -\varepsilon$, we obtain

$$|s_0(z - z_\nu)\exp[-\eta(|z_\nu|^2 - (z_\nu e^{-i\theta})^2)]| \\ \leq C\exp[-\varepsilon(x - x'_\nu)^2 + 3\varepsilon {y'_\nu}^2 - 2\eta {y'_\nu}^2] \leq C\exp(-\varepsilon|z - z_\nu|^2).$$

And finally

$$\Sigma_2(z) < CP(z)\sum \frac{c_\nu}{P(z_\nu)} e^{-\varepsilon|z-z_\nu|^2}.$$

Let

$$a(z) = \begin{cases} |c_\nu|/P(z), & |z - z_\nu| \leq d_0, \\ 0, & \operatorname{dist}(z, Z) > d_0. \end{cases}$$

The function $a(z)P(z)$ is square integrable. The estimate for Σ_2 assumes the form

$$|\Sigma_2(z)| < \operatorname{Const} P(z)\iint_{\mathbb{C}} a(\zeta)e^{-\varepsilon|\zeta-z|^2}\,dm_\zeta.$$

Let the quantity ε in our construction be chosen so that $\delta < \varepsilon < \eta/2$, where δ is the quantity from Definition 1. Then taking (3) into account

$$|\Sigma_2(z)| < \operatorname{Const}\iint_{\mathbb{C}} a(\zeta)P(\zeta)e^{-(\delta-\varepsilon)|\zeta-z|^2}\,dm_\zeta.$$

Relation (14) for $j = 2$ now follows immediately by the triangle inequality, and Theorem 1 is proved.

§3. Proof of Theorems 2 and 3

a. Our construction is based on the reproducing property of the function $e_w(z) = \exp(z\overline{w}/2)$ in the space $\mathscr{B}$, i.e., $f(w) = \langle f, e_w\rangle$, $f \in \mathscr{B}$. The proof of Theorem 2 is quite standard. Let $\gamma \in (0, 1]$, $S \in \mathscr{S}_\gamma$. Let $Z = \{z_\nu\}$ be the zero set of S and $\mathscr{E}(Z) = \{\exp(zz_\nu/2)\}$ be the corresponding exponential system. Denote $\overline{S}(z) = \overline{S(\overline{z})}$. The minimality of $\mathscr{E}(Z)$ is a consequence of the existence of the biorthogonal system of functions:

$$f_\nu(z) = \frac{\overline{S}(z)}{\overline{S}'(\overline{z}_\nu)(z - \overline{z}_\nu)} \in \mathscr{B}, \qquad \langle f_\nu, e_{z_\mu}\rangle = f_\nu(\overline{z}_\mu) = \delta_{\mu\nu}. \tag{15}$$

If any function $f \in \mathscr{B}$ is orthogonal to $\mathscr{E}(Z)$, then $f(\overline{z}_\nu) = 0$ for all ν's and a simple estimate shows that the entire function $f(z)/\overline{S}(z)$ vanishes when z tends to infinity. So $f/\overline{S}$ (and consequently f) is identically zero. This implies the completeness of the system $\mathscr{E}(Z)$ in $\mathscr{B}$.

b. The proof of Theorem 3 is a modification to the space $\mathscr{B}$ of the reasoning of the proofs of Theorems 4.1 and 4.2 in [5]. We must prove the relation

$$\mathfrak{S}_R f \to f, \qquad R \to \infty,\ f \in \mathscr{B}, \tag{16}$$

where the operator $\mathfrak{S}_R$ is defined by

$$\mathfrak{S}_R\colon f \to \mathfrak{S}_R f(z) = \sum \alpha(|z_\nu|/R)c_\nu(f)e^{z_\nu z/2}.$$

Relation (16) is evidently true when f belongs to the linear span of $\mathscr{E}(Z)$ which is dense in $\mathscr{B}$. So we need only prove that $\sup\{\|\mathfrak{S}_R\|;\ R > 0\} < \infty$, or, which is stronger,

$$\mathfrak{S}_R^* f \to f, \qquad R \to \infty,\ f \in \mathscr{B}. \tag{17}$$

Here the adjoint operators $\mathfrak{S}_R^*$ may be expressed directly by means of the biorthogonal system (15):

$$\mathfrak{S}_R^* f(z) = \sum \alpha\left(\frac{|z_\nu|}{R}\right) f(\overline{z}_\nu) \frac{\overline{S}(z)}{\overline{S}'(\overline{z}_\nu)(z - \overline{z}_\nu)}.$$

Thus (17) gives a summability method for the Lagrange interpolation series in the space $\mathscr{B}$.

c. We need an analog of the residue theorem.

LEMMA 2. *Let $\omega \subset \mathbb{C}$ be a bounded domain with smooth boundary, and let a C^1-function $\rho(x, y)$ in its closure and a meromorphic function h in ω with only simple poles $\{\zeta_k\}$ and continuous on $\partial\omega$ be given. Then*

$$\frac{1}{2i\pi} \int_{\partial\omega} \rho(\zeta) h(\zeta)\, d\zeta = \frac{1}{\pi} \iint_\omega h \frac{\partial \rho}{\partial \overline{\zeta}}\, dm_\zeta + \sum \rho(\zeta_k) \operatorname{Res}_{\zeta_k} h. \tag{18}$$

This statement is a direct consequence of the Green formula.

Let $z \in \mathbb{C}$ be fixed and $\omega = \{\zeta : |\zeta| < 3R\}$, $\rho(\zeta) = \alpha_R(\zeta) = \alpha(\zeta/R)$, $h(\zeta) = f(\zeta)/[\overline{S}(\zeta)(\zeta - z)]$. Then (18) gives

$$\begin{aligned} 0 = \frac{1}{\pi} \iint_{\mathbb{C}} &\frac{f(\zeta)}{\overline{S}(\zeta)(z - \zeta)} \frac{\partial}{\partial \overline{\zeta}} \alpha_R(\zeta)\, dm_\zeta \\ &+ \sum_\nu \alpha_R(\overline{z}_\nu) \frac{f(\overline{z}_\nu)}{\overline{S}'(\overline{z}_\nu)(z - \overline{z}_\nu)} - \frac{f(z)}{\overline{S}(z)} \alpha_R(z). \end{aligned}$$

Multiplication by $\overline{S}(z)$ yields

$$\begin{aligned} \mathfrak{S}_R^* f(z) &= f(z) + f(z)(\alpha_R(z) - 1) - \frac{\overline{S}(z)}{\pi} \iint_{\mathbb{C}} \frac{f(\zeta)}{\overline{S}(\zeta)(z - \zeta)} \frac{\partial}{\partial \overline{\zeta}} \alpha_R(\zeta)\, dm_\zeta \\ &= f(z) + A_{1R}(z) + A_{2R}(z). \end{aligned} \tag{19}$$

d. Let us introduce some additional notation. For any Hilbert space $\mathfrak{H}$ let $B(\mathfrak{H})$ be its unit ball. For a given domain $G \in \mathbb{C}$ and a measurable function $\rho: G \to [0, +\infty]$ we denote by $L^2(G, \rho)$ the space of functions $\Phi: G \to \mathbb{C}$ such that

$$\|\Phi\|^2_{L^2(G,\rho)} = \iint_G |\Phi(z)|^2 \rho(z)\, dm_z < \infty.$$

$L^2(G)$ will denote the space $L^2(G, 1)$.

By virtue of (19), in order to obtain (17) we must prove

$$\|A_{jR}\|_{L^2(\mathbb{C}, \exp(-|z|^2/2))} \to 0, \qquad R \to \infty,\ j = 1, 2. \tag{20}$$

This is clear for $j=1$. If any function $\Phi \in L^2(\mathbb{C}, \exp(-|z|^2/2))$ is given, let

$$J_R(\Phi) = \langle A_{2R}, \Phi\rangle_{L^2(\mathbb{C}, \exp(-|z|^2/2))}. \tag{21}$$

Relation (20) for $j=2$ is now equivalent to

$$\sup\{|J_R(\Phi)|;\ \Phi \in B(L^2(\mathbb{C}, \exp(-|z|^2/2)))\} \to 0, \qquad R \to \infty.$$

Direct calculation gives

$$\frac{\partial}{\partial \bar{\zeta}} \alpha_R(\zeta) = \frac{\zeta}{2R|\zeta|} \alpha'\left(\frac{|\zeta|}{R}\right).$$

Let $\varphi(z) = \Phi(z)S(z)\exp(-|z|^2/2)$. Relation (4) implies

$$\iint_{\mathbb{C}} |\varphi(z)|^2 |z|^{2\gamma} dm_z < C.$$

Here the constant C does not depend on $\Phi \in B(L^2(\mathbb{C}, \exp(-|z|^2/2)))$. By substituting these relations into (21), we get an explicit expression for $J_R(\Phi)$:

$$J_R(\Phi) = \frac{1}{2\pi R} \iint_{\mathbb{C}} \frac{f(\zeta)}{\overline{S}(\zeta)} \frac{\zeta}{|\zeta|} \alpha'\left(\frac{|\zeta|}{R}\right) \iint_{\mathbb{C}} \frac{\varphi(z)}{\zeta - z} dm_z dm_\zeta.$$

Let $\tilde{\varphi}(\tilde{z}) = R^{1+\gamma}\varphi(R\tilde{z})$. Clearly $\tilde{\varphi} \in CB(L^2(\mathbb{C}, |\tilde{z}|^{2\gamma}))$ and the substitution $z \to \tilde{z}R$, $\zeta \to \tilde{\zeta}R$ gives

$$J_R(\Phi) = \frac{1}{2\pi} R^{1-\gamma} \iint_{\mathbb{C}} \frac{f(R\tilde{\zeta})}{\overline{S}(R\tilde{\zeta})} \frac{\tilde{\zeta}}{|\tilde{\zeta}|} \alpha'(|\tilde{\zeta}|) \iint_{\mathbb{C}} \frac{\tilde{\varphi}(\tilde{\zeta})}{\tilde{\zeta} - \tilde{z}} dm_{\tilde{z}} dm_{\tilde{\zeta}}.$$

In what follows the sign $\tilde{}$ will be omitted.

LEMMA 3. *The operator*

$$T\colon \varphi \to \iint_{\mathbb{C}} \frac{\varphi(\zeta)}{\zeta - z} dm_z \tag{22}$$

considered as an operator from $L^2(\mathbb{C}, |z|^{2\gamma})$ *to* $L^2(\{\zeta : 1 < |\zeta| < 2\})$ *is bounded.*

PROOF. For $\varphi \in L^2(\mathbb{C}, |z|^{2\gamma})$, let

$$\varphi_1(z) = \begin{cases} \varphi(z), & 1/2 < |z| < 3, \\ 0, & \text{otherwise}, \end{cases} \qquad \varphi_2(z) = \varphi(z) - \varphi_1(z).$$

The boundedness of the norm of $T\varphi_1$ is a consequence of a well-known fact on weak-singularity integral estimates. The estimate of $T\varphi_2$ is obtained directly from the Schwarz inequality.

So the expression for $J_R(\Phi)$ may be reduced to the form

$$J_R(\Phi) = \frac{1}{2\pi} R^{1-\gamma} \iint_{\mathbb{C}} \frac{f(R\zeta)}{\overline{S}(R\zeta)} \frac{\zeta}{|\zeta|} \alpha'(|\zeta|)(T\varphi)(\zeta)\, dm_\zeta. \tag{23}$$

It follows from the definition of α that integration in (23) is performed over the annulus $K(1,2) = \{\zeta : 1 < |\zeta| < 2\}$ only.

e. Let $\{\zeta_1, \zeta_2, \dots, \zeta_{N(R)}\} = R^{-1}\overline{Z} \cap K(1, 2)$, and let $\varepsilon > 0$ be small enough for the disks $\{\zeta : |\zeta - \zeta_j| < 10\varepsilon/R\}$ to be pairwise disjoint. Also let $D_j = \{\zeta : |\zeta - \zeta_j| < 5\varepsilon/R\}$ and $E_R = \bigcup D_j$. Put $\omega(\zeta) = (\zeta/|\zeta|)\alpha'(|\zeta|)$; this function is smooth in $K(1, 2)$. We have

$$\begin{aligned} J_R(\Phi) &= \frac{1}{2\pi} R^{1-\gamma}\left(\iint_{K(1,2)\setminus E_R} + \iint_{E_R}\right) \frac{f(R\zeta)}{\overline{S}(R\zeta)}\, \omega(\zeta)(T\varphi)(\zeta)\, dm_\zeta \\ &= I_{1R}(\varphi) + I_{2R}(\varphi). \end{aligned}$$

We shall prove that

$$\sup\{|I_{mR}(\varphi)| : \varphi \in B(L^2(\mathbb{C}, |z|^{2\gamma}))\} \to 0, \qquad R \to \infty,\ m = 1, 2. \tag{24}$$

It follows from (4) that

$$\begin{aligned} |I_{1R}(\varphi)| &< \text{Const} \iint_{K(1,2)} R|f(R\zeta)| e^{-|R\zeta|^2/4} |T\varphi(\zeta)|\, dm_\zeta \\ &< \text{Const}\left(\iint_{K(1,2)} |f(R\zeta)|^2 e^{-|R\zeta|^2/2} R^2\, dm_\zeta\right)^{1/2} \\ &\quad \times \left(\iint_{K(1,2)} |T\varphi(\zeta)|^2\, dm_\zeta\right)^{1/2} \\ &= \text{Const}\left(\iint_{RK(1,2)} |f(\zeta)| e^{-|\zeta|^2/2}\, dm_\zeta\right)^{1/2} \\ &\quad \times \left(\iint_{K(1,2)} |T\varphi(\zeta)|^2\, dm_\zeta\right)^{1/2}. \end{aligned}$$

By virtue of Lemma 3, the last factor in the rightmost part is bounded uniformly with respect to $\varphi \in B(L^2(\mathbb{C}, |z|^{2\gamma}))$, while the first factor tends to zero as $R \to \infty$. The validity of (24) for $m = 1$ is proved.

f. The representation of $I_{2R}(\varphi)$ may be written in the form

$$\begin{aligned} I_{2R}(\varphi) &= \frac{1}{2\pi} R^{1-\gamma} \sum_j \iint_{D_j} \left[\frac{f(R\zeta)}{\overline{S}(R\zeta)} - \frac{f(R\zeta_j)}{\overline{S}'(R\zeta_j)R(\zeta - \zeta_j)}\right] \omega(\zeta) T\varphi(\zeta)\, dm_z \\ &\quad + \sum_j \frac{1}{2\pi R^\gamma} \frac{f(R\zeta_j)}{\overline{S}'(R\zeta_j)} \iint_{D_j} \frac{\omega(\zeta)}{\zeta - \zeta_j} T\varphi(\zeta)\, dm_\zeta \\ &= \sum_j A_{1j}(\varphi, R) + \sum_j A_{2j}(\varphi, R). \end{aligned}$$

We shall prove that for $l = 1, 2$

$$\sup\left\{\left|\sum_j A_{lj}(\varphi, R)\right|;\ \varphi \in B(L^2(\mathbb{C}, |z|^{2\gamma}))\right\} \to 0, \qquad R \to \infty. \tag{25}$$

For the case $l = 1$ we use the representation

$$\frac{f(R\zeta)}{\overline{S}(R\zeta)} - \frac{f(R\zeta_j)}{\overline{S}'(R\zeta_j)R(\zeta - \zeta_j)} = \frac{1}{2i\pi} \int_{|w - \zeta_j| = l} \frac{f(Rw)}{\overline{S}(Rw)} \frac{dw}{w - \zeta}, \qquad \zeta \in D_j,$$

which is valid for $5\varepsilon R^{-1} < t < 10\varepsilon R^{-1}$.

Let $K_j = \{\zeta : 8\varepsilon R^{-1} < |\zeta - \zeta_j| < 10\varepsilon R^{-1}\}$. Integration of both sides of the last equality over t from $8\varepsilon R^{-1}$ to $10\varepsilon R^{-1}$ gives

$$\frac{f(R\zeta)}{\overline{S}(R\zeta)} - \frac{f(R\zeta_j)}{\overline{S}'(R\zeta_j)R(\zeta-\zeta_j)} = \frac{R}{4\varepsilon i\pi}\iint_{K_j}\frac{f(Rw)}{\overline{S}(Rw)}\frac{w-\zeta_j}{|w-\zeta_l|}\frac{dm_w}{w-\zeta}.$$

By substituting this into the expression for $A_{1j}(\varphi, R)$ and changing the order of integration, we obtain

$$A_{1j}(\varphi, R) = \frac{R^{2-\gamma}}{4\varepsilon i\pi}\iint_{K_j}\frac{f(Rw)}{\overline{S}(Rw)}\frac{w-\zeta_j}{|w-\zeta_j|}\iint_{D_j}\frac{\omega(\zeta)}{w-\zeta}T\varphi(\zeta)\,dm_\zeta dm_w.$$

To estimate the inner integral, let us notice that for $\zeta \in D_j$, $w \in K_j$ we have $|w-\zeta| > 3\varepsilon/R$ and

$$\begin{aligned}\left|\iint_{D_j}\frac{\omega(\zeta)}{w-\zeta}T\varphi(\zeta)\,dm_\zeta\right| &< \mathrm{Const}\,\frac{R}{3\varepsilon}\left(\iint_{D_j}dm_\zeta\right)^{1/2}\left(\iint_{D_j}|T\varphi(\zeta)|^2\,dm_\zeta\right)^{1/2}\\ &< \mathrm{Const}\left(\iint_{D_j}|T\varphi(\zeta)|^2\,dm_\zeta\right)^{1/2}.\end{aligned}$$

After returning to $A_{1j}(\varphi, R)$, we obtain by using (4)

$$\begin{aligned}&|A_{1j}(\varphi, R)|\\ &\quad< \mathrm{Const}\left(\iint_{D_j}|T\varphi(\zeta)|^2\,dm_\zeta\right)^{1/2}R^2\iint_{K_j}|f(Rw)|e^{-|Rw|^2/4}\,dm_w\\ &\quad< \mathrm{Const}\left(\iint_{D_j}|T\varphi(\zeta)|^2\,dm_\zeta\right)^{1/2}\left(\iint_{RK_j}|f(w)|^2e^{-|w|^2/2}\,dm_w\right)^{1/2},\end{aligned}$$

and finally

$$\begin{aligned}\sum_j|A_{1j}(\varphi, R)| &< \mathrm{Const}\left(\sum_j\iint_{RK_j}|f(w)|^2e^{-|w|^2/2}\,dm_w\right)^{1/2}\\ &\qquad\times\left(\sum_j\iint_{D_j}|T\varphi(\zeta)|^2\,dm_\zeta\right)^{1/2}\\ &< \mathrm{Const}\left(\iint_{R<|w|<2R}|f(w)|^2e^{-|w|^2/2}\,dm_w\right)^{1/2}\\ &\qquad\times\left(\iint_{K(1,2)}|T\varphi(\zeta)|^2\,dm_\zeta\right)^{1/2} \rightrightarrows 0\end{aligned}$$

as R tends to infinity. Estimate (25) for $l = 1$ is established.

g. Consider (25) for the case $l = 2$. We have

$$\sum_j |A_{1j}(\varphi, R)| < \left(\sum_j \left|\frac{f(R\zeta_j)}{R^\gamma \overline{S}'(R\zeta_j)}\right|^2\right)^{1/2} \left(\sum_j \left|\iint_{D_j} \frac{\omega(\zeta)}{\zeta - \zeta_j} T\varphi(\zeta)\, dm_\zeta\right|^2\right)^{1/2}. \qquad (26)$$

The first factor in the right-hand side can easily be estimated

$$\begin{aligned}\sum_j \left|\frac{f(R\zeta_j)}{R^\gamma \overline{S}'(R\zeta_j)}\right|^2 &< \text{Const} \sum_j |f(R\zeta_j)|^2 e^{-|R\zeta_j|^2/2} \\ &= \text{Const} \sum_{R<|z_\nu|<2R} |f(z_\nu)|^2 e^{-|z_\nu|^2/2} \\ &< \text{Const} \iint_{R<|z|<2R} |f(z)|^2 e^{-|z|^2/2} dm_z \to 0, \qquad R \to \infty.\end{aligned}$$

It is enough to prove now that the second factor in the right-hand side of (26) is bounded uniformly with respect to $\varphi \in B(L^2(\mathbb{C}, |z|^{2\gamma}))$. Or in other words

$$\sum_j \left|\iint_{D_j} \frac{\omega(\zeta) - \omega(\zeta_j)}{\zeta - \zeta_j} T\varphi(\zeta)\, dm_\zeta\right|^2 < C, \qquad (27)$$

$$\sum_j \left|\omega(\zeta_j) \iint_{D_j} \frac{T\varphi(\zeta)}{\zeta - \zeta_j} dm_\zeta\right|^2 < C, \qquad (28)$$

C being independent of the choice of the function $\varphi \in B(L^2(\mathbb{C}, |z|^2 2\gamma))$. Since ω is smooth, there exists a $c > 0$ such that $|\omega(\zeta_j)| < c$ and

$$|[\omega(\zeta) - \omega(\zeta_j)]/(\zeta - \zeta_j)| < c.$$

Therefore

$$\sum_j \left|\iint_{D_j} \frac{\omega(\zeta) - \omega(\zeta_j)}{\zeta - \zeta_j} T\varphi(\zeta)\, dm_\zeta\right|^2 < \text{Const} \frac{1}{R^2} \sum_j \iint_{D_j} |T\varphi(\zeta)|^2\, dm_\zeta.$$

This and Lemma 3 imply (28).

h. To prove estimate (28), consider:

the maximal function

$$M\varphi(z) = \sup_{r>0} \frac{1}{\pi r^2} \iint_{|\zeta - z|<r} |\varphi(\zeta)|\, dm_\zeta;$$

the truncated singular operator

$$H_\delta \varphi(z) = \iint_{|z-\zeta|>\delta} \frac{\varphi(\zeta)}{(z-\zeta)^2} dm_\zeta;$$

and the maximal singular operator $H^*\varphi(z) = \sup_{\delta>0}\{|H_\delta \varphi(z)|\}$.

LEMMA 4. *Operators* M, H_δ, *and* H^* *considered as operators from the space* $L^2(\mathbb{C}, |z|^{2\gamma})$ *to the space* $L^2(K(1, 2))$ *are bounded.*

The proof of this lemma is trivial.

LEMMA 5. *A constant* C *exists such that for every* $\delta > 0$, *each function* $\varphi \in B(L^2(\mathbb{C}, |z|^2 2\gamma))$ *and points* $\zeta, \tilde{\zeta}, \hat{\zeta} \in K(1, 2)$ *satisfying* $|\zeta - \tilde{\zeta}| < \delta$ *and* $|\zeta - \hat{\zeta}| < \delta$ *the following inequality is fulfilled*

$$|H_{5\delta}\varphi(\zeta) - H_{5\delta}\varphi(\hat{\zeta})| < CM\varphi(\tilde{\zeta}). \tag{29}$$

The proof of this well-known lemma is given, e.g., in [5].

i. By substituting the explicit expression (22) for the operator T into (28) and changing the order of integration, we obtain

$$\iint_{D_j} \frac{T\varphi(\zeta)}{\zeta - \zeta_j}\, dm_j = \iint_{\mathbb{C}} \varphi(\lambda) \iint_{D_j} \frac{dm_\zeta}{(\zeta - \zeta_j)(\lambda - \zeta)}\, dm_\lambda.$$

The inner integral in the right-hand side can be calculated directly

$$\iint_{D_j} \frac{dm_\zeta}{(\zeta - \zeta_j)(\lambda - \zeta)} = \begin{cases} 25\varepsilon^2 R^{-2}(\lambda - \zeta_j)^{-2}, & \lambda \notin D_j, \\ |\lambda - \zeta_j|^2(\lambda - \zeta_j)^{-2}, & \lambda \in D_j. \end{cases}$$

Returning to (28) we obtain

$$\iint_{D_j} \frac{T\varphi(\zeta)}{\zeta - \zeta_j}\, dm_\zeta = \iint_{D_j} \varphi(\lambda) \frac{|\lambda - \zeta_j|^2}{(\lambda - \zeta_j)^2}\, dm_\lambda + \frac{25\varepsilon^2}{R^2} \iint_{\mathbb{C}\setminus D_j} \frac{\varphi(\lambda)}{(\lambda - \zeta_j)^2}\, dm_\lambda.$$

The estimate

$$\sum_j \left| \iint_{D_j} \varphi(\lambda) \frac{|\lambda - \zeta_j|^2}{(\lambda - \zeta_j)^2}\, dm_\lambda \right|^2 < C$$

may be obtained by using the Schwarz inequality directly. To prove (29) and thus to end the proof of the theorem, we need only establish the following inequality

$$\sum_j |R^{-2} H_{5\delta}\varphi(\zeta_j)|^2 < C, \qquad \delta = \varepsilon/R. \tag{30}$$

Let the points $\tilde{\zeta}_j, \hat{\zeta}_j \in \{\zeta : |\zeta - \zeta_j| < \delta\}$ be chosen so that

$$\begin{aligned} |H_{5\delta}\varphi(\hat{\zeta}_j)| &< \max\left\{2 \inf_{|\zeta - \zeta_j| < \delta} \{|H_{5\delta}\varphi(\zeta_j)|\}, N(R)^{-1}\right\}, \\ |M\varphi(\tilde{\zeta}_j)| &< \max\left\{2 \inf_{|\zeta - \zeta_j| < \delta} \{|M\varphi(\zeta_j)|\}, N(R)^{-1}\right\}. \end{aligned} \tag{31}$$

Here, as before, $N(R)$ is the number of points $\{\zeta_j\}$ in the annulus $K(1, 2)$. Because of (31), the proof of (30) may be reduced to checking the inequalities

$$\sum_j |R^{-2} H_{5\delta}\varphi(\hat{\zeta}_j)|^2 < C, \qquad \sum_j |R^{-2} M\varphi(\tilde{\zeta}_j)|^2 < C.$$

We shall restrict ourselves to the first inequality. Without loss of generality one can assume that the maximum in the right-hand sides of (31) is achieved on the first quantity. Therefore,

$$\begin{aligned}\sum_j |R^{-2} H_{5\delta}\varphi(\hat{\zeta}_j)|^2 &< \sum_j \left(\frac{2\varepsilon^2}{\pi} \iint_{|\zeta-\zeta_j|<\delta} |H_{5\delta}\varphi(\zeta)|\, dm_\zeta\right)^2 \\ &< \text{Const}\, \frac{1}{R^2} \sum_j \iint_{|\zeta-\zeta_j|<\delta} |H_{5\delta}\varphi(\zeta)|^2\, dm_\zeta \\ &< \text{Const}\, \frac{1}{R^2} \iint_{K(1,2)} |H_{5\delta}\varphi(\zeta)|^2\, dm_\zeta.\end{aligned}$$

To finish the proof it remains to use Lemma 4.

References

1. I. Daubechies and A. Grossmann, *Frames in the Bargmann space of entire functions*, Comm. Pure Appl. Math. **41** (1988), no. 2, 151–164.

2. B. Ya. Levin, *Distribution of zeros of entire functions*, GITTL, Moscow, 1956; English transl., Amer. Math. Soc., Providence, R.I., 1964.

3. ——, *On bases of exponentials in* $L^2(0, \pi)$, Zap. Mekh. Mat. Fak. Kharkov. Gos. Univ. i Kharkov. Mat. Obshch. (4) **27** (1961), no. 4, 29–37. (Russian)

4. Yu. I. Lyubarskiĭ and M. L. Sodin, *Analogues of the sine-type functions for convex domains*, Preprint no. 17, Institute for Low Temperature Physics and Engineering, Academy of Sciences of the Ukrain. SSR, Kharkov, 1986. (Russian)

5. Yu. I. Lyubarskiĭ, *Series of exponentials in Smirnov spaces and interpolation by entire functions of exponential type*, Izv. Akad. Nauk SSSR Ser. Mat. **52** (1988), no. 3, 559–590; English transl. in Math. USSR-Izv. **32** (1988) .

6. A. Hurwitz, *Algemeine Funktionentheorie und elliptische Funktionen*, Springer-Verlag, Berlin, 1964.

7. K. Seip and R. Wallstein, *Sampling and interpolating in the Bargmann-Fock space*, Report Institute Mittag-Leffler (1990/91), no. 4.

Institute for Low Temperature Physics & Engineering, 47 Lenin pr., Kharkov 310164 Ukraine

Translated by THE AUTHOR

ADVANCES IN SOVIET MATHEMATICS
Volume 11, 1992

Subharmonic Functions of Finite Degree Bounded on Subsets of the "Real Hyperplane"

B. YA. LEVIN, V. N. LOGVINENKO, AND M. L. SODIN

Introduction

In 1953 A. C. Schaeffer [1] established the following important fact.

Let E *be a closed subset of the real axis such that for some values* $L < \infty$ *and* $\delta > 0$ *the Lebesgue measure of the intersection of* E *and any interval of length* $2L$ *is not less than* δ. *Then there exists a unique function* $v(z)$ *which is harmonic in the domain* $\mathbb{C} \setminus E$ *and has the following properties*:

a) *a.e. on* E *the function* $v(z)$ *has zero limit values*;

b) *for every* $z \in \mathbb{C}$, $0 \leqslant v(z) - |\operatorname{Im} z| \leqslant \pi L^2/(2\delta)$.

A. C. Schaeffer also proved that every entire function $f(z)$ of finite degree[1] σ and modulus on E not larger than 1 satisfies the inequality

$$\log|f(z)| \leqslant \sigma v(z) \tag{1}$$

in the whole complex plane. Hence $\sup\{\log|f(x)| : x \in \mathbb{R}\} \leqslant C(L, \delta)\sigma$, where $C(L, \delta) = \pi L^2/(2\delta)$.

Applying the Phragmen-Lindelöf principle, one can show (see [2, 3]) that inequality (1) remains valid for all subharmonic functions with nonpositive restriction on E, having finite degree not exceeding σ. We denote by $K_\sigma(E)$ the class of such functions. For every function $u \in K_\sigma(E)$, the following analog of inequality (1) $u(z) \leqslant \sigma v(z)$ remains valid everywhere on the complex plane.

Since $\sigma v(z) \in K_\sigma(E)$, the function $\sigma v(z)$ is the upper envelope of the class $K_\sigma(E)$. In other words, $\sigma v(z)$ is the subharmonic majorant of this class. It follows from the above-mentioned result of A. C. Schaeffer that for every function $u(z)$ subharmonic in the complex plane and of finite degree σ the inequality

$$\sup\{u(x) : x \in \mathbb{R}\} \leqslant C(L, \sigma)\sigma + \sup\{u(\xi) : \xi \in E\} \tag{2}$$

1991 *Mathematics Subject Classification.* Primary 30D20, 31A05.

[1]This means that $\sigma = \limsup_{|z|\to\infty} |z|^{-1} \log|f(z)| < \infty$.

is valid. Recently, A. E. Fryntov [4] and A. Baernstein [5] proved independently that the exact value of $C(L, \delta)$ is $\log\cot(\pi\delta/(4L))$.

Inequality (2) is the starting point for our investigations. If $u(z)$ is a plurisubharmonic function defined in $\mathbb{C}^n$ and of finite degree at most σ, and E is a relatively dense subset of $\mathbb{R}^n$, then an estimate similar to (2) was reported [3] by one of the authors. Recall that a set E is relatively dense if for some $L < \infty$ and $\delta > 0$ the Lebesgue measure of $E \cap K(x, L)$ is not less than δ for any $x \in \mathbb{R}^n$. Here

$$K(x, L) = \{\xi = (\xi_1, \ldots, \xi_n) \in \mathbb{R}^n : |x_j - \xi_j| < L, \ j = 1, \ldots, n\}.$$

V. E. Katsnelson [6] showed that this analog holds also for the wider class of finite degree functions polysubharmonic in $\mathbb{C}^n$. A similar inequality was proved by M. Benedicks [7] for functions subharmonic in $\mathbb{R}^{n+1}$ and for relatively dense subsets of $\mathbb{R}^n$. In the recent paper [8], a result including the theorems of Katsnelson and M. Benedicks and strengthening both of them was established.

Upper envelopes of classes of subharmonic in $\mathbb{R}^n$ functions of finite degree not larger than σ and nonpositive on the set $E \subset \mathbb{R}^n$ were studied in [2, 3, 7, 9]. We preserve the notation $K_\sigma(E)$ for such classes. If the set E is regular with respect to the Dirichlet problem in $\mathbb{R}^{n+1}$ and the subharmonic majorant of the class $K_\sigma(E)$ is finite, then this majorant belongs itself to this class, equals zero on E and is a harmonic function outside E.

Positive harmonic functions in unbounded domains which are equal to zero on the boundary of these domains were investigated in many papers. In the case when the domain has the form $\mathbb{R}^{n+1} \setminus E$, $E \subset \mathbb{R}^n$, the investigations began in [10] in the late forties, were continued in [11, 1, 2] and developed and systematized in [7] and [9] (see also [12–14]). This theory has a number of applications in the case $n = 1$ when subharmonic majorants are closely connected with special conformal mappings (see [2, 9]). From our viewpoint, the main applications of this theory pertain to extremal problems in various classes of subharmonic functions [9], to estimates of derivatives of analytic functions on the boundary of their domain of analyticity [2], to the spectral theory of differential operators (see references in [15]) and to problems of quasianalyticity [3, 12, 16, 17].

Now, before stating the main results of this paper, let us introduce some notations.

$\mathbb{R}^{n+1} = \{z = (x, y) : (x_1, \ldots, x_n) \in \mathbb{R}^n, \ y \in \mathbb{R}\}$;

$\mathbb{R}^{n+1}_+ = \{z = (x, y) : y > 0\}$ is the upper half-plane;

$\mathbb{R}^{n+1}_- = \{z = (x, y) : y < 0\}$ is the lower half-plane;

$\{z = (x, 0)\}$ is the "real hyperplane"; we identify this set with $\mathbb{R}^n$ and write x instead of $(x, 0)$;

$\|z\| = \{\|x\|^2 + y^2\}^{1/2} = \{x_1^2 + \cdots + x_n^2 + y^2\}^{1/2}$;

$|z| = \max\{|x|, |y|\} = \max\{|x_1|, \ldots, |x_n|, |y|\}$;

$B(z, \rho) = \{w \in \mathbb{R}^{n+1} : \|w - z\| < \rho\}$ is a ball in $\mathbb{R}^{n+1}$;
$D(x, \rho) = \{u \in \mathbb{R}^n : \|u - x\| < \rho\} = B(x, \rho) \cap \mathbb{R}^n$;
$S^n = \partial B(0, 1)$ is the unit sphere in $\mathbb{R}^{n+1}$;
$Q(z, \rho) = \{w \in \mathbb{R}^{n+1} : |w - z| < \rho\}$ is a hypercube in $\mathbb{R}^{n+1}$;
$K(x, \rho) = \{u \in \mathbb{R}^n : |u - x| < \rho\} = Q(x, \rho) \cap \mathbb{R}^n$.
We say that a subharmonic function $u(z)$ in $\mathbb{R}^{n+1}$ has finite degree σ, if

$$\limsup_{\|z\|\to\infty} u(z)/\|z\| = \sigma.$$

The paper consists of four sections. In §1 which is mainly systematic in character, we shall establish some properties of functions subharmonic in the half-space $\mathbb{R}^{n+1}_+$ and having a nonnegative harmonic majorant in this half-space. Analyzing these properties, we obtain the following result.

A subharmonic in $\mathbb{R}^{n+1}$ *function* $u(z)$ *has nonnegative harmonic majorants in upper and lower half-spaces if and only if this function has finite degree and satisfies the Cartwright condition*

$$\int_{\mathbb{R}^n} (1 + \|\xi\|^2)^{-(n+1)/2} u_+(\xi)\, d\xi < \infty.$$

If $n = 1$ and $u(z) = \log|f(z)|$, $f(z)$ being an entire function, this theorem was proved by M. G. Kreĭn [18]. Apparently, this theorem is known in the general case too. Nevertheless, for the convenience of the reader, we give its proof in §1.

Section 2 is devoted to subharmonic majorants

$$v(z) = v(z, K_\sigma(E)) = \sup\{u(z) : u \in K_\sigma(E)\}.$$

These majorants are either finite everywhere on $\mathbb{R}^{n+1}$ or equal to $+\infty$ on $\mathbb{R}^{n+1} \setminus E$. The set E is said to be *of type* (α) in the first case, and *of type* (β) in the second.

For any $R > 0$ we consider the set

$$E_R = E \cup \{x \in \mathbb{R}^n : \|x\| \geqslant R\}$$

and define the functions $v_R(z) = v(z, K_\sigma(E_R))$. These functions are bounded on $\mathbb{R}^n$ (possibly, by different constants), and for any point $z \in \mathbb{R}^{n+1}$ the function $R \mapsto v_R(z)$ does not decrease as R increases. In §2 we prove the equality $\lim_{R\to\infty} v_R(z) = v(z)$, $z \in \mathbb{R}^{n+1}$, and the statement:

A set $E \subset \mathbb{R}^n$ *has type* (α) *if and only if the function* $R \mapsto v_R(z)$ *is bounded at some point* $z \in \mathbb{R}^{n+1} \setminus E$.

For the case $n = 1$ this criterion was obtained in [2]. In the general case the proof uses the same outline.

A set $E \subset \mathbb{R}^n$ is called *relatively dense with respect to capacity* if there exist values $L < \infty$ and $\delta > 0$ such that for any $x \in \mathbb{R}^n$ the estimate

$$\operatorname{cap}(E \cap Q(x, L)) \geqslant \delta$$

is fulfilled. Here cap(A) is the Wiener capacity of a set $A \subset \mathbb{R}^{n+1}$ for $n \geqslant 2$ and the logarithmic capacity for $n = 1$.

The main result of the paper is proved in §3. Here is a particular case:

Let E be a closed subset of $\mathbb{R}^n$. The following statements are equivalent:

a) *the set E is relatively dense with respect to capacity;*

b) *there exists a value $C < \infty$ such that for any subharmonic in $\mathbb{R}^{n+1}$ function $u(z)$ of finite degree σ the inequality*

$$\sup\{u(z) : x \in \mathbb{R}^n\} \leqslant C\sigma + \sup\{u(\xi) : \xi \in E\}$$

is valid.

It follows from standard estimates of capacity (see, for example, [19, 5.4]) that a subset of $\mathbb{R}^n$ relatively dense with respect to Lebesgue measure is relatively dense with respect to capacity as well. Therefore, the above-mentioned theorem of M. Benedicks is an immediate corollary of the implication a) $\Rightarrow$ b). The criterion stated above is part of a more general theorem which is proved in §3. In this theorem a necessary and sufficient condition for the growth of every function of $K_\sigma(E)$ along the "real hyperplane" to be slower than $|x|^\alpha$, $0 \leqslant \alpha < 1$, is given. This criterion is formulated in terms of capacity characteristics of the set $E \subset \mathbb{R}^n$.

In the last §4 we give some remarks, supplements and generalizations of the main theorem of §3.

We would like to mention here that the method applied in this paper is based on simple properties of subharmonic functions and on elementary facts from potential theory only. However, this allows us to give simple proofs of the Benedicks [7] and Carleson [20] estimates of harmonic measure and even slightly refine these estimates. We hope to return to this subject in a future paper.

§1. Subharmonic functions that have a nonnegative harmonic majorant in $\mathbb{R}^{n+1}_+$

We shall use the following classical representation of functions subharmonic in $\mathbb{R}^{n+1}_+$ (see, for example, [22, Theorem 16.1.7]).

THEOREM A. *Let $u(z)$ be a subharmonic function in $\mathbb{R}^{n+1}_+$ having a nonnegative harmonic majorant in the half-space. Then*

$$u(z) = \sigma y + \frac{2y}{|S^n|}\int_{\mathbb{R}^n} \frac{d\nu(\xi)}{(\|x-\xi\|^2 + y^2)^{(n+1)/2}} + G(z), \qquad z = (x, y). \tag{3}$$

Here σ is a real number, the charge $d\nu(\xi)$ is the weak limit as $\eta \to 0$ (in the dual space to the space of continuous functions with finite support) of the charges $u(\xi, \eta)\,d\xi$,

$$\int_{\mathbb{R}^n} (1 + \|\xi\|^2)^{-(n+1)/2} d|\nu|(\xi) < \infty$$

and $G(z)$ is the Green potential of the measure $d\mu(\xi)$ "Riesz associated" with the function $u(z)$, i.e.,

$$G(z) = \int_{\mathbb{R}_+^{n+1}} \{\|(\xi, -\eta) - z\|^{1-n} - \|(\xi, \eta) - z\|^{1-n}\} \, d\mu(\xi).$$

Further we shall use the following theorem[2] of V. S. Azarin:

THEOREM B. *Suppose that a function $u(z)$ satisfies the conditions of Theorem A. Then everywhere on $\mathbb{R}_+^{n+1}$ except, maybe, at the points of an exceptional set*

$$\mathscr{E} = \bigcup_j B(z^{(j)}, \delta_j),$$

such that the following series converges

$$\sum_j \{\delta_j / \|z^{(j)}\|\}^{n-1} < \infty,$$

the asymptotic estimate $u(z) = \sigma y + o(\|z\|)$, $\|z\| \to \infty$, is valid. (Here σ is the value from representation (3).)

We shall also need the following version of Phragmen-Lindelöf principle[3].

THEOREM 1. *Let $u(z)$ be a subharmonic in $\mathbb{R}^{n+1}$ function of finite degree not larger than σ, and*

$$\int_{\mathbb{R}^n} (1 + \|\xi\|^2)^{-(n+1)/2} u_+(\xi, 0) \, d\xi < \infty.$$

Then for every point $(x, y) \in \mathbb{R}$, $y \neq 0$, the estimate

$$u(x, y) \leqslant \sigma|y| + \frac{2|y|}{|S^n|} \int_{\mathbb{R}^n} \frac{u(\xi, 0)\, d\xi}{(\|x - \xi\|^2 + y^2)^{(n+1)/2}} \tag{4}$$

is valid. Besides, for every $x \in \mathbb{R}^n$

$$u(x, 0) \leqslant \inf \left\{ \sigma|y| + \frac{2|y|}{|S^n|} \int_{\mathbb{R}^n} \frac{u(\xi, 0)\, d\xi}{(\|x - \xi\|^2 + y^2)^{(n+1)/2}} : y \neq 0 \right\}. \tag{5}$$

PROOF. To begin with we consider the particular but most important case $u(z) \geqslant 0$. Define the nonnegative function v harmonic in $\mathbb{R}_+^{n+1}$:

$$v(x, y) = \frac{2|y|}{|S^n|} \int_{\mathbb{R}^n} \frac{u(\xi, 0)\, d\xi}{(\|x - \xi\|^2 + y^2)^{(n+1)/2}}.$$

The boundary values of $v(z)$ on the hyperplane $\mathbb{R}^n = \partial\mathbb{R}_+^{n+1}$ coincide a.e. with $u(x, 0)$. The difference $w(z) = u(z) - v(z)$ is subharmonic in $\mathbb{R}_+^{n+1}$ and for every $x \in \mathbb{R}^n$ we have $\limsup_{z \to (x, 0),\, z \in \mathbb{R}_+^{n+1}} w(z) \leqslant 0$.

---

[2]The general case of this theorem pertains to functions subharmonic in a cone of the space $\mathbb{R}^{n+1}$. In the half-plane and half-space cases, it gives a more precise estimate of the size of an exceptional set than earlier theorems of W. Hayman and J. Lelong-Ferrand.

[3]A similar statement was proved by M. Benedicks [17, Lemma 6].

For $z = (x, y)$, we set $\overline{z} = (x, -y)$. Define the function

$$\widetilde{w}(\zeta) = \begin{cases} w(\zeta), & \zeta \in S^n_+, \\ -w(\overline{\zeta}), & \zeta \in S^n_-, \end{cases}$$

on the sphere S^n. Here $S^n_\pm = S^n \cap \mathbb{R}^{n+1}_\pm$ are the "upper" and "lower" hemispheres. The harmonic continuation $\varkappa(z)$ of this function to the ball $B(0, 1)$ has the form

$$\begin{aligned}\varkappa &= \frac{1-\|z\|^2}{|S^n|}\int_{S^n}\widetilde{w}(\zeta)\frac{d\lambda(\zeta)}{(1-2\langle z, \zeta\rangle + \|z\|^2)^{(n+1)/2}} \\ &= \frac{1-\|z\|^2}{|S^n|}\int_{S^n_+}\widetilde{w}(\zeta)\left\{\frac{1}{(1-2\langle z, \zeta\rangle + \|z\|^2)^{(n+1)/2}}\right. \\ &\qquad\qquad \left. - \frac{1}{(1-2\langle z, \overline{\zeta}\rangle + \|z\|^2)^{(n+1)/2}}\right\} d\lambda(\zeta),\end{aligned}$$

where λ is the area of S^n. In particular, it follows that $\varkappa(x, 0) = 0$. Thus $\varkappa(z)$ is the harmonic majorant of the function $w(z)$ in the upper half-ball $B_+(0, 1)$.

Set $w_R(z) = w(Rz)$, $R > 0$, and define the function $\varkappa_R(z)$ by $w_R(z)$ in the same way as we have already defined the function $\varkappa(z)$ using $w(z)$. Then we have $w_R(z) \leqslant \varkappa_R(z)$ everywhere in $B_+(0, 1)$. Fix a point $z = (x, y) \in \mathbb{R}^{n+1}_+$. For $R > \|z\|$, we have

$$\begin{aligned} w(z) &= w_R(z/R) \leqslant \varkappa_R(z/R) \\ &= \frac{1-\|z/R\|^2}{|S^n|}\int_{S^n_+} w(R\zeta)\left\{\frac{1}{(1-2\langle z/R, \zeta\rangle + \|z\|^2)^{(n+1)/2}}\right. \\ &\qquad\qquad \left. - \frac{1}{(1-2\langle z/R, \overline{\zeta}\rangle + \|z\|^2)^{(n+1)/2}}\right\} d\lambda(\zeta) \\ &= \frac{1-\|z/R\|^2}{|S^n|}(n+1)\int_{S^n_+}\frac{w(R\zeta)}{\zeta}\,d\lambda(\zeta) \\ &\qquad\qquad \times\{-\langle z, \overline{\zeta}\rangle + \langle z, \zeta\rangle + O(1/R)\}, \qquad R\to\infty. \end{aligned}$$

Sending R to infinity, we deduce

$$w(z) \leqslant \frac{2\sigma y(n+1)}{|S^n|}\int_{S^n_+}\eta\, d\lambda(\xi, \eta) = \sigma y.$$

Hence inequality (4) is proved for the upper half-space under the assumption $u(z) \geqslant 0$. For the lower half-space $\mathbb{R}^{n+1}_-$ the proof is similar.

To deduce the estimate (5), one can apply inequality (4) twice and use the semigroup property of the Poisson kernel for a half-space (see [24]). Indeed, if, for example, $y > 0$, then

$$u(x, 0) = u\left(x, \frac{y}{2} - \frac{y}{2}\right) \leqslant \sigma\frac{y}{2} + \frac{y}{|S^n|}\int_{\mathbb{R}^n}\frac{u(\xi, y/2)\, d\xi}{(\|x-\xi\|^2 + y^2/4)^{(n+1)/2}}$$

$$\leqslant \sigma\frac{y}{2} + \frac{y}{|S^n|}\int_{\mathbb{R}^n}\left\{\frac{\sigma y}{2} + \frac{y}{|S^n|}\int_{\mathbb{R}^n}\frac{u(\nu, 0)\,d\nu}{(\|\xi-\nu\|^2+y^2/4)^{(n+1)/2}}\right\} \times \frac{d\xi}{(\|x-\xi\|^2+y^2/4)^{(n+1)/2}}$$

$$= \sigma y + \frac{2y}{|S^n|}\int_{\mathbb{R}^n}\frac{u(\nu, 0)\,d\nu}{(\|\xi-\nu\|^2+y^2)^{(n+1)/2}}.$$

A similar inequality is valid for $y < 0$.

Thus we have proved Theorem 1 for nonnegative functions $u(z)$. In the general case, consider the functions $u_m(z) = m + \max(-m, u(z))$, $m \in \mathbb{N}$, apply the particular case just proved to these functions and send m to infinity. The theorem is proved.

Now we are able to prove the following analog of M. G. Kreĭn's theorem for a subharmonic function in $\mathbb{R}^{n+1}$.

THEOREM 2. *Necessary and sufficient conditions for a subharmonic in $\mathbb{R}^{n+1}$ function $u(z)$ to have a nonnegative harmonic majorant in the "upper" and "lower" half-spaces are that $u(z)$ has a finite degree and satisfies the Cartwright condition*

$$\int_{\mathbb{R}^n}(1+\|x\|^2)^{-(n+1)/2}u_+(x)\,dx < \infty. \tag{6}$$

PROOF. *Sufficiency.* Suppose that the degree of the function $u(z)$ is equal to σ. According to Theorem 1, the function

$$\sigma|y| + \frac{2|y|}{|S^n|}\int_{\mathbb{R}^n}\frac{u_+(\xi)\,d\xi}{(\|x-\xi\|^2+y^2)^{(n+1)/2}}$$

is a positive harmonic majorant of $u(z)$ on each of the half-spaces $\mathbb{R}^{n+1}_{\pm}$.

Necessity. Let $v(z)$ be a nonnegative harmonic majorant of the function $u(z)$ in the half-space $\mathbb{R}^{n+1}_{+}$, and let $d\nu(\xi)$ be the nonnegative boundary measure corresponding to $v(z)$ according to the representation (3). Since $v(z)$ is a majorant of the function $u_+(z)$, there is a majorization relation between the measures: $u_+(\xi)\,d\xi \leqslant d\nu(\xi)$. Hence

$$\int_{\mathbb{R}^n}\frac{u_+(x)\,dx}{(1+\|x\|^2)^{(n+1)/2}} \leqslant \int_{\mathbb{R}^n}\frac{d\nu(x)}{(1+\|x\|^2)^{(n+1)/2}} < \infty.$$

The convergence of the integral (6) is proved.

By Theorem B there exist values σ_+ and σ_- such that

$$u(z) = \sigma_+ y + o(\|z\|), \qquad \|z\| \to \infty, \quad z \in \mathbb{R}^{n+1}_+ \setminus \mathscr{E},$$
$$u(z) = \sigma_- y + o(\|z\|), \qquad \|z\| \to \infty, \quad z \in \mathbb{R}^{n+1}_- \setminus \mathscr{E}.$$

By the maximum principle, the estimate

$$u(z) \leqslant \max(\sigma_+, \sigma_-)|y| + o(\|z\|), \qquad \|z\| \to \infty,$$

holds everywhere on $\mathbb{R}^{n+1}$. Consequently $u(z)$ has the finite degree $\sigma = \max(\sigma_+, \sigma_-)$. Theorem 2 is proved.

§2. Majorants of classes $K_\sigma(E)$ and positive harmonic functions vanishing on the boundary

Let E be a closed regular subset of the hyperplane $\mathbb{R}^n$, let $K_\sigma(E)$ be the class of subharmonic functions in $\mathbb{R}^{n+1}$ of finite degree not larger than σ, which are nonpositive on E, and let

$$v(z) = v(z, K_\sigma(E)) = \sup\{u(z) : u \in K_\sigma(E)\}$$

be the subharmonic majorant[4] of the class $K_\sigma(E)$. As we mentioned in the Introduction, either the majorant is a nonnegative subharmonic function in $\mathbb{R}^{n+1}$ vanishing on E and harmonic on $\mathbb{R}^{n+1} \setminus E$ (case (α)), or $v(z) \equiv +\infty$ on $\mathbb{R}^{n+1} \setminus E$ (case (β)). In view of the equality

$$v(z, K_\tau(E)) = (\tau/\sigma)v(z, K_\sigma(E)),$$

further we set $\sigma = 1$ and consider only the class $K(E) = K_1(E)$. Later we shall use another version of the Phragmen-Lindelöf principle[5].

LEMMA 1. *Let $w(z) \in K(E)$ be a positive harmonic function in $\mathbb{R}^{n+1} \setminus E$ that satisfies the estimate $w(z) \geqslant |y|$, $z \in \mathbb{R}^{n+1}$. Then $w(z) = v(z, K(E))$, and consequently, the majorant $v(z, K(E))$ is finite.*

PROOF. Let $u(z)$ be an arbitrary function from the class $K(E)$. To prove our lemma, we must show that $u(z) \leqslant w(z)$, $z \in \mathbb{R}^{n+1}$.

Fix $\varepsilon > 0$ and consider the function $u_\varepsilon(z) = u(z) - (1+\varepsilon)w(z)$, which is subharmonic in $\mathbb{R}^{n+1} \setminus E$. By Theorem B of §1, this function tends to $-\infty$ as $\|z\| \to \infty$ uniformly on every double-based cone with vertex at the origin embedded compactly into $\mathbb{R}^{n+1} \setminus \mathbb{R}^n$. In particular, this cone can be chosen so wide that the cone

$$C = \{z = (x, y) \in \mathbb{R}^{n+1} : \|x\|^2 \leqslant ny^2\}$$

is embedded compactly into it. Given $\delta > 0$, consider the function

$$u_{\varepsilon,\delta}(z) = u_\varepsilon(z) - \delta(\|x\|^2 - ny^2).$$

This function is subharmonic in $\mathbb{R}^{n+1} \setminus E$, nonpositive in E. It tends to $-\infty$ as $\|z\| \to \infty$ on the set $\operatorname{clos}(\mathbb{R}^{n+1} \setminus C)$. By the maximum principle we have $u_{\varepsilon,\delta}(z) \leqslant \max_{\zeta \in \partial C}[u(\zeta) - (1+\varepsilon)w(\zeta)]_+$ for all $z \in \mathbb{R}^{n+1} \setminus C$. The maximum on the right-hand side of this inequality does not depend on δ. Passing to

[4]We do not use the upper regularization in the definition of the subharmonic majorants since the set E is regular.

[5]For $n = 1$ this lemma was proved in [2]. For the general case it was proved by M. Benedicks [7, Theorem 3]. The proof given below is a repetition of the proof from [2] for the multidimensional case.

the limit as $\delta \to 0$, we see that the subharmonic in $\mathbb{R}^{n+1} \setminus E$ function $u_\varepsilon(z)$ is bounded from above.

Now consider the function $h_\varepsilon(z) = (u_\varepsilon)_+(z)$. It is easy to show that this function is subharmonic in the entire space $\mathbb{R}^{n+1}$ and is bounded from above. At the same time, if R is large enough, then $u_\varepsilon(z) \equiv 0$ for all $z \in C \setminus B(0, R)$. It follows (see, for example, [19, Theorem 3.21]) that $h_\varepsilon(z) \equiv 0$ everywhere on $\mathbb{R}^{n+1}$. Sending ε to zero, we conclude that $u(z) \leqslant w(z)$ for all $z \in \mathbb{R}^{n+1}$. The lemma is proved.

Along with the set E, we consider the family of sets

$$E_R = E \cup \{x \in \mathbb{R}^n : \|x\| \geqslant R\}, \qquad R > 0.$$

Denote by $v_R(z) = v(z, K(E_R))$ the subharmonic majorant of the class $K(E_R)$.

LEMMA 2. *For every $R < \infty$, the majorant $v_R(z)$ is finite and bounded on every layer $\{z = (x, y) : -H \leqslant y \leqslant H\}$, $H < \infty$.*

PROOF. Any function $u(z) \in K(E_R)$ is bounded from above on the hyperplane $\mathbb{R}^n$. Hence it satisfies the conditions of Theorem 1. By this theorem

$$\begin{aligned} u(x, 0) &\leqslant R + \frac{2R}{|S^n|} \int_{\mathbb{R}^n} \frac{u(\xi, 0)\, d\xi}{(\|x - \xi\|^2 + R^2)^{(n+1)/2}} \\ &\leqslant R + \frac{2R}{|S^n|} \int_{\|\xi\| \leqslant R} \frac{u(\xi, 0)\, d\xi}{(\|x - \xi\|^2 + R^2)^{(n+1)/2}}. \end{aligned}$$

Set $M = \sup\{u(z) : x \in \mathbb{R}^n\}$. Then

$$M \leqslant R + M \frac{2R}{|S^n|} \int_{\|\xi\| \leqslant R} \frac{d\xi}{(\|x - \xi\|^2 + R^2)^{(n+1)/2}} \leqslant R + (1 - \gamma) M,$$

where γ ($0 < \gamma < 1$) does not depend on R. The γ coincides up to a constant factor with the angular measure of the cone with vertex at $(0, R) \in \mathbb{R}^{n+1}$ and base $\mathbb{R}^n \setminus D(0, R)$. Thus $M \leqslant R/\gamma$. On the other hand, $|y| \in K(E_R)$. Therefore, $|y| \leqslant v_R(z) \leqslant |y| + R/\gamma$ for all $z \in \mathbb{R}^{n+1}$. Lemma 2 is proved.

THEOREM 3. *The case (α) holds if and only if the set of values $\{v_R(z)\}$, $R > 0$, is bounded at some point $z \in \mathbb{R}^{n+1} \setminus E$. Further $v(z) = \lim_{R \to \infty} v_R(z)$ for all $z \in \mathbb{R}^{n+1}$.*

PROOF. Suppose that the case (α) holds. Then, as R grows, the majorants $v_R(z)$ do not decrease and are bounded by the function $v(z)$. By Harnack's Theorem, the set of functions $\{v_R(z)\}$ converges uniformly on every compact set in $\mathbb{R}^{n+1} \setminus E$ to a positive harmonic function $w(z)$ as $R \to \infty$. (In fact, as it follows from Theorem 1, the convergence is uniform on every compact set in $\mathbb{R}^{n+1}$.) The nonnegative harmonic function $w(z)$ belongs to the class

$K(E)$ and satisfies the estimate $w(z) \geqslant |y|$ for all $z = (x, y) \in \mathbb{R}^{n+1}$. By Lemma 1, $v(z) \equiv w(z)$.

Conversely, if the set $\{v_R(z)\}$, $R > 0$, is bounded at least at one point $z \in \mathbb{R}^{n+1} \setminus E$, then the limit $w(z) = \lim_{R\to\infty} v_R(z)$ exists everywhere on $\mathbb{R}^{n+1}$. As before, $w(z)$ is a subharmonic function in $\mathbb{R}^{n+1}$ that vanishes on E and is positive and harmonic outside E. In particular, by Theorem 2, the function $w(z)$ has finite degree $\tau \geqslant 1$. Thus $w(z) \geqslant \tau |y|$ for all $z = (x, y) \in \mathbb{R}^{n+1}$. By Lemma 1, the majorant $v(z)$ is finite and $\tau v(z) \equiv w(z)$. Applying part of the theorem already established, we conclude that $\tau = 1$. The theorem is proved.

REMARK 1. In the case (α), the cone $P(E)$ of positive functions harmonic outside E and vanishing on E is two-dimensional.

For $n = 1$, this result was stated by one of the authors in [3] (see also [9]). In the general case, it was proved by M. Benedicks, but the proof given below seems to be more elementary.

Indeed, the majorant $v(z, K(E))$ belongs to the cone $P(E)$. If a function $\tilde{v}(z) \in P(E)$ has coinciding degrees in the upper and lower half-spaces, then by Lemma 1 $v(z) \equiv \tilde{v}(z)$. If one can find a function $v_0(z)$ of zero degree in the cone $P(E)$, then by Lemma 1 $v_0(z) < v(z)$, $z \in \mathbb{R}^{n+1} \setminus E$. Applying this lemma again, we conclude that $v(z) - v_0(z) \equiv v(z)$, thus $v_0(z) \equiv 0$.

Note now that, along with each majorant $v(z, K_\sigma(E))$, every function of the form $v(z, K_\sigma(E)) + \varkappa y$, $-\sigma \leqslant \varkappa \leqslant \sigma$, belongs to the cone $P(E)$. There are no other functions in this cone. In fact, every function $w(z) \in P(E)$ has positive degree in both half-spaces $\mathbb{R}^{n+1}_\pm$. By adding an appropriate function of the form $\varkappa y$, these degrees can be made equal. Hence the function $w(z) + \varkappa y$ coincides with $v(z, K_\sigma(E))$ for some σ.

REMARK 2. In the case (β), the cone $P(E)$ is one-dimensional. The known proofs of this fact either use estimates of the harmonic measure going back to Carleman (see [9]) or theorems proved by using such estimates (for example, the proof in [7] relies on the theorem of S. Friedland and W. Hayman on the number of tracts of subharmonic functions).

REMARK 3. Other criteria of finiteness of the majorant $v(z, K_\sigma(E))$ may be found in the papers of M. Benedicks [7] and P. P. Kargaev [12].

§3. Boundedness and slow growth along the "real hyperplane" of subharmonic functions of finite degree

The following theorem is the main result of this paper.

THEOREM 4. *Let E be a closed regular subset of the "real" hyperplane $\mathbb{R}^n$, and λ be a fixed number from the interval $[0, 1)$. The following statements are equivalent:*

a) *there exist values $L < \infty$ and $\delta > 0$ such that*

$$\operatorname{cap}(E \cap Q(x, L(1 + |x|^\lambda))) \geqslant \delta |x|^{\lambda(n-1)}$$

for all $x \in \mathbb{R}^n$;

b) *there exists a value* $C = C(L, \delta, n, \lambda)$ *such that for every* $\sigma > 0$ *and for every function* $u(z) \in K_\sigma(E)$ *the estimate*

$$u(z) \leqslant C\sigma(1 + |x|^\lambda)$$

holds for all $x \in \mathbb{R}^n$.

The proof of this theorem is based on the well-known simple estimates of the harmonic measure (see, for example, [20]). For the convenience of the reader, we give these estimates with proofs.

LEMMA 3. *Let* Q *be a cube in* $\mathbb{R}^{n+1}$ *(with unit edges) symmetric with respect to the "real hyperplane"* $\mathbb{R}^n$, $K = Q \cap \mathbb{R}^n$, *and let* $\widehat{Q}$ *be the cube concentric with* Q *whose linear size is twice as small. Then for an arbitrary compact set* $F \subset \widehat{Q}$ *we have*

$$\omega(z, F, Q \setminus F) \geqslant \gamma_1 \operatorname{cap}(F), \qquad z \in \widehat{Q}, \tag{7}$$

$$\int_K \omega(x, F, Q \setminus F)\, dx \leqslant \gamma_2 \operatorname{cap}(F), \tag{8}$$

where $\omega(\cdot, F, Q \setminus F)$ *is the harmonic measure of* F *with respect to the domain* $Q \setminus F$, *and the positive constants* γ_1 *and* γ_2 *depend only on* n.

PROOF. We consider only the case $n > 1$. The right-hand sides of the inequalities (7) and (8) are slightly different. For the case of logarithmic capacity the proof is similar. Assume that the compact set F is regular. Denote by ν the equilibrium measure of the compact set F, and by $U(z)$ its Newton potential:

$$U(z) = \int_F \|z - \zeta\|^{1-n}\, d\nu(\zeta).$$

Assume that $\|\nu\| = \operatorname{cap}(F)$, where $\|\nu\|$ is the total variation of the measure ν. Then $U(z) \equiv 1$ as $z \in F$.

First we prove estimate (7). For every measure μ supported by F we have

$$\|\mu\| = \int_F U(z)\, d\mu(z) = \int_F \left\{ \int_F \|z - \zeta\|^{1-n}\, d\mu(z) \right\} d\nu(\zeta).$$

The function $\|z - \zeta\|^{1-n}$ is not less than the Green function $G(z, \zeta)$ of the cube Q everywhere in this cube. Thus

$$\|\mu\| \geqslant \int_F \left\{ \int_F G(z, \zeta)\, d\mu(z) \right\} d\nu(\zeta).$$

Now we apply this inequality to the measure μ from the representation

$$\omega(\zeta, F, Q \setminus F) = \int_F G(z, \zeta)\, d\mu(z),$$

i.e., to the Riesz measure of the superharmonic function $\omega(\zeta, F, Q \setminus F)$ in

Q, extended by the unit value on F. We get

$$\|\mu\| \geqslant \int_F \omega(\zeta, F, Q \setminus F)\, d\nu(\zeta) = \int_F d\nu(\zeta) = \operatorname{cap}(F).$$

Since $\zeta \in \widehat{Q}$, the Green function $G(z, \zeta)$ of the cube Q is bounded away from zero by a positive value γ_1 depending on the dimension n only. Hence, as $\zeta \in \widehat{Q}$

$$\omega(\zeta, F, Q \setminus F) = \int_F G(z, \zeta)\, d\mu(\zeta) \geqslant \gamma_1 \|\mu\| \geqslant \gamma_1 \operatorname{cap}(E).$$

Inequality (7) is proved.

To prove (8), observe that the inequality $\omega(z, F, Q \setminus F) \leqslant U(z)$, where $U(z)$ is the equilibrium Newton potential of the compact set F as before, is valid everywhere on Q. Thus

$$\int_K \omega(\zeta, F, Q \setminus F)\, d\nu(\zeta) \leqslant \int_K U(\zeta)\, d\nu(\zeta) = \int_F d\nu(\zeta) \left\{ \int_K \|x - \zeta\|^{1-n}\, dx \right\}.$$

At every point $\xi \in K$, the inner integral does not exceed the value

$$\gamma_3 \int_0^{\sqrt{n}} r^{1-n} r^{n-1}\, dr = \gamma_2.$$

By the maximum principle this estimate is valid everywhere on $\mathbb{R}^{n+1}$. Therefore, the estimate (8) is proved.

PROOF OF THEOREM 4. We consider only the case $n > 1$. As before, we set $\sigma = 1$. First we prove the implication a) $\Rightarrow$ b).

Set $\varepsilon = \gamma_1 \delta$, where γ_1 is the constant from Lemma 3, and δ is the number from the statement a). Define the function

$$w(z, y) = \int_{\mathbb{R}^{n+1}} P_{\pm}(x - t, y)(1 + |t|^{\lambda})\, dt,$$

harmonic in each of the half-spaces $\mathbb{R}^{n+1}_{\pm}$. Here $P_{\pm}(x, t)$ are the Poisson kernels for these half-spaces. For a chosen ε, one can find a value $\rho < \infty$ such that for any $x \in \mathbb{R}^n$, $|x| \geqslant \rho$, we have the estimate

$$\sup\{\omega(x', y) : |x - x'| \leqslant L|x|^{\lambda},\ |y| \leqslant L|x|^{\lambda}\} \leqslant (1 + \varepsilon)(1 + |x|^{\lambda}). \tag{9}$$

Without loss of generality one can assume that E contains the cube $K(0, \rho)$. Fix $R < \infty$ and define as in §2 the set E_R, the class $K(E_R)$ and the majorant $v_R(z)$. Each of the functions $v_R(z)$ is bounded on the hyperplane $\mathbb{R}^n$. Therefore the value $c_R = \sup\{v_R(x)/(1 + |x|^{\lambda}) : x \in \mathbb{R}^n\}$ is finite for any $R < \infty$. We shall show that there exists a value C that does not depend on R and such that for any $R < \infty$ the inequality $c_R \leqslant C$ holds.

Let $x^* = x^*(R)$ be a point where the supremum c_R is reached, and let Q^* be the cube $Q^* = Q(x^*, L(1+|x^*|^\lambda))$. Then

$$\begin{aligned} c_R(1+|x^*|^\lambda) = v_R(x^*) &\leqslant \omega(x^*, \partial Q^*, Q^* \setminus E) \sup\{v_R(z) : z \in \partial Q^*\} \\ &\leqslant (1-\varepsilon) \sup\{v_R(z) : z \in \partial Q^*\} \\ &\leqslant (1-\varepsilon)\{L(1+|x^*|^\lambda) + c_R \sup\{w(z) : z \in \partial Q^*\}\} \\ &\leqslant (1-\varepsilon)(L + c_R(1+\varepsilon))(1+|x^*|^\lambda). \end{aligned} \tag{10}$$

The first inequality uses the maximum principle, the second uses Lemma 3, condition a) of our theorem and the fact that after a t-times linear expansion of a set its capacity is multiplied by t^{n-1}. The third inequality is based on the estimate $v_R(z) \leqslant c_R w(z) + |y|$ and, finally, the fourth one is based on the estimate (9) for the function $w(z)$. Now we conclude from (10) that $c_R \leqslant (1-\varepsilon)\varepsilon^{-2} L$. We obtain

$$v_R(z) \leqslant C(1+|x|^\lambda) \tag{11}$$

for any $R < \infty$ and any $x \in \mathbb{R}^n$, where $C = (1-\varepsilon)\varepsilon^{-2} L$. Theorem 3 implies that the majorant $v(z, K(E))$ is finite and satisfies the same estimate (11). Hence the condition b) of Theorem 4 is fulfilled.

Now we prove the converse implication b) $\Rightarrow$ a). The negation of condition a) means that

$$\begin{gathered} (\exists \{x^{(m)}\}_{m=1}^\infty : \|x^{(m)}\| \to \infty) \quad (\exists L_m \to \infty) \quad (\exists \delta_m \searrow 0) \\ \operatorname{cap}(Q(x^{(m)}, L_m |x^{(m)}|^\lambda) \cap E) \leqslant \delta_m |x^{(m)}|^{\lambda(n-1)}. \end{gathered} \tag{12}$$

By virtue of condition b), the subharmonic majorant $v(z, K(E))$ exists, hence it belongs to $K(E)$ and satisfies b). We shall show that (12) implies that

$$\sup\{v(x, K(E))/(1+|x|^\lambda) : x \in \mathbb{R}^n\} = \infty. \tag{13}$$

This contradiction means that the implication b) $\Rightarrow$ a) is correct.

Set $Q_m = Q(x^{(m)}, L|x^{(m)}|^\lambda)$, $K_m = Q_m \cap \mathbb{R}^n$. Let $h(z)$ be a harmonic function in the cube Q_m that equals $|y| = L_m|x^{(m)}|^\lambda$ on the two faces of the cube parallel to the "real hyperplane" $\mathbb{R}^n$, and equals zero on other faces. Applying the maximum principle, we get

$$\begin{aligned} v(z) &\geqslant h(z) - \int_{Q_m \cap E} h(\xi)\omega(z, d\zeta, Q_m \setminus E) \\ &\geqslant h(z) - L_m|x^{(m)}|^\lambda \int_{Q_m \cap E} \omega(z, d\zeta, Q_m \setminus E) \\ &= L_m|x^{(m)}|^\lambda \left\{ \frac{h(z)}{L|x^{(m)}|^\lambda} - \omega(z, E, Q_m \setminus E) \right\}, \qquad z \in Q_m. \end{aligned}$$

Integrating this estimate over K_m, we get

$$\begin{aligned}\frac{1}{\operatorname{meas}_n(K_m)}\int_{K_m} v(x)\,dx \\ \geqslant L_m|x^{(m)}|^\lambda\Bigg\{\frac{1}{\operatorname{meas}_n(K_m)}\int_{K_m}\frac{h(x)\,dx}{L_m|x^{(m)}|^\lambda} \\ - \frac{1}{\operatorname{meas}_n(K_m)}\int_{K_m}\omega(x, E, Q_m\setminus E)\,dx\Bigg\}. \quad (14)\end{aligned}$$

By similarity reasons, the first summand in the braces on the right-hand side of (14) is bounded away from zero by a positive value depending on n only. By Lemma 3 and (12) the second (negative) summand in these braces tends to zero as $m\to\infty$. Thus

$$\frac{1}{\operatorname{mes}_n(K_m)}\int_{K_m} v(x)\,dx \geqslant cL_m|x^{(m)}|^\lambda,$$

where the value $c>0$ depends on n only. Of course, this implies (13). The theorem is proved.

§4. Remarks and supplements to Theorem 4

1. In the cases $\lambda=0$ or $\lambda>0$, $n=1$, suitable sufficient conditions of boundedness of functions of class $K_\sigma(E)$ on the real hyperplane may be deduced from a standard lower estimate of capacity (see, for example, [19, Chapter 5]):

$$\operatorname{cap}(F)\geqslant \gamma m_s(E), \qquad s>n-1. \qquad (15)$$

Here F is a compact set contained in some cube of the space $\mathbb{R}^{n+1}$ with sides of length $2L$, γ depends on n, s and L, but not on F, $m_s(F)$ is the s-contents of the set F, i.e., $m_s(F)=\inf\sum r_j^s$, where the infimum is taken over all coverings of F by balls, the r_j being the radii of these balls.

2. In the case $\lambda>0$, $n>1$ the direct application of estimate (15) gives rough results, at least, if the set E is "sparse". More precise conditions on E may be obtained by using the following proposition.

LEMMA 4. *Let E be a closed regular subset of the "real" hyperplane $\mathbb{R}^n$ such that for some $\nu\in(0,1/n)$ the condition*

$$(\exists l<\infty)\,(\exists\delta>0)\,(\forall x\in\mathbb{R}^n)\quad \operatorname{cap}(Q(x, l|x|^\nu)\cap E)\geqslant\delta \qquad (16)$$

is fulfilled. Then condition a) *of Theorem* 4 *holds for $\lambda=n\nu$.*

PROOF. Devide the cube $K(x, |x|^\lambda)$ into congruent cubes with edges parallel to the coordinate axes and equal to $l|x|^\nu$. (Without loss of generality, we can suppose that such a decomposition is possible.) Consider a subset of "small cubes" such that any two cubes of this subset have no common boundary points, and the number $N=N_x$ of small cubes of this subset is maximal. It is easy to see that $N_x \approx |x|^{(\lambda-\nu)n}$ (i.e., the ratio of these values is bounded away from zero and infinity).

Denote by E_j the intersection of E with the jth small cube of the chosen subset, and by μ_j the uniform probability measure of E_j. Then for $z \in E_j$, we have

$$U^{\mu_j}(z) = \int_{E_j} \|z-\zeta\|^{1-n}\, d\mu_j(\zeta) = \frac{1}{\operatorname{cap}(E_j)} \leqslant \frac{1}{\delta}, \qquad j = 1, \dots, N. \tag{17}$$

Set

$$\mu = \frac{1}{N}\sum_{j=1}^{N} \mu_j$$

and estimate the Newton potential of this probability measure from above on the set

$$E^* = \bigcup_{j=1}^{N} E_j.$$

For any $z \in E_j$

$$\begin{aligned} U^{\mu}(z) &= \frac{1}{N}\sum_{k=1}^{N} \int_{E_k} \|z-\zeta\|^{1-n}\, d\mu_k(\zeta) \\ &= \frac{1}{N} U^{\mu_j}(z) + \frac{1}{N}\sum_{k\neq j} \int_{E_k} \|z-\zeta\|^{1-n}\, d\mu_k(\zeta). \end{aligned} \tag{18}$$

Assuming, as before, that $z \in E_j$, we estimate from above the sum on the right-hand side of (18):

$$\begin{aligned} &\frac{1}{N}\sum_{k\neq j} \int_{E_k} \|z-\zeta\|^{1-n}\, d\mu_k(\zeta) \\ &\leqslant \frac{C}{N_x} \sum_{p=1}^{[N_x^{1/n}]+1} (p|x|^{\nu})^{1-n}\{(2p+1)^n - (2p-1)^n\} \\ &\leqslant \frac{C}{N_x}|x|^{\nu(1-n)} N_x^{1/n} \leqslant \frac{C}{N_x}|x|^{\nu(1-n)+\lambda-\nu} = \frac{C}{N_x}. \end{aligned} \tag{19}$$

Substituting (17) and (19) into (18), we obtain $U^{\mu}(z) \leqslant C N_x^{-1}(\delta^{-1}+1)$, $z \in E^*$, whence

$$\begin{aligned} \operatorname{cap}(E \cap Q(x, L|x|^{\lambda})) &\geqslant \operatorname{cap}(E^*) \geqslant C^{-1}(1+\delta)^{-1} N_x \delta \\ &\geqslant C^{-1}|x|^{(\lambda-\nu)n} = C^{-1}|x|^{\lambda(n-1)}. \end{aligned}$$

(We denote by C various positive constants which do not depend on x.) Therefore, Lemma 4 is proved.

3. Lemma 4 and estimate (15) imply the following

COROLLARY. *Suppose that $l < \infty$ and $\delta > 0$ are such that for any $x \in \mathbb{R}^n$*

$$\operatorname{meas}_n(E \cap Q(x, l(1+|x|^{\lambda}))) \geqslant \delta.$$

Then

$$u(x) \leqslant C(1+|x|^{\lambda}), \qquad \lambda = n\nu,$$

for every $u \in K(E)$ *and every* $x \in \mathbb{R}^n$.

A similar statement was proved by M. Benedicks [7, Corollary 2], but his method gave a less precise estimate ($\lambda = (3n+1)\nu$ only). Note that the condition $\nu < 1/n$ in Lemma 4 and in the last Corollary cannot be improved. Namely, repeating the arguments which proved the implication b) $\Rightarrow$ a) in Theorem 4, one can prove the following statement.

Let γ *be a number from the open interval* $(1/n, 1)$ *and let* E *be a closed regular subset of the "real hyperplane"* $\mathbb{R}^n$ *such that*

$$\operatorname{cap}(Q(x, 1+|x|^{\gamma}) \cap E) \leqslant \text{const}$$

for every $x \in \mathbb{R}^n$. *Then case* (β) *holds for the set* E, *i.e., the majorant* $v(z, K(E))$ *does not exist.*

4. Let us mention a generalization.

The requirement that E *is a subset of the real hyperplane can be replaced in Theorem 4, as well as in the above remarks, by the requirement that* E *is contained in a parabolic layer*

$$\{z \in \mathbb{R}^{n+1} : |y| \leqslant L(1+|x|^{\lambda})\}.$$

5. *In the statement* b) *of Theorem* 4 *we can assume that the function* $u(z)$ *is subharmonic only in* $\mathbb{R}^{n+1} \setminus E$.

Indeed, if $u(z)$ is a function subharmonic in $\mathbb{R}^{n+1} \setminus E$ with nonnegative boundary values on E, then the function $u_+(z)$ is subharmonic everywhere in $\mathbb{R}^{n+1}$. Thus $u_+(z) \in K(E)$ and the implication b) $\Rightarrow$ a) proved previously can be applied.

References

1. A. C. Schaeffer, *Entire functions and trigonometric polynomials*, Duke Math. J. **20** (1953), 77–88.

2. N. I. Akhiezer and B. Ya. Levin, *Generalizations of the Bernstein inequalities for derivatives of entire functions*, Studies in Modern Problems of the Theory of Functions of Complex Variable, Fizmatgiz, Moscow, 1960, pp. 125–183. (Russian)

3. B. Ya. Levin, *Subharmonic majorants and their applications*, Proc. All-Union Conference in Function Theory, Kharkov, 1971, pp. 117–120. (Russian)

4. A. E. Fryntov, *One extremal problem of potential theory*, Dokl. Akad. Nauk SSSR **300** (1988), no. 4; English transl. in Soviet Math. Dokl. **37** (1988), 754–755.

5. A. Baernstein, *An extremal problem for certain subharmonic functions in the plane*, Rev. Mat. Iberoamericana **4** (1988), no. 2, 199–218.

6. V. E. Katsnelson, *Equivalent norms in spaces of entire functions of exponential type*, Mat. Sb. **92 (134)** (1973), 34–54; English transl. in Math. USSR-Sb. **21** (1973).

7. M. Benedicks, *Positive harmonic functions vanishing on the boundary of certain domains in* $\mathbb{R}^n$, Ark. Math. **18** (1980), 53–72.

8. B. Ya. Levin and V. N. Logvinenko, *On the classes of functions subharmonic in* $\mathbb{R}^n$ *and bounded on certain sets*, Zap. Nauchn. Sem. Leningrad. Otdel. Mat. Inst. Steklov. (LOMI) **170** (1989), 157–175; English transl. in J. Soviet Math. (to appear).

9. B. Ya. Levin, *Majorants in classes of subharmonic functions*. I, Teor. Funktsiĭ, Funktsional. Anal i Prilozhen. **51** (1989), 3–17; II **52** (1989), 3–33; English transl. in J. Soviet Math. (to appear).

10. B. Kjellberg, *On certain integral and harmonic functions*, Thesis (1948), Uppsala.

11. ______, *On growth of minimal positive functions in a plane domain*, Ark. Math. **1** (1951), 347–351.

12. P. P. Kargaev, *Existence of the Phragmen-Lindelöf function and some conditions of quasianalyticity*, Zap. Nauchn. Sem. Leningrad. Otdel. Mat. Inst. Steklov. (LOMI) **126** (1983), 97–108; English transl. in J. Soviet Math. **27** (1984), no. 1.

13. S. Segawa, *Martin boundaries of Denjoy domains*, Proc. Amer. Math. Soc. **103** (1988), 177–183.

14. S. Gardiner, *Minimal harmonic functions on Denjoy domains*, Proc. Amer. Math. Soc. **107** (1989), 963–975.

15. B. Ya. Levin, *Subharmonic majorants and some applications*, Complex Analysis (J. Hersh and A. Huber, eds.), Birkhäuser, Basel, 1988, pp. 181–190.

16. ______, *Completeness of systems of functions, quasianalyticity and subharmonic majorants*, Zap. Nauchn. Sem. Leningrad. Otdel. Mat. Inst. Steklov. (LOMI) **170** (1989), 102–156; English transl. in J. Soviet Math. (to appear).

17. M. Benedicks, *The support of functions and distributions with a spectral gap*, Math. Scand. **55** (1984), 285–309.

18. M. G. Kreĭn, *To the theory of entire functions of exponential type*, Izv. Akad. Nauk SSSR Ser. Mat. **11** (1947), 309–326. (Russian)

19. W. K. Hayman and P. B. Kennedy, *Subharmonic functions*, 1, Academic Press, New York, 1976.

20. L. Carleson, *Estimates of harmonic measures*, Ann. Acad. Sci. Fenn. Ser. A **7** (1982), no. 1, 25–32.

21. P. Koosis, *The logarithmic integral*, 1, Cambridge University Press, Cambridge, 1988.

22. L. Hörmander, *The analysis of linear partial differential operators*, II, Springer-Verlag, Berlin and New York, 1983.

23. V. S. Azarin, *Generalization of a theorem of Hayman on subharmonic functions in an n-dimensional cone*, Mat. Sb. **66** (1965), no. 2, 248–264; English transl. in Amer. Math. Soc. Transl. (2) **80** (1969).

24. E. Stein, *Singular integrals and differentiability properties of functions*, Princeton Univ. Press, Princeton, 1970.

Translated by M. L. SODIN

ADVANCES IN SOVIET MATHEMATICS
Volume 11, 1992

Uniqueness Theorems for Algebraic Functions

A. A. GOLDBERG AND V. A. PYANA

Let $P(z, w)$ be an irreducible polynomial in z and w, $\deg_w P = m \geqslant 1$, $\deg_z P = n \geqslant 1$. The equation

$$P(z, w) \equiv p_m(z)w^m + p_{m-1}(z)w^{m-1} + \cdots + p_0(z) = 0 \tag{1}$$

determines an algebraic function $w = w(z)$ such that $P(z, w(z)) \equiv 0$. It is known that the algebraic function $w = w(z)$ enables one to determine the polynomial $P(z, w)$ up to a constant factor. Let $\mathscr{A} \in \mathbb{C}$. The roots of the equation $P(z, \mathscr{A}) = 0$ are called *$\mathscr{A}$-points* of the algebraic function $w = w(z)$; if $\deg_z P(z, \mathscr{A}) = n - p$, $p \geqslant 1$, then $w = w(z)$ is considered to have an $\mathscr{A}$-point of order p at an infinity. In other words, $w = w(z)$ has an $\mathscr{A}$-point of order p at infinity if $z^n P(1/z, \mathscr{A})$ has a zero of order p at the point $z = 0$. If $\mathscr{A} = \infty$, then the roots of the equation $w^m P(z, 1/w)|_{w=0} = 0$ or, which is the same, roots of the equation $p_m(z) = 0$ are called *∞-points* (or *poles*) of the function $w = w(z)$. By $E(\mathscr{A}) = E(\mathscr{A}, P)$, we shall denote the set of $\mathscr{A}$-points in $\overline{\mathbb{C}}$ of the algebraic function given by equation (1) (orders of $\mathscr{A}$-points are considered here). If we also take into consideration the orders of $\mathscr{A}$-points, then we shall obtain a multiset[1] $\mathscr{M}E(\mathscr{A}) = \mathscr{M}E(\mathscr{A}, P)$. Accordingly, we shall distinguish between $\operatorname{card} E(\mathscr{A})$ and $\operatorname{card} \mathscr{M}E(\mathscr{A})$.

In what follows we shall consider the numbers $\mathscr{A}_j \in \overline{\mathbb{C}}$ as different. The algebraic function determined by equation (1) is said to be uniquely determined by q multisets $\mathscr{M}E(\mathscr{A}_j)$, $1 \leqslant j \leqslant q$, if for any algebraic function $w = \widehat{w}(z)$ defined by the equation

$$Q(z, w) = 0, \qquad \deg_z Q = n, \ \deg_w Q = m,$$

the relation $\mathscr{M}E(\mathscr{A}_j, P) = \mathscr{M}E(\mathscr{A}_j, Q)$ implies that $P = cQ$, where c is some nonzero constant. The statement: a function is uniquely determined by sets $E(\mathscr{A}_j)$, $1 \leqslant j \leqslant q$, is defined in similar fashion.

1991 *Mathematics Subject Classification.* Primary 14H05.

[1]We have not come across the term "multiset" in publications on the theory of functions, but in discrete mathematics, in particular, in combinatorics, this term is widely used (e.g., see [1]).

We have trivial statement that for $m = 1$ the algebraic function determined by equation (1) (which is a rational function in this case) is uniquely determined by two multisets $\mathscr{M}E(\mathscr{A}_1)$, $\mathscr{M}E(\mathscr{A}_2)$ and one set $E(\mathscr{A}_3)$. It is equally obvious that the function $w(z)$ is not uniquely determined by the multisets $\mathscr{M}E(\mathscr{A}_1)$ and $\mathscr{M}E(\mathscr{A}_2)$. For the case $m = 1$ Adams and Straus [2, Theorem 3] proved that four sets $E(\mathscr{A}_j)$, $1 \leqslant j \leqslant 4$, do determine a rational function uniquely, while Pizer [3] constructed an example showing that three sets $E(\mathscr{A}_j)$ do not suffice.

The following question arises: what is the minimal number of the multisets $\mathscr{M}E(\mathscr{A}_j)$ or sets $E(\mathscr{A}_j)$ needed to uniquely determine an algebraic function for an arbitrary $m \geqslant 2$. Here we shall present some estimates and construct examples showing that in some cases the estimates cannot be improved.

THEOREM 1. *$2m$ multisets $\mathscr{M}E(\mathscr{A}_j)$, $1 \leqslant j \leqslant 2m$, together with one set $E(\mathscr{A}_{2m+1})$ are sufficient to determine an m-valued algebraic function. Further $L = 4m + 1 + [-2m/n]$ sets $E(\mathscr{A}_j)$, $1 \leqslant j \leqslant L$, are sufficient to determine an m-valued algebraic function whose inverse function is n-valued.*

PROOF. Let us suppose that the multisets $\mathscr{M}E(\mathscr{A}_j)$, $1 \leqslant j \leqslant s$, and the sets $E(\mathscr{A}_j)$, $s+1 \leqslant j \leqslant s+T$, coincide for the algebraic function determined by equation (1) and for the algebraic function determined by the equation

$$\widehat{P}(z, w) \equiv \widehat{p}_m(z)w^m + \widehat{p}_{m-1}(z)w^{m-1} + \cdots + \widehat{p}_0(z), \tag{2}$$

$\deg_w \widehat{P} = m$, $\deg_z \widehat{P} = n$, $\widehat{P} \not\equiv cP$, $c = \text{const} \neq 0$. Without loss of generality, one can assume that all the $\mathscr{A}_j$ are finite and the $E(\mathscr{A}_j)$ do not contain ∞. Let $R(P, \widehat{P})$ be the resultant of the polynomials P and $\widehat{P}$ regarded as polynomials in w (e.g., see [4, §§34, 35]). For the resultant we have $R(P, \widehat{P}) \not\equiv 0$. Assuming the contrary, we see that in the neighborhood of some point z_0 two one-valued branches $w_j(z)$ and $\widehat{w}_k(z)$ of the functions $w(z)$ and $\widehat{w}(z)$ are identically equal. The identical equality will be preserved under any analytic continuation of these branches. So

$$P(z, w) = h_1(z, w)g(z, w), \qquad \widehat{P}(z, w) = h_2(z, w)g(z, w),$$

where $\deg_w g \geqslant 1$. Since $\widehat{P} \not\equiv cP$, we have $\deg_z h_j + \deg_w h_j \geqslant 1$, $j = 1, 2$. This contradicts the condition that P and $\widehat{P}$ are irreducible.

Since the resultant is a symmetric function of the roots, it is easily established that

$$\sum_{j=1}^{s} \operatorname{card} \mathscr{M}E(\mathscr{A}_j) + \sum_{j=s+1}^{s+T} \operatorname{card} E(\mathscr{A}_j) \tag{3}$$

does not exceed the sum of orders of the zeros in the resultant $R(P, \widehat{P})$. In our case the resultant is a determinant of order $2m$ whose elements are either zeros or $p_j(z)$ or $\widehat{p}_k(z)$. Therefore, the resultant is a polynomial of degree not exceeding $2mn$, and the sum (3) does not exceed $2mn$.

The algebraic function $w = w(z)$ determined by equation (1) maps an m-sheeted compact Riemann surface $\mathscr{R}_z$ which covers the z-plane bijectively onto an n-sheeted compact Riemann surface $\mathscr{R}_w$ which covers the w-plane. The sum (3) is evidently not less that $sn+T$. Let us consider the case when $s = 2m$, $T = 1$. We shall get $2mn \geqslant 2mn + 1$. This contradiction proves the first assertion of Theorem 1.

Now let $s = 0$, $T = L$. Then the number $\sum_{j=1}^{L} \operatorname{card} E(\mathscr{A}_j)$ equals the number of points $\mathscr{R}_w$ lying above $\mathscr{A}_1, \dots, \mathscr{A}_L$. Let

$$P(z, w) = q_n(w)z^n + q_{n-1}(w)z^{n-1} + \cdots + q_0(w).$$

The sum of orders of algebraic branching points of the Riemann surface $\mathscr{R}_w$ does not exceed the number of zeros (with order taken into account) of the discriminant for P regarded as a polynomial in z ([4, §35]). The discriminant is a determinant of order $2n-1$, whose first column contains 1 (in the first row) and n (in the nth row) with all other elements of the column being equal to zero. In the other $2n-2$ columns there are either zeros or $q_j(w)$ or $jq_j(w)$. That is why the discriminant as a polynomial in w has degree not exceeding $m(2n-2)$. Hence

$$\sum_{j=1}^{L} \operatorname{card} E(\mathscr{A}_j) \geqslant nL - m(2n-2).$$

We have obtained $2mn \geqslant nL - m(2n-2)$ and $L \leqslant 4m - 2m/n$, which contradicts the definition of L. This proves the second assertion of Theorem 1.

A natural question is how exact are the estimates of the number of sets and multisets in Theorem 1. We can present examples showing that the estimates cannot be improved, but not for all possible cases.

EXAMPLE 1. For $n = 1$ the multisets $\mathscr{M}E(\mathscr{A}_j)$ and the sets $E(\mathscr{A}_j)$ coincide, hence both assertions of the theorem coincide as well and it assumes the following form: $2m+1$ sets $E(\mathscr{A}_j)$, $1 \leqslant j \leqslant 2m+1$, are sufficient to uniquely determine an m-valued algebraic function whose inverse function is a rational function of degree m. One can show that $2m$ sets do not suffice. Let

$$P(z, w) = w^m - z, \qquad Q(z, w) = zw^m - 1,$$
$$\mathscr{A}_j = \exp(i2\pi j/m), \qquad 1 \leqslant j \leqslant m.$$

We have $\mathscr{A}_{m+j} = \exp(i\pi(2j-1)/m)$, $1 \leqslant j \leqslant m$. Then

$$\left.\begin{aligned} E(\mathscr{A}_j, P) &= E(\mathscr{A}_j, Q) = \{1\}, \\ E(\mathscr{A}_{m+j}, P) &= E(\mathscr{A}_{m+j}, Q) = \{-1\}, \end{aligned}\right\} \quad 1 \leqslant j \leqslant m.$$

EXAMPLE 2. For $n = 2$, knowing $E(\mathscr{A})$, one can uniquely determine $\mathscr{M}E(\mathscr{A})$. The first assertion of the theorem states that $2m+1$ sets $E(\mathscr{A}_j)$, $1 \leqslant j \leqslant 2m+1$, uniquely determine an m-valued algebraic function, while the second assertion of the theorem in this case yields $L = 3m + 1$. We shall show that $2m$ sets do not suffice to uniquely determine the algebraic function at least for even m. Let $m = 2k$, $k \in \mathbb{N}$,

$$P(z, w) = (-3z^2 + 2z - 5)w^{2k} + 14(z^2 + 1)w^k + (5z^2 - 2z + 7),$$
$$Q(z, w) = (-3z^2 - 2z - 1)w^{2k} + 14(z^2 + 1)w^k + (5z^2 + 2z + 3).$$

Denote $\omega_{jk} = \exp(i2\pi j/k)$, $\eta_{jk} = \exp(i(2j-1)\pi/k)$, $1 \leqslant j \leqslant k$. Then

$$\mathscr{M}E(\omega_{jk}, P) = \mathscr{M}E(\omega_{jk}, Q) = \mathscr{M}E(\eta_{jk}, P) = \mathscr{M}E(\eta_{jk}, Q) = \{i, -i\},$$
$$\mathscr{M}E(5\omega_{jk}, P) = \mathscr{M}E(5\omega_{jk}, Q) = \mathscr{M}E(\eta_{jk}/3, P)$$
$$= \mathscr{M}E(\eta_{jk}/3, Q) = \{1, \infty\}, \qquad 1 \leqslant j \leqslant k.$$

All in all $4k = 2m$ multisets have been used. Let us show that P and Q are irreducible polynomials. This can be easily proved by using the fact that the algebraic functions determined by the equations $P(z, w) = 0$ and $Q(z, w) = 0$ are expressed in radicals. After the substitution $w^k = u$, $P(z, w)$ will become a quadratic trinomial with discriminant equal to $16(16z^4 - 4z^3 + 37z^2 - 6z + 21)$. Having assumed that this is a square of a quadratic trinomial, we conclude that the discriminant must have the form $16(4z^2 + az + \sqrt{21})^2$. Comparing the coefficients at z^3 and z from both expressions, we arrive at a contradiction. It is clear now that the equation $P(z, w) = 0$ defines a complete analytic function, i.e., the polynomial $P(z, w)$ is irreducible. The irreducibility of the polynomial Q is proved in a similar way.

EXAMPLE 3. Let $m = 2k$, $n \geqslant 3$. Let

$$P(z, w) = (-3z^n + 2z^{n-1} - 2z^{n-2} - 3)w^{2k}$$
$$+ 14(z^n + 1)w^k + (5z^n - 2z^{n-1} + 2z^{n-2} + 5),$$
$$Q(z, w) = (-3z^n - 2z^{n-1} + 2z^{n-2} - 3)w^{2k}$$
$$+ 14(z^n + 1)w^k + (5z^n + 2z^{n-1} - 2z^{n-2} + 5).$$

Let ω_{jk} and η_{jk} mean the same as in the previous example. Then ($1 \leqslant j \leqslant k$)

$$\mathscr{M}E(\omega_{jk}, P) = \mathscr{M}E(\omega_{jk}, Q) = \mathscr{M}E(\eta_{jk}, P)$$
$$= \mathscr{M}E(\eta_{jk}, Q) = \{\eta_{sn} : 1 \leqslant s \leqslant n\},$$
$$\mathscr{M}E(\eta_{jk}/3, P) = \mathscr{M}E(\eta_{jk}/3, Q) = \mathscr{M}E(5\omega_{jk}, P)$$
$$= \mathscr{M}E(5\omega_{jk}, Q) = \{0, \dots, 0, 1, \infty\}.$$

Here we come across $4k = 2m$ multisets $\mathscr{M}E(\mathscr{A}_j)$. The irreducibility is analyzed as in Example 2. Indeed, after the substitution $w^k = u$, we obtain quadratic trinomials whose discriminants are respectively equal to

$$16(16z^{2n} - 4z^{2n-1} + 5z^{2n-2} - 2z^{2n-3} + z^{2n-4} + 32z^n - 4z^{n-1} + 4z^{n-2} + 16),$$
$$16(16z^{2n} + 4z^{2n-1} - 3z^{2n-2} - 2z^{2n-3} + z^{2n-4} + 32z^n + 4z^{n-1} - 4z^{n-2} + 16).$$

If one of these trinomials is reducible, then the corresponding discriminant must be equal to $16(4z^n + az^{n-1} + bz^{n-2} + 4)^2$. Comparing the coefficients at z^{2n-1}, z^{2n-2}, z^{n-2}, we arrive at a contradiction.

EXAMPLE 4. Let $m = 2$, $n = 3$. Then $L = 7$. We show that six sets $E(\mathscr{A}_j)$ do not always determine an algebraic function uniquely. Let

$$\begin{aligned} P(z, w) = {} & \{20z^3 - (63 + 5\sqrt{3}i)z^2 + 16(3 + i\sqrt{3})z - 8(1 + i\sqrt{3})\}w^2 \\ & - \{20z^3 - (207 + 5\sqrt{3}i)z^2 + 64(3 + i\sqrt{3})z - 32(1 + i\sqrt{3})\}, \\ Q(z, w) = {} & \{80z^3 - 40(3 + i\sqrt{3})z^2 + (30 + 62\sqrt{3}i)z + 16(1 - i\sqrt{3})\}w^2 \\ & - \{80z^3 - 40(3 + i\sqrt{3})z^2 + (30 + 158\sqrt{3}i)z + 64(1 - i\sqrt{3})\}. \end{aligned}$$

An elementary check gives

$$\begin{aligned} \mathscr{M}E(\pm 1, P) &= \{(3 + i\sqrt{3})/6, (3 + i\sqrt{3})/6, \infty\}, \\ \mathscr{M}E(\pm 2, P) &= \{0, 0, (3 + i\sqrt{3})/4\}, \\ \mathscr{M}E(\pm 3, P) &= \{1, 1, (1 + i\sqrt{3})/4\}, \\ \mathscr{M}E(\pm 1, Q) &= \{(3 + i\sqrt{3})/6, \infty, \infty\}, \\ \mathscr{M}E(\pm 2, Q) &= \{0, (3 + i\sqrt{3})/4, (3 + i\sqrt{3})/4\}, \\ \mathscr{M}E(\pm 3, Q) &= \{1, (1 + i\sqrt{3})/4, (1 + i\sqrt{3})/4\}. \end{aligned}$$

Therefore,

$$E(\pm 1, P) = E(\pm 1, Q), \quad E(\pm 2, P) = E(\pm 2, Q), \quad E(\pm 3, P) = E(\pm 3, Q).$$

The irreducibility of the polynomials P and Q follows from the fact that the algebraic functions that they determine can be presented as the square roots of the quotient of two irreducible polynomials of odd degree.

The algebraic function determined by equation (1) with $p_m(z) \equiv 1$ will be called an *entire algebraic function*. In other words, an entire algebraic function is an algebraic function whose only pole is at $z = \infty$.

THEOREM 2. *$2m - 1$ multisets $\mathscr{M}E(\mathscr{A}_j)$, $\mathscr{A}_j \neq \infty$, $1 \leqslant j \leqslant 2m - 1$, and one set $E(\mathscr{A}_{2m})$, $\mathscr{A}_{2m} \neq \infty$ are sufficient to uniquely determine an m-valued entire algebraic function. In order to uniquely determine an m-valued entire algebraic function whose inverse function is n-valued, $L_1 = 4m - 1 + [-(2m - 1)/n]$ sets $E(\mathscr{A}_j)$ are sufficient with $\mathscr{A}_j \neq \infty$, $1 \leqslant j \leqslant L_1$.*

PROOF. The first assertion of Theorem 2 is an immediate corollary of the first assertion of Theorem 1. Indeed, under the assumptions of Theorem 2,

$\mathcal{M}E(\infty) = \{\infty, \infty, \dots, \infty\}$, i.e., actually one additional multiset is given. The second assertion of Theorem 2 is proved as the corresponding assertion of Theorem 1 with the following corrections. The resultant $R(P, \widehat{P})$ is a determinant of order $2m$ whose first column contains units in the first and $(m+1)$th rows, the remaining elements being zeros. Other columns contain either zeros or $p_j(z)$ or $\widehat{p}_j(z)$. Thus the resultant is a polynomial of degree not exceeding $n(2m-1)$.

The sum of orders of algebraic branching points of the Riemann surface $\mathcal{R}_w$ does not exceed, as before, $m(2n-2)$, but an algebraic branching point of order $n-1$ lies above $w = \infty$, so the sum of orders of algebraic branching points of $\mathcal{R}_w$ lying above the finite plane does not exceed

$$m(2n-2) - (n-1) = (2m-1)(n-1).$$

As in the proof of Theorem 1, we obtain

$$n(2m-1) \geqslant nL_1 - (2m-1)(n-1),$$

and as before, we arrive at a contradiction.

REMARK. It is easy to check that $L - L_1 \leqslant 2$, and the equality can be achieved (e.g., when $n = 2$).

EXAMPLE 5. For $m = 1$, we have $L_1 = 2$. This is the well-known unimprovable theorem on uniqueness for polynomials (see, e.g., [2]).

EXAMPLE 6. Let $m = 2$, $n = 1$. Then $L_1 = 4$. Three sets $E(\mathcal{A}_j)$, $\mathcal{A}_j \neq \infty$, $1 \leqslant j \leqslant 3$, do not, generally speaking, determine an entire algebraic function uniquely. Let

$$P(z, w) = w^2 + 2(z+1)w - (z+2),$$
$$\widehat{P}(z, w) = w^2 + 3(z+1)w - (z+2).$$

In this case

$$E(0, P) = \mathcal{M}E(0, P) = E(0, \widehat{P}) = \mathcal{M}E(0, \widehat{P}) = \{-2\},$$
$$E(\pm 1, P) = \mathcal{M}E(\pm 1, P) = E(\pm 1, \widehat{P}) = \mathcal{M}E(\pm 1, \widehat{P}) = \{-1\}.$$

The polynomials P and $\widehat{P}$ are obviously irreducible.

REFERENCES

1. V. I. Baranov and B. S. Stechkin, *Extremal combinatorial problems and their applications*, "Nauka", Moscow, 1989. (Russian)

2. W. W. Adams and E. G. Straus, *Non-Archimedian analytic functions taking the same values at the same points*, Illinois J. Math. **15** (1971), no. 3, 418–424.

3. A. K. Pizer, *A problem on rational functions*, Amer. Math. Monthly **80** (1973), no. 5, 552–553.

4. B. L. Van der Waerden, *Algebra* I, Springer-Verlag, Berlin, Heidelberg, and New York, 1967.

LVOV STATE UNIVERSITY, UL. UNIVERSITETSKAYA 1, LVOV 290000 UKRAINE

Translated by V. A. TKACHENKO

ADVANCES IN SOVIET MATHEMATICS
Volume 11, 1992

Subharmonic Functions and $\cos(\pi\lambda)$-Theorems for Entire Functions Represented by Gap Series

A. E. FRYNTOV

Let $f(z)$ be an entire function of finite order ρ. We define its characteristics

$$M(r, f) = \max_{|z|=r} |f(z)|, \qquad \mu(r, f) = \min_{|z|=r} |f(z)|.$$

The classical theorem of Wiman [1] and Valiron [2] states that if $\rho < 1$ then

$$\limsup_{r\to\infty} \frac{\ln \mu(r, f)}{\ln M(r, f)} \geqslant \cos(\pi\rho).$$

At the present time there are many papers related to this theorem. We do not cite these works here, but the reader can find a survey of them in the book by Essen [3]. We only mention the paper of Kjellberg [4] in which the analog of the theorem of Wiman and Valiron for entire functions of finite lower order was proved.

In order to characterize a subharmonic function $u(z)$, we shall use the following characteristics

$$B(r, u) = \max_{|z|=r} u(z), \qquad L(r, u) = \min_{|z|=r} u(z).$$

Let $u(z) = \ln|f(z)|$, where $f(z)$ is an entire function. The characteristics of the functions $u(z)$ and $f(z)$ are related by the following equalities

$$B(r, u) = \ln M(r, f), \qquad L(r, u) = \ln \mu(r, f).$$

Now we mention an analog of the theorem of Wiman and Valiron for subharmonic functions on the complex plane.

THEOREM 0. *Let $u(z)$ be a subharmonic function in the complex plane $\mathbb{C}$ which satisfies the following conditions*

1) $0 \leqslant B(r, u) \leqslant r^{\lambda}$, $\forall r \in (0, \infty)$;
2) $B(1, u) = 1$.

1991 *Mathematics Subject Classification*. Primary 30D20, 31A02.

If $\lambda < 1$, *then*

$$\sup_{r>0} \frac{L(r, u)}{B(r, u)} \geqslant \cos(\pi\lambda).$$

One can find the proof of this theorem, for example, in [10].

The aim of the present paper is to refine the relation between $\mu(r, f)$ and $M(r, f)$ in Wiman's and Valiron's theorem in the case when the expansion of the entire function $f(z)$ into power series has sufficiently many zero terms.

We shall say that an entire function $f(z)$ is represented by a *gap series of density* Δ if the following expansion is valid:

$$f(z) = \sum_{k \in A} a_k z^k,$$

where A is a subset of nonnegative integers with maximal density Δ. Note that the maximal density of A is the greatest lower bound of densities of sets B that contain A and have the usual density. Note also that a set B has the density iff the following limit exists: $\mathrm{dens}(B) = \lim_{r\to\infty} n_B(r)/r$. In the last relation, $n_B(r)$ is the number of points belonging to $B \cap [0, r]$. In this case, $\mathrm{dens}(B)$ is called the density of B. In other words, the *maximal density* of a set A is $\mathrm{dens}^*(A) = \inf\{\mathrm{dens}(B) : A \subseteq B\}$. (The maximal density and the upper density should not be confused. The upper density of a set A is defined by $\lim_{r\to\infty} \sup n_A(r)/r$ and may differ from the maximal density. In this paper the notion of upper density is not used.)

The following example allows us to hope that the theorem of Wiman and Valiron may be refined for gap series.

Let $f(z)$ be an entire function which has the representation $f(z) = g(z^d)$, where $g(z)$ is an entire function. If the lower order of $f(z)$ is equal to λ, then the lower order of $g(z)$ is equal to λ/d. Since $\lambda/d < 1$, we can apply the Kjellberg theorem to the function $g(z)$. It follows from this theorem that

$$\limsup_{r\to\infty} \frac{\ln \mu(r, f)}{\ln M(r, f)} \geqslant \cos(\pi\lambda/d).$$

Note that the function $f(z)$ may be represented by a gap series with density $\Delta = 1/d$. We can ask the following question. Suppose that the values λ and Δ are connected by the relation $\lambda\Delta < 1$. Is it true that

$$\limsup_{r\to\infty} \frac{\ln \mu(r, f)}{\ln M(r, f)} \geqslant \cos(\pi\lambda\Delta) \tag{1}$$

for an arbitrary entire function represented by a gap power series with density Δ and lower order λ?[1]

This conjecture was stated in the problem book by Hayman [6]. A short review of this problem can be found in [6]. The results of our paper allow us to solve some problems which are connected with the asymptotic relations

[1] This question was asked by Kövari. A survey of problems related to this question may be found in the paper [11].

between the characteristics $M(r, f)$ and $\mu(r, f)$ for entire functions represented by gap power series. These questions are connected with the analogs of Theorem 0 for a special class of subharmonic functions. In particular, this special class contains the limit functions of the family of subharmonic functions

$$\ln|f(rz)|/A \tag{2}$$

as $A \to \infty$ and $r \to \infty$, where $f(z)$ is a function represented by a gap power series.

In this work we also construct an example of a subharmonic function which shows that if we slightly extend the class of limit functions of the family (2), then the conjecture (1) may fail. Further on we shall define this class exactly (it will be denoted by $\mathscr{A}(\Delta)$). It seems plausible that the class $\mathscr{A}(\Delta)$ contains only the limit functions represented by (2). Now we pass to the main results of this paper.

§1. Subharmonic function of classes $\mathscr{A}(\Delta)$ and $\mathscr{B}(\Delta)$

The main tool for asymptotic estimations in this paper is the following theorem.

THEOREM 1. *Let $u(z)$ be a subharmonic function in the domain*

$$D = \{z : |\arg(z)| < \varphi_0,\ |z| \in (r_1, r_2)\}, \qquad \varphi_0 \leqslant \pi.$$

Then $\widetilde{u}(z)$, $\widetilde{u}(z) = \widetilde{u}(re^{i\theta}) = \sup_{|\varphi|\leqslant|\theta|} u(re^{i\varphi})$, is subharmonic in D. If the function $u(z)$ is subharmonic in the annulus $\Pi = \{z : |z| \in (r_1, r_2)\}$, then $\widetilde{u}(z)$ is subharmonic in this annulus as well.

PROOF. First let us suppose that the function $u(z)$ is continuous in the domain D and show that in this case $\widetilde{u}(z)$ is a continuous subharmonic function in the domain $D^+ = \{z : 0 < \arg(z) < \varphi_0,\ |z| \in (r_1, r_2)\}$. Indeed, the inequality

$$\begin{aligned}\widetilde{u}(r_1e^{i\theta_1}) - \widetilde{u}(r_2e^{i\theta_2}) &= \sup_{|\varphi|\leqslant|\theta_1|} u(r_1e^{i\varphi}) - \sup_{|\varphi|\leqslant|\theta_2|} u(r_2e^{i\varphi})\\ &= u(r_1e^{i\varphi_1}) - \sup_{|\varphi|\leqslant|\theta_2|} u(r_2e^{i\varphi})\\ &\leqslant u(r_1e^{i\varphi_1}) - u(r_2e^{i\varphi_2}),\end{aligned}$$

where φ_1 and φ_2 satisfy the condition $|\varphi_1 - \varphi_2| \leqslant |\theta_1 - \theta_2|$, implies the continuity of the function $\widetilde{u}(z)$ in the domain D. In order to prove the subharmonicity of $\widetilde{u}(z)$ in D^+, we shall show that for every $z \in D^+$ there exists a positive number $\rho(z)$ such that for an arbitrary $\rho \in (0, \rho(z))$ the following inequality is true

$$\frac{1}{2\pi}\int_0^{2\pi} \widetilde{u}(z + \rho^{i\psi})\, d\psi \geqslant \widetilde{u}(z).$$

Now we write an identity which is valid for every measurable function $g(z)$ on the circle $\{z = re^{i\omega} + \rho e^{i\psi} : \psi \in [0, 2\pi)\}$:

$$\begin{aligned}
&\int_0^{2\pi} g(re^{i\omega} + \rho e^{i\psi})\,d\psi \\
&\quad = \int_0^{\pi} g(re^{i\omega} + \rho e^{i(\omega+\psi)})d\psi + \int_0^{\pi} g(re^{i\omega} + \rho e^{i(\omega-\psi)})\,d\psi \\
&\quad = \int_0^{\pi} \{g(r(\psi)e^{i\omega+\alpha(\psi)}) + g(r(\psi)e^{i\omega-\alpha(\psi)})\}\,d\psi,
\end{aligned}$$

where $r(\psi) = |r + \rho e^{i\psi}|$, $\alpha(\psi) = \arg(r + \rho e^{i\psi})$. (We note that $\alpha(\psi)$ tends to zero uniformly as $\rho \to 0$.) It follows from the subharmonicity of $u(z)$ in D that

$$\begin{aligned}
u(re^{i\varphi}) &\leqslant \frac{1}{2\pi}\int_0^{2\pi} u(re^{i\varphi} + \rho e^{i\psi})d\psi \\
&= \frac{1}{2\pi}\int_0^{\pi} \{u(r(\psi)e^{i\varphi+\alpha(\psi)}) + u(r(\psi)e^{i\varphi-\alpha(\psi)})\}d\psi.
\end{aligned}$$

This inequality holds for all values $\rho > 0$ such that $r(\psi)e^{i\varphi+\alpha(\psi)}$ and $r(\psi)e^{i\varphi-\alpha(\psi)}$ belong to D for every ψ contained in $[0, 2\pi)$. Hence, if $z = re^{i\theta} \in D$, then

$$\widetilde{u}(re^{i\theta}) = \sup_{|\varphi|\leqslant|\theta|} u(re^{i\varphi}) \leqslant \frac{1}{2\pi}\int_0^{\pi} \sup_{|\varphi|\leqslant\theta} \{u(r(\psi)e^{i\varphi+\alpha(\psi)}) + u(r(\psi)e^{i\varphi-\alpha(\psi)})\}d\psi.$$

Since

$$\min\{|\varphi-\alpha|, |\varphi+\alpha|\} \leqslant \theta - \alpha, \qquad \max\{|\varphi-\alpha|, |\varphi+\alpha|\} \leqslant \theta + \alpha$$

for every φ and α such that $|\varphi| < \theta$ and $|\alpha| < \theta$, we have

$$\begin{aligned}
&\sup_{|\varphi|\leqslant\theta} \{u(r(\psi)e^{i\varphi+\alpha(\psi)}) + u(r(\psi)e^{i\varphi-\alpha(\psi)})\} \\
&\qquad\qquad \leqslant \sup_{|\varphi|\leqslant\theta+\alpha} u(r(\psi)e^{i\varphi}) + \sup_{|\varphi|\leqslant\theta-\alpha} u(r(\psi)e^{i\varphi})
\end{aligned}$$

for every $|\alpha| < \theta$. If $\rho \to 0$, then $\max|\alpha(\psi)| \to 0$; hence, there is a positive number $\rho(re^{i\theta})$ such that for all $\rho \leqslant \rho(re^{i\theta})$ the following inequality is true

$$\begin{aligned}
\widetilde{u}(re^{i\theta}) &\leqslant \frac{1}{2\pi}\int_0^{\pi} \{\widetilde{u}(r(\psi)e^{i\theta+\alpha(\psi)}) + \widetilde{u}(r(\psi)e^{i\theta-\alpha(\psi)})\}d\psi \\
&= \frac{1}{2\pi}\int_0^{2\pi} \widetilde{u}(re^{i\theta} + \rho e^{i\psi})d\psi.
\end{aligned}$$

We have shown that the function $\widetilde{u}(z)$ is subharmonic and continuous in the domain D^+ if the function $u(z)$ is subharmonic and continuous in D. Now let us show that the function $u(\widetilde{z})$ is subharmonic in the domain D. Indeed, $\widetilde{u}(z)$ is subharmonic in the domain

$$D^- = \{z : 0 > \arg(z) > -\varphi_0,\ |z| \in (r_1, r_2)\}$$

because $\widetilde{u}(z) = \widetilde{u}(\overline{z})$ and for $\arg(z) = 0$ we have the inequality

$$\widetilde{u}(r) = u(r) \leqslant \frac{1}{2\pi}\int_0^{2\pi} u(r+\rho e^{i\psi})d\psi \leqslant \frac{1}{2\pi}\int_0^{2\pi} \widetilde{u}(r+\rho e^{i\psi})d\psi, \qquad \rho < \rho(r).$$

If the function $u(z)$ is continuous and subharmonic in the annulus Π, then, as was proved above, the function $\widetilde{u}(z)$ is subharmonic in the domain $\Pi \setminus (-r_1, -r_2)$ and, obviously, is continuous in the whole set Π. We shall show that $\widetilde{u}(z)$ will be a subharmonic function in the whole annulus Π.

To prove this, we define the function $u_\varepsilon(z)$ for $\varepsilon > 0$ by the equation $u_\varepsilon(z) = u(z^{1+\varepsilon})$ in the domain $\{z : |\arg(z)| < \pi,\ |z| \in (r_1^{1/(1+\varepsilon)}, r_2^{1/(1+\varepsilon)})\}$. The function $\widetilde{u}_\varepsilon(z)$ is subharmonic in this domain and, moreover, the equality

$$\widetilde{u}_\varepsilon(z) = \max_{|\zeta|=|z^{1+\varepsilon}|} u(\zeta) = B(|z|^{1+\varepsilon}, u).$$

is valid in the domain

$$\{z : |\arg(z) - \pi| < \pi(1 - 1/(1+\varepsilon)),\ |z| \in (r_1^{1/(1+\varepsilon)}, r_2^{1/(1+\varepsilon)})\}.$$

Hence the function $u_\varepsilon(z)$ is subharmonic in the whole annulus

$$\{z : |z| \in (r_1^{1/(1+\varepsilon)}, r_2^{1/(1+\varepsilon)})\}.$$

It now follows from the uniformly compact convergence $u_\varepsilon(z)$ to $u(z)$ as $\varepsilon \to 0$ on $\Pi = \{z : |z| \in (r_1, r_2)\}$ that the function $u(z)$ is subharmonic and continuous in Π.

Now let us get rid of the assumption of continuity of the function $u(z)$ in the domain D. We consider the function

$$(A_\delta u)(z) = \frac{1}{\pi\delta^2}\iint_{|\xi|\leqslant\delta} u(z+\xi)\,d\xi d\overline{\xi}.$$

This function is continuous and subharmonic in the ring

$$\{|z| \in (r_1+\delta, r_2-\delta)\},$$

and if $\delta \searrow 0$, then

$$\left(A_\delta u\right)(z) \searrow u(z). \tag{3}$$

If we showed that $(\widetilde{A_\delta u})(z)$ increases and tends to $\widetilde{u}(z)$ as $\delta \searrow 0$ for every fixed z, then the proof of Theorem 1 would be completed. Indeed, the limit of an increasing sequence of subharmonic functions is a subharmonic function. Obviously, for every fixed z, the function $(\widetilde{A_\delta u})(z)$ is an increasing function with respect to δ because the function $(A_\delta u)(z)$ has the same property. We have $(z = re^{i\theta})$

$$\begin{aligned} 0 \leqslant (\widetilde{A_\delta u})(z) - \widetilde{u}(z) &= \sup_{|\varphi|\leqslant\theta} (A_\delta u)(re^{i\varphi}) - \sup_{|\varphi|\leqslant\theta} u(re^{i\varphi}) \\ &\leqslant \sup_{|\varphi|\leqslant\theta,\,|\zeta|\leqslant\delta} u(re^{i\varphi}+\zeta) - \sup_{|\varphi|\leqslant\theta} u(re^{i\varphi}). \end{aligned}$$

Since the function $u(z)$ is upper semicontinuous, $\widetilde{(A_\delta u)}(z)$ tends to $\widetilde{u}(z)$ for every fixed z. Theorem 1 is proved.

Now we shall define the classes $\mathscr{A}(\Delta)$ and $\mathscr{B}(\Delta)$. We say that a subharmonic function $u(z)$ belongs to the class $\mathscr{A}(\Delta)$, if

$$B(r, u) = \sup u(re^{i\varphi}) = B_I(r, u) = \sup_{\varphi\in I} u(re^{i\varphi})$$

for every interval I of length $2\pi\Delta$. We define $\mathscr{B}(\Delta)$ as the set of subharmonic functions which satisfy the following condition

$$B(r, u) = \sup_{\varphi} u(re^{i\varphi}) = \sup_{\varphi\in[-\pi\Delta, \pi\Delta]} u(re^{i\varphi}).$$

These classes of subharmonic functions are interesting due to the following properties:

a) the set of limit functions of the family $A^{-1}\ln|f(rz)|$ as $A\to\infty$ and $r\to\infty$, where f is an entire function represented by a gap series with maximal density Δ of indices, is contained in $\mathscr{A}(\Delta)$;

b) the set of limit functions of the family $A^{-1}\ln|f(rz)|$ as $A\to\infty$ and $r\to\infty$, where f is a real entire function with maximal density Δ of changes of signs in its power expansion, is contained in $\mathscr{B}(\Delta)$.

These properties will be proved soon, but first we shall prove some theorems related to these classes. According to the definitions of $\mathscr{A}(\Delta)$ and $\mathscr{B}(\Delta)$, we have $\mathscr{A}(\Delta)\subseteq\mathscr{B}(\Delta)$. The following theorems are analogs of the $\cos\pi\lambda$-theorem for subharmonic functions of the classes $\mathscr{A}(\Delta)$ and $\mathscr{B}(\Delta)$.

THEOREM 2. *Let* $u(z)\in\mathscr{B}(\Delta)$, $u\not\equiv\text{const}$, *and moreover,*

1) $0 < B(r, u)\leqslant r^\lambda$, $\forall r\in(0,\infty)$;

2) $B(1, u) = 1$.

If $\lambda\Delta < 1$, *then*

$$\sup_{r>0}\frac{u(r)}{B(r, u)}\geqslant\cos(\pi\lambda\Delta).$$

PROOF OF THEOREM 2. We consider the function $\widetilde{u}(z)$ in the domain $\mathbb{C}\setminus\{0\}$. According to Theorem 1, this function is subharmonic. Let

$$\widetilde{u}(0) = \limsup_{z\to 0} u(z).$$

Since $\widetilde{u}(z) < \infty$, this function is subharmonic in the whole complex plane $\mathbb{C}$. If $\Delta = 1$, then the conclusion of Theorem 2 follows from Theorem 0. If $\Delta < 1$, then for every $\varepsilon > 0$ such that $\Delta+\varepsilon < 1$, the function $v_\varepsilon(z) = \widetilde{u}(z^{\Delta+\varepsilon})$ is subharmonic in the domain $\{|\arg(z)|\neq\pi\}$. The definition of $\mathscr{B}(\Delta)$ implies

$$\widetilde{u}(z) = B(|z|, u)$$

for $|\arg(z)| > \pi\Delta$; hence, the function $v_\varepsilon(z)$ is subharmonic in the whole complex plane. Since $\varepsilon > 0$ is arbitrary, we conclude that the function

$v_0(z) = \lim_{\varepsilon\to 0} v_\varepsilon(z)$ is subharmonic in the whole complex plane too. Hence we have the following properties of $v_0(z)$:

1) $0 \leqslant B(r, v_0) \leqslant r^{\lambda\Delta}$;
2) $B(1, v_0) = 1$.

Since $B(r, v_0) = B(r^\Delta, u)$, $v_0(r) = u(r^\Delta)$, Theorem 0 implies the inequality

$$\sup_{r>0} \frac{u(r)}{B(r, u)} \geqslant \cos(\pi\lambda\Delta).$$

Theorem 2 is proved.

THEOREM 3. *Let $u(z)$ be a subharmonic function belonging to $\mathscr{A}(\Delta)$ and the following conditions be fulfilled*

1) $0 \leqslant B(r, u) \leqslant r^\lambda$, $\forall r \in (0, \infty)$;
2) $B(1, u) = 1$.

If $\lambda(\Delta + \delta) < 1$, $\delta \geqslant 0$, then

$$\sup_{r>0} \frac{L_\delta(r, u)}{B(r, u)} \geqslant \cos(\pi\lambda(\Delta + \delta)),$$

where $L_\delta(r, u) = \inf_{|\varphi|\leqslant\delta} u(re^{i\varphi})$.

PROOF OF THEOREM 3. Suppose that $\Delta + \delta < 1$; otherwise the conclusion of the theorem is a simple corollary of Theorem 0. We construct the function

$$w(z) = \begin{cases} u(z), & |\arg(z)| \leqslant \delta, \\ w_\delta(z), & \delta < \arg(z) \leqslant \pi, \\ w_{-\delta}(z), & -\pi < \arg(z) \leqslant -\delta, \end{cases}$$

where $w_{\pm\delta}(re^{i\theta\pm i\delta}) = \sup_{|\varphi\pm\delta|\leqslant\theta} u(re^{i\varphi})$. Theorem 1 implies that the function $w(z)$ is subharmonic for $\{\arg(z) = \pm\delta\}$ and for $\{\arg(z) = -\pi\}$. We shall show that this function is subharmonic in the whole complex plane $\mathbb{C}$. Indeed, its subharmonicity follows from that of the function $w(z)$ inside the angle $\{|\arg(z) - \pi| \leqslant \pi(1 - \Delta - \delta)\}$ and from the relation

$$w(z) = B(|z|, u), \qquad \forall z \in \{|\arg(z) - \pi| \leqslant \pi(1 - \Delta - \delta)\}$$

outside this angle. Subharmonicity on $\{\arg(z) = \pm\delta\}$ follows from that of $u(z)$ on this set and from the inequality $w_\delta(z) \geqslant u(z)$.

Now, as in the proof of Theorem 2, we construct the functions

$$v_\varepsilon(z) = w(z^{\Delta+\delta+\varepsilon}), \qquad v_0(z) = \lim_{z\to 0} v_\varepsilon(z),$$

which are subharmonic in the whole complex plane $\mathbb{C}$ too. The function $v_0(z)$ satisfies the following relations:

1) $0 \leqslant B(r, v_0) \leqslant r^{\lambda(\Delta+\delta)}$;
2) $B(1, v_0) = 1$.

Moreover, $B(r, v_0) = B(r^{\Delta+\delta}, u)$, $L(r, v_0) = L_\delta(r^{\Delta+\delta}, u)$. It follows from Theorem 0 that

$$\sup_{r>0} \frac{L_\delta(r, u)}{B(r, u)} \geqslant \cos(\pi\lambda(\Delta+\delta)).$$

Theorem 3 is proved.

§2. Limit sets of entire functions represented by gap series and limit sets of real functions having power expansions with a small number of sign changes. Relationship between limit sets of such functions and the classes $\mathscr{A}(\Delta)$ and $\mathscr{B}(\Delta)$

We shall find relationships between the classes $\mathscr{A}(\Delta)$ and $\mathscr{B}(\Delta)$ on the one hand and entire functions representable by power gap series on the other hand.

Let $f(z)$ be an entire function. We say that a subharmonic function $u(z)$ belongs to the limit set of $\ln|f(z)|$ if there exist sequences $r_k \to +\infty$ and $A_k \to +\infty$ such that $u(z) = \lim_{k\to\infty} \ln|f(r_k z)|/A_k$. Here the limit is understood in the sense of the space of distributions $\mathscr{D}'(\mathbb{C})$, the space $\mathscr{D}(\mathbb{C})$ being the space of infinitely differentiable functions with compact supports in $\mathbb{C}$.

Now we mention an important property of convergence of subharmonic functions in the topology of $\mathscr{D}'(\mathbb{C})$, widely used in what follows. Its proof can be found, for example, in the paper by Eremenko, Sodin, and Shea [5].

This property is the following. If $v_n(z) \to v(z)$ in $\mathscr{D}'(\Omega)$, and the circle $R = \{\zeta \in \mathbb{C} : |\zeta - z| = \rho\}$ is contained in Ω, then $v_n(z) \to v(z)$ in the topology of $L^1(R)$. In particular, the convergence $v_n(z) \to v(z)$ in the topology of $L^1(R)$ and semicontinuity of subharmonic functions imply that

$$\max_{z\in R} v_n(z) \to \max_{z\in R} v(z).$$

The connection between limit sets of entire functions which are represented by gap series with density Δ and the class $\mathscr{A}(\Delta)$ is established by the following theorem.

THEOREM 4. *If the entire function $f(z)$ is represented by a power series with the density Δ^* of nonzero indices, then every limit function of the family $A^{-1}\ln|f(rz)|$, $A \to +\infty$, $r \to +\infty$, belongs to $\mathscr{A}(\Delta^*)$.*

PROOF OF THEOREM 4. Let

$$f(z) = c_0 + \sum_{k\in A} c_k z^k.$$

Without loss of generality we can assume what the set A has the usual density $\Delta > \Delta^*$. In this case we shall show that every limit function of the family $A^{-1}\ln|f(rz)|$, $A \to +\infty$, $r \to +\infty$, belongs to $\mathscr{A}(\Delta)$.

We consider the function

$$\phi(z) = z \prod_{k \in A} (1 - z^2/k^2).$$

This function is of exponential type. We shall make use of some results from the monograph by B. Ya. Levin [7]. The function $\phi(z)$ has completely regular growth in the sense of Levin-Pfluger, and its indicator diagram is the interval of the imaginary axis $[-i\pi\Delta, i\pi\Delta]$. Zeros of this function form an R-set, because the disks $\{z : |z-k| < 1/2\}$ do not intersect for distinct $k \in A$. It follows from a theorem due to B. Ya. Levin [7, p. 125] that outside the set

$$R = \bigcup_{k \in A} \{|z-k| < 1/2\}$$

there exists the limit

$$\lim_{\substack{r \to \infty \\ re^{i\psi} \notin R}} \ln|\phi(re^{i\psi})|/r = \pi\Delta \sin|\psi| \tag{4}$$

uniformly with respect to ψ. In particular, it follows now that

$$\lim_{\substack{k \to \infty \\ k \in A}} \frac{\theta^k}{|\phi'(k)|} = 0, \qquad \forall\theta, \ |\theta| < 1.$$

Hence for all θ, $|\theta| < 1$, the function

$$\phi(z, \theta) = \sum_{k \in A} \frac{\phi(z)\theta^k}{(z-k)\phi'(k)}$$

is a function of exponential type with the indicator diagram contained in $[-\pi\Delta, \pi\Delta]$. In other words,

$$\limsup_{|z| \to \infty} \ln|\phi(z, \theta)|/|z| \leqslant \pi\Delta \sin|\arg(z)|.$$

For every fixed θ, we denote by $\varphi(\zeta, \theta)$ the Borel transform of the function $\phi(z, \theta)$. It is represented by

$$\varphi(\zeta, \theta) = \int_0^\infty \phi(z, \theta) e^{-\zeta z}\, dz,$$

where the integral is taken over an arbitrary ray for which $\operatorname{Re}\zeta z > 0$. According to the inversion formula for the Borel transform, we have

$$\phi(z, \theta) = \frac{1}{2\pi i} \int_\Omega \varphi(\zeta, \theta) e^{\zeta z}\, d\zeta,$$

where Ω is any closed path containing the indicator diagram of $\phi(z, \theta)$. (Note that the indicator diagram of $\phi(z, \theta)$ is the interval of the imaginary axis $[-i\pi\Delta, i\pi\Delta]$.) From the definition of the function $\phi(z, \theta)$ we have

$$\theta^k = \phi(k, \theta) = \frac{1}{2\pi i} \int_\Omega \varphi(\zeta, \theta) e^{\zeta k}\, d\zeta, \qquad k \in A.$$

In this way we obtain

$$\sum_{k\in A} c_k(z\theta)^k = \frac{1}{2\pi i}\int_\Omega \varphi(\zeta, \theta)\sum_{k\in A} c_k z^k e^{k\zeta}\, d\zeta$$

or, equivalently,

$$f(z\theta) = \frac{1}{2\pi i}\int_\Omega \varphi(\zeta, \theta) f(ze^\zeta)\, d\zeta.$$

Using the latter relation, we have the estimate

$$|f(z\theta)| \leqslant C(\theta, \Omega)\max_{\zeta\in\Omega} \ln|f(ze^\zeta)|,$$

where $C(\theta, \Omega) = \max_{\zeta\in\Omega}|\varphi(\zeta, \theta)|$, hence

$$\ln|f(z\theta)| \leqslant \ln C(\theta, \Omega) + \max_{\zeta\in\Omega}\ln|f(ze^\zeta)|. \tag{5}$$

Using $\lim \ln|f(r_k z)|/A_k = u(z)$ we obtain the following estimate from inequality (5)

$$u(\theta z) \leqslant \max_{\zeta\in\Omega} u(ze^\zeta)$$

for every θ, $|\theta| < 1$, and hence for every φ_0 we have

$$\max_{\varphi\in[-\pi, \pi]} u(|\theta| re^{i\varphi}) \leqslant \max_{\zeta\in\Omega} u(re^{i\varphi_0+\zeta}), \qquad z = re^{i\varphi_0}.$$

Since the last inequality holds for every $|\theta| < 1$ and every Ω which contains the interval $[-i\pi\Delta, i\pi\Delta]$, we can choose Ω coinciding with the boundary of the δ-neighborhood of $[-i\pi\Delta, i\pi\Delta]$. Taking the limit as δ tends to zero and using the lower semicontinuity of $u(z)$, we get

$$B(|\theta| r, u) \leqslant \max_{|\varphi|\leqslant\pi\Delta} u(re^{i\varphi_0+\varphi}).$$

Now sending $|\theta|$ to 1 and using the continuity of $B(r, u)$ with respect to r, we obtain

$$B(r, u) \leqslant \max_{|\varphi|\leqslant\pi\Delta} u(re^{i\varphi_0+\varphi}) = B_I(r, u)$$

for every interval I such that $|I| = 2\pi\Delta$. The theorem is proved.

Thus we have established the required connection between entire functions represented by gap series and subharmonic functions belonging to the class $\mathscr{A}(\Delta)$. The following theorem gives us a similar connection between entire functions represented by a real power series with few changes of signs and subharmonic functions belonging to $\mathscr{B}(\Delta)$. Now we formulate an exact definition.

Let $f(z)$ be an entire function represented by the series

$$f(z) = c_0 + \sum_{k\in\mathbb{N}} c_k z^k,$$

with the real coefficients c_k. A set $A \subseteq \mathbb{N}$ is said to be the *set of sign changes* of the sequence c_k, if for some continuous real function $h(x)$ which changes its sign only on A, the sequence $c_k h(k)$ does not change its sign for any integer k. In other words, either $c_k h(k) \geqslant 0$ for every $k \in \mathbb{N}$ or $c_k h(k) \leqslant 0$ for every $k \in \mathbb{N}$.

The following theorem characterizes the limit set for an entire function represented by a real power series whose set of sign changes has maximal density not larger than Δ.

THEOREM 5. *If an entire real function $f(z)$ is represented by a power series whose set of sign changes has the maximal density Δ^*, then every limit function of the family $A^{-1} \ln |f(rz)|$, as A and r tend to infinity, belongs to $\mathscr{B}(\Delta^*)$.*

PROOF OF THEOREM 5. Let

$$f(z) = c_0 + \sum_{k \in \mathbb{N}} c_k z^k.$$

According to the conditions of Theorem 5, the maximal density of A, the set of sign changes of c_k, is not larger than Δ^*. Hence for every $\Delta > \Delta^*$ one can find a set $B \subset \{\lambda + 1/2 : \lambda \in \mathbb{N}\}$ which has density Δ and contains A.

Now we shall construct a set C which will satisfy the conditions:

a) the set $C \subset \{\lambda \pm 1/3 : l \in \mathbb{N}\}$ is measurable with respect to density and its density is equal to Δ;

b) the function

$$\phi(z) = z \prod_{\lambda \in C} (1 - z^2/\lambda^2) \tag{6}$$

is such that the sequence $\phi(k)c_k$ does not change sign for $k \in \mathbb{N}$.

In order to construct the set C, we consider $B \backslash A$ and note that according to the inequality $\operatorname{dens}(B) > \Delta$ we have $\lim_{r\to\infty} n_{B\backslash A}(r) = +\infty$. Now we divide the set $B \setminus A$ into parts D_1 and D_2, where

$$D_1 = \bigcup \{D \subseteq B \setminus A : n_D(r) \leqslant (1/2) n_{B\backslash A}(r),\ \forall r > 0\},$$

and $D_2 = B \setminus A$. It is obvious that $n_{D_2}(r) - n_{D_1}(r)$ is bounded and hence the set C defined by $C = \{\lambda + 1/6 : \lambda \in A\} \cup \{\lambda + 1/6 : \lambda \in D_1\}$ has density Δ. Moreover, the sequence $\phi(k)c_k$ does not change its sign for $k \in \mathbb{N}$. The function $\phi(z)$ is a function of exponential type and has completely regular growth in the sense of B. Ya. Levin. The set C is an R-set because the disks $\{z : |z - \lambda| < 1/3\}$ do not intersect for different $\lambda \in C$. The indicator diagram of ϕ is contained in $[-i\pi\Delta, i\pi\Delta]$. Hence, according to the theorem of B. Ya. Levin [7, p. 125], outside the set

$$R = \bigcup_{\lambda \in C} \{|z - \lambda| < 1/3\},$$

the limit

$$\lim_{\substack{r\to\infty \\ re^{i\psi} \notin R}} \ln |re^{i\psi}|/r = \pi\Delta \sin|\psi| \tag{7}$$

exists uniformly with respect to ψ.

In particular, the last relation yields

$$\lim_{\substack{k\to\infty \\ k\in\mathbb{N}}} \frac{\theta^k}{|\phi(k)|} = 0, \qquad \forall\theta,\ \theta\in(0,1).$$

It follows from the previous relation that for every fixed $\theta\in(0,1)$ there exists a constant $C(\theta)>0$ such that for all $k\in\mathbb{N}$ the inequality

$$|\phi(k)| \geqslant C(\theta)\theta^k \tag{8}$$

is fulfilled. As in the proof of Theorem 4, we denote by $\varphi(\zeta)$ the Borel transform of $\phi(z)$

$$\varphi(\zeta) = \int_0^\infty \phi(z)e^{-\zeta z}\,dz.$$

According to the inversion formula for the Borel transform, we have

$$\phi(z) = \frac{1}{2\pi i}\int_\Omega \varphi(\zeta)e^{\zeta z}\,d\zeta,$$

where Ω is any closed path not intersecting the indicator diagram of $\phi(z)$. (In our case the indicator diagram is $[-i\pi\Delta, i\pi\Delta]$.) Setting $z=k$ and using the power expansion of ϕ, we have

$$\phi(k) = \frac{1}{2\pi i}\int_\Omega \varphi(\zeta)e^{\zeta k}\,d\zeta, \qquad k\in\mathbb{N},$$

$$\sum_{k\in\mathbb{N}} c_k\phi(k)z^k = \frac{1}{2\pi i}\int_\Omega \varphi(\zeta)\sum_{k\in\mathbb{N}} c_k z^k e^{k\zeta}\,d\zeta = \frac{1}{2\pi i}\int_\Omega \varphi(\zeta)f(ze^\zeta)\,d\zeta.$$

The sequence $c_k\phi(k)$ does not change sign, so we may assume that $c_k\phi(k)\geqslant 0$. It follows from (8) and from the previous formula that

$$M(f, r\theta) \leqslant C^{-1}(\theta)\sum_{k\in\mathbb{N}} c_k\phi(k)z^k = \frac{1}{2\pi i}\int_\Omega \varphi(\zeta,\theta)f(ze^\zeta)\,d\zeta.$$

Now we have the estimate $M(f, r\theta) \leqslant C(\theta,\Omega)\max_{\zeta\in\Omega}\ln|f(re^\zeta)|$, where $C(\theta,\Omega) = C^{-1}(\theta)\max_{\zeta\in\Omega}|\varphi(\theta,\zeta)|$ or

$$M(f, r\theta) \leqslant \ln C(\theta,\Omega) + \max_{\zeta\in\Omega}\ln|f(re^\zeta)|, \tag{9}$$

for every $\theta\in(0,1)$.

Now repeating the proof of Theorem 5, we conclude that for

$$u(z) = \lim \ln|f(r_k z)|/A_k$$

the inequality $B(r,u) \leqslant \max_{|\varphi|\leqslant\pi\Delta} u(re^{i\varphi})$ is fulfilled, and, therefore, the function $u(z)$ belongs to $\mathscr{B}(\Delta)$ for every $\Delta>\Delta^*$. Theorem 5 is proved.

Before proceeding with our exposition, we recall the definition of Polya peaks. Let $K(r)$, $r>0$, be a nonnegative unbounded increasing function.

The sequence r_k is a sequence of *Polya peaks of order* $\mu \geqslant 0$ for the function $K(r)$ if there exists a sequence $\delta_k \to 0$, $\delta_k > 0$, such that

$$K(rr_k) \leqslant (1+\delta_k)K(r_k)r^{\mu}, \qquad r \in [\delta_k, \delta_k^{-1}].$$

The supremum and infimum of orders of Polya peaks are called the *upper* and *lower Polya orders of* $K(r)$, respectively. We shall denote them by ρ^* and λ^*, respectively. It is known (see for example [9]) that the set of orders of Polya peaks is either the segment $[\lambda^*, \rho^*]$ or is empty. If λ and ρ are the usual lower and upper orders, respectively, then

$$[\lambda, \rho] \subseteq [\lambda^*, \rho^*]. \tag{10}$$

THEOREM 6. *Let $f(z)$ be an entire function represented by a gap power series which has maximal density Δ of nonzero indices. Let the function $\ln|M(r, f)|$ have Polya peaks of order μ. If $\mu(\Delta+\delta) < 1$ for some $\delta \geqslant 0$, then*

$$\limsup \frac{\ln \mu_\delta(r, f)}{\ln M(r, f)} \geqslant \cos \pi\mu(\Delta+\delta),$$

where $\mu_\delta(r, f) = \min_{|\varphi|\leqslant\delta} |f(re^{i\varphi})|$.

The following theorem is an analog of Theorem 6 for real entire functions represented by power series which have few sign changes.

THEOREM 7. *Let $f(z)$ be a real entire function represented by the series*

$$f(z) = c_0 + \sum_{k\in\mathbb{N}} c_k z^k.$$

Let the set of sign changes of coefficients of this series has maximal density not larger than Δ. If the function $\ln|f(z)|$ has Polya peaks of finite order μ and $\mu\Delta < 1$, then the following inequality holds

$$\limsup \frac{\ln|f(r)|}{\ln M(r, f)} \geqslant \cos(\pi\mu\Delta).$$

We shall obtain Theorems 6 and 7 as corollaries of Theorems 2 and 3. We need the following lemma.

LEMMA 1. *Let $v_n(r)$ be a sequence of functions subharmonic in an annulus $\{|z| \in (r_1, r_2)\}$ which tends to some subharmonic function in the topology of $\mathscr{D}'$. If for every $r \in (r_1, r_2)$ the inequality*

$$L(r, v_n) = \inf_{\varphi} v_n(re^{i\varphi}) \leqslant \gamma B(r, v_n)$$

is true, then for the limit function $v(z)$ the inequality $L(r, v) \leqslant \gamma B(r, v)$ is true as well.

In order to prove this lemma we use the following theorem of A. Baernstein [8].

THEOREM (A. Baernstein). *Let $u(z)$ be a δ-subharmonic function in the annulus $|z| \in (r_1, r_2)$ and $u(z) = u_+(z) - u_-(z)$ be one of its representations as the difference of two subharmonic functions. Let*

$$u^*(re^{i\theta}) = \sup\left\{ \int_E u(re^{i\varphi})\, d\varphi \,:\, \operatorname{meas}(E) = 2\theta \right\},$$

where the supremum is taken over all measurable sets which have Lebesgue measure 2θ, and let

$$N(r, u_-) = \int_{-\pi}^{+\pi} u_-(re^{i\varphi})\, d\varphi.$$

Under these conditions the function $u^*(re^{i\varphi}) + N(r, u_-)$ of the variable $\zeta = re^{i\theta}$ is continuous for $r \in (r_1, r_2)$, $\theta \in [0, \pi]$, and is subharmonic for $r \in (r_1, r_2)$, $\theta \in (0, \pi)$.

PROOF OF LEMMA 1. We consider the function $u_n(z) = v_n(z) - \gamma B(r, v_n)$. Baernstein's theorem implies that the function $u_n^*(re^{i\theta}) + 2\pi|\gamma| B(r, v_n)$ is a subharmonic function of the variable $\zeta = re^{i\theta}$ in the domain $r \in (r_1, r_2)$, $\theta \in (0, 2\pi)$, and, moreover, is continuous for $r \in (r_1, r_2)$, $\theta \in [0, 2\pi]$. It follows from the definition of $u_n(re^{i\varphi})$ and from the conditions of Lemma 1 that $\inf_\varphi u_n(re^{i\varphi}) \leqslant 0$ and hence, the definition of $u_n^*(re^{i\theta})$ immediately implies

$$\lim_{\theta \to -\pi} (\pi - \theta)^{-1} (u_n^*(re^{i\theta}) - u_n^*(re^{i\pi})) \geqslant 0, \qquad \lim_{\theta \to 0} \theta^{-1} (u_n^*(re^{i\theta}) - u_n^*(r)) \geqslant 0$$

(note that according to the conditions of Lemma 1, we have $\gamma < 1$). From the latter inequalities for u_n^*, it follows that the function

$$u_n^*(re^{i\theta}) + 2\pi|\gamma| B(r, v_n),$$

defined in the half-annulus $r \in (r_1, r_2)$, $\theta \in [0, 2\pi]$, can be continued to the whole annulus by $u_n^*(re^{i\theta}) = u_n^*(re^{-i\theta})$ for $\theta \in (-\pi, 0)$.

The new function is continuous and subharmonic in the annulus $r \in (r_1, r_2)$. In order to complete the proof, it is sufficient to show that if $v_n(z)$ converges to $u(z)$ in the topology of $\mathscr{D}'$ in the annulus $r \in (r_1, r_2)$, then $u_n^*(re^{i\theta}) \to u^*(re^{i\theta})$ and $B(r, v_n) \to B(r, v)$ at each point of the domain $r \in (r_1, r_2)$, $\theta \in [0, 2\pi]$. Indeed, if it is proved, then

$$\lim (\pi - \theta)^{-1} (u^*(re^{i\theta}) - u^*(re^{i\pi})) \geqslant 0,$$

and hence

$$\inf_\varphi u(re^{i\varphi}) = \inf_\varphi v(z) - \gamma B(r, v) \leqslant 0.$$

From the convergence $v_n(re^{i\varphi}) \to v(re^{i\varphi})$ in the topology of $\mathscr{D}'$, it follows that $\lim B(r, v_n) = B(r, v)$. It is sufficient to prove that

$$v_n^*(re^{i\theta}) \to v^*(re^{i\theta}), \qquad \forall r \in (r_1, r_2),\ \forall \theta \in [0, \pi].$$

As we have mentioned above, the convergence $v_n(re^{i\varphi}) \to v(re^{i\varphi})$ in the topology of $\mathscr{D}'$ implies that

$$\lim \int_{-\pi}^{\pi} |v_n(re^{i\varphi}) - v(re^{i\varphi})|\, d\varphi = 0$$

for all $r \in (r_1, r_2)$. The definition of the $*$-operator implies

$$\begin{aligned} v_n^*(re^{i\theta}) - v^*(re^{i\theta}) &\leqslant \sup\left\{ \int_E (v_n(re^{i\varphi}) - v(re^{i\varphi}))\, d\varphi \,:\, \mathrm{meas}(E) = 2\theta \right\} \\ &\leqslant \int_{-\pi}^{\pi} |v_n(re^{i\varphi}) - v(re^{i\varphi})|\, d\varphi = o(1), \qquad n \to \infty. \end{aligned}$$

A similar estimate is true for $v^*(re^{i\theta}) - v_n^*(re^{i\theta})$ as well. Lemma 1 is proved.

PROOF OF THEOREMS 6 AND 7. Let μ be an order of Polya peaks for $\ln|M(r, f)|$. We consider the family of subharmonic functions

$$v_n(z) = \frac{\ln|f(z, r_n)|}{\ln M(r_n, f)},$$

where r_n is the sequence of Polya peaks for $\ln|M(r, f)|$. This family is precompact and for every limit function $u(z)$ of this family the following relations hold:

a) $0 \leqslant B(r, u) \leqslant r^{\mu}$,

b) $B(1, r) = 1$.

We shall prove Theorem 6 only (the proof of Theorem 7 is analogous). Suppose that the assertion of Theorem 6 is false. In this case for some $\varepsilon > 0$ and all sufficiently large r we have

$$\frac{\ln \mu_\delta(r, f)}{\ln M(r, f)} \leqslant \cos \pi\mu(\Delta + \delta) - \varepsilon.$$

It follows from Lemma 1 for the limit function $u(z) = \lim v_n(z)$ that the inequality

$$L_\delta(r, u) \leqslant B(r, u)(-\varepsilon + \cos \pi\mu(\Delta + \delta)) \tag{11}$$

is fulfilled. For the proof it is sufficient to apply Lemma 1 for the functions

$$w_n(z) = \{v_n(z) - (\cos \pi\mu(\delta + \Delta) + \varepsilon)\, B(r_0, v_n)\}^+$$

in the annulus $(r_0 - \alpha, r_0 + \alpha)$ for all sufficiently small α. According to Theorem 4, $u(z)$ belongs to $\mathscr{A}(\Delta)$, consequently inequality (11) contradicts the assertion of Theorem 2.

§3. An example of a subharmonic function belonging to $\mathscr{A}(\Delta)$

Let $\mathscr{A}^*(\Delta)$ be the family of all subharmonic functions $u(z)$ satisfying the following condition: there exists an entire function $f(z)$ represented by a gap power series with maximal density Δ such that for some sequences A_k and r_k the limit $A_k^{-1} \ln|f(r_k z)|$ is equal to $u(z)$. It follows from the present work that $\mathscr{A}^*(\Delta) \subseteq \mathscr{A}(\Delta)$. Now we consider functions which belong to $\mathscr{A}(\Delta)$ and $\mathscr{A}^*(\Delta)$ and are bounded from above by $|z|^\mu$. In other words,

$$0 \leqslant B(r, u) \leqslant r^\mu \tag{12}$$

for some $\mu > 0$. We denote these classes by $\mathscr{A}^*(\Delta, \mu)$ and $\mathscr{A}(\Delta, \mu)$, respectively. We have $\mathscr{A}^*(\Delta, \mu) \subseteq \mathscr{A}(\Delta, \mu)$. It seems that $\mathscr{A}^*(\Delta, \mu) = \mathscr{A}(\Delta, \mu)$, but we cannot prove it.

One of the questions from [6] is the following:

Let $f(z)$ be an entire function of lower order λ represented by a gap series with a density of nonzero indices Δ and let $\lambda < 1/\Delta$. Is it true that

$$\limsup_{r\to\infty} \frac{\ln \mu(r, f)}{\ln M(r, f)} \geqslant \cos(\pi\lambda\Delta)\,?$$

It is natural to suppose that a subharmonic analog of this question has the following form:

Let $u(z)$ be a nontrivial subharmonic function which belongs to the class $\mathscr{A}(\Delta)$ and satisfies inequality (12) with $\mu < 1/\Delta$. Is it true that for some $r > 0$ the inequality

$$L(r, u) \geqslant \cos(\pi\mu\Delta)B(r, u) \tag{13}$$

holds?

The following example constructed in Theorem 8 demonstrates that the answer is negative.

THEOREM 8. *For every $\alpha \in (0, 1)$ and every $A \in (-\infty, +\infty)$ one can find $\mu > 0$ and $\Delta = \alpha/\mu$ such that there exists a subharmonic function $u(z)$ which satisfies the following conditions*

a) $u(z) \in \mathscr{A}(\Delta)$;
b) $B(r, u) = r^\mu$;
c) $L(r, u) \leqslant AB(r, u)$, $\forall r \geqslant 0$.

PROOF OF THEOREM 8. First, we construct a function $v(z)$ which satisfies the following conditions for some $q > 1$:

1) $v(qz) = q^\alpha v(z)$;
2) $v(-r) = B(r, v) = r^\alpha$, $r \geqslant 0$;
3) $v(1) = -\infty$.

The basic idea for constructing this function is taken from the work of A. Eremenko, M. Sodin, and Shea [5]. We define the point measure γ on the sequence $\{q^k\}_{k=-\infty}^{\infty}$ by the equations $\gamma(\{q^k\}) = q^{k\alpha}$. Now we consider a subharmonic function defined by the formula

$$w_q(z) = C_q \int_0^{+\infty} \ln|1 - z^2/t^2|\, d\gamma(t),$$

where C_q is a normalizing coefficient determined by the condition $w_q(i) = 1$. It follows from the definition of γ that

$$w_q(qz) = q^\alpha w_q(z). \tag{14}$$

If $q \to 1$, then, evidently, $w_q(re^{i\varphi}) \to w_1(re^{i\varphi}) = r^\alpha \cos(\alpha(\varphi - \pi/2))$ for $\varphi \in (0, \pi)$. Moreover, this convergence is uniform on every compact subset of the upper half-plane. The function $\widetilde{w}_q(z)$, by Theorem 1, is subharmonic in the whole complex plane because the function $B(w_q, r) = w_q(re^{\pm\pi/2})$ satisfies the condition

$$\widetilde{w}_q(re^{i\varphi}) = B(w_q, r), \qquad \forall\varphi, \ |\varphi - \pi/2| < \pi/2.$$

From the uniform convergence of $B(w_q, r)$ to r^α on every compact set of the ray $(0, +\infty)$ and from relation (14), it follows that

$$(1 - A_q)r^\alpha \leqslant B(w_q, r) \leqslant (1 + A_q)r^\alpha, \tag{15}$$

where A_q depends on q only, and $A_q \to 1$ as $q \to 1$. We choose q to satisfy the inequality

$$(1 - A_q) \geqslant (1 + A_q)\cos(\pi\alpha/2) \tag{16}$$

and define the function $v(z)$ for such a value of q by

$$v(re^{i\varphi}) = \begin{cases} \max\left\{r^\alpha \cos(\alpha(\pi - \varphi)), \dfrac{1 - A_q}{1 + A_q}\widetilde{w}_q(re^{i\varphi})\right\}, & |\pi - \varphi| \leqslant \pi/2, \\ \dfrac{1 - A_q}{1 + A_q}\widetilde{w}_q(re^{i\varphi}), & |\varphi| < \pi/2. \end{cases}$$

It follows from relations (15) and (16) that this function is subharmonic in the entire complex plane and satisfies conditions 1)–3).

Now we pass to the construction of the function $u(z)$. Since the function $v(z)$ is upper semicontinuous and $v(1) = -\infty$, the inequality $v(r) \leqslant Ar^\alpha$ is valid in any neighborhood of $\{r : |r - 1| < \delta\}$. According to 1), for the function $v(z)$ we have

$$v(r) \leqslant Ar^\alpha, \qquad r \in \bigcup_{k=-\infty}^{\infty} \{r : |r - q^k| < \delta q^k\}.$$

From the latter relation and property 1) of $v(z)$ it follows that for some $\theta > 1$

$$\min\{v_k(r) : k \in \mathbb{Z}, \ 0 \leqslant k \leqslant k_0\} \leqslant Ar^\alpha, \qquad r > 0, \ k_0 \geqslant \frac{\ln q}{\ln\theta + 1}, \tag{17}$$

where $v_k(z) = \theta^{\alpha k} v(\theta^k z)$.

In order to define the function $u(z)$, we set $\Delta = 1/k_0$, $\mu = \alpha/\Delta = \alpha k_0$, $u(z) = v_k(z^{k_0})$ as $z \in \{z = re^{i\varphi} : |\varphi - 2\pi/k| \leqslant \pi/k_0\}$, $k = 0, \dots, k_0$. This function is subharmonic of class $\mathscr{A}(\Delta)$ and $B(r, u) = r^\mu$, $\mu = \alpha/\Delta$; further (17) implies $L(r, u) \leqslant AB(r, u) = Ar^\mu$, $r > 0$. Theorem 8 is proved.

To conclude, I would like to thank M. L. Sodin, who read this paper and made several useful remarks.

References

1. A. Wiman, *Uber eine Eigenschaft der ganzen Funktionen von der Hohe Null*, Math. Ann. **76** (1915), 197–211.

2. G. Valiron, *Sur le minimum du module des fonctions entières d'ordre inférieur à un*, Mathematica **11** (1935), 264–269.

3. M. Essen, *The* $\cos \pi\lambda$*-theorem*, Lecture Notes in Math., vol. 467, Springer-Verlag, Berlin and New York, 1975.

4. B. Kjellberg, *On the minimum modulus of entire functions of lower order less than one*, Math. Scand. **8** (1960), 189–197.

5. A. E. Eremenko, M. L. Sodin, and D. F. Shea, *On the minimum of the modulus of an entire function on the Polya peak sequence*, Teor. Funktsiĭ, Funktsional. Anal. i Prilozhen. **45** (1986), 26–40; English transl. in J. Soviet Math. **48** (1990), no. 4.

6. W. K. Hayman, *Research problems in function theory*, The Athlone Press, London, 1967.

7. B. Ya. Levin, *Distribution of zeros of entire functions*, GITTL, Moscow, 1956; English transl., Amer. Math. Soc., Providence, R.I., 1964.

8. A. Baernstein, *Integral means, univalent functions and circular symmetrization*, Acta Math. **133** (1974), 139–169.

9. D. Drasin and D. Shea, *Polya peaks and the oscillation of positive functions*, Proc. Amer. Math. Soc. **34** (1972), no. 2, 403–411.

10. W. K. Hayman and P. B. Kennedy, *Subharmonic functions*, 2, Academic Press, New York, 1990.

11. T. Murai, *Gap series*, Analytic Function Theory of One Complex Variable, Pitman Res. Notes Math. Ser., vol. 212, Longman Sci. Tech., Harlow, 1989, pp. 149–177.

Institute for Low Temperature Physics & Engineering, 47 Lenin pr., Kharkov 310164 Ukraine

Translated by THE AUTHOR

ADVANCES IN SOVIET MATHEMATICS
Volume 11, 1992

Lacunary Series and Fourier Integrals of Functions of Several Variables

V. N. LOGVINENKO AND S. YU. FAVOROV

Recently M. Benedicks [1] has proved the following uniqueness theorem:

THEOREM A. *Let* $f \in L^1(\mathbb{R}^d)$ *and let* $\hat{f}$ *be its Fourier transform. Let*

$$\operatorname{meas}_d\{x \in \mathbb{R}^d : f(x) \neq 0\} < \infty,$$
$$\operatorname{meas}_d\{t \in \mathbb{R}^d : \hat{f}(t) \neq 0\} < \infty,$$

where meas_d *denotes the Lebesgue measure on* $\mathbb{R}^d$. *Then* $f(x) = 0$ *a.e.*[1]

The simplicity of the author's proof is striking. M. Benedicks shows that for almost all $t \in \mathbb{R}^d$ the function

$$g(t, x) = \sum_{k \in \mathbb{Z}^d} f(x - 2\pi k) \exp\{-i\langle t, x - 2\pi k\rangle\}, \tag{1}$$

with $\langle t, x\rangle = t_1x_1 + \cdots + t_dx_d$, is a trigonometric polynomial with respect to x. Without loss of generality we can assume that

$$\operatorname{meas}_d\{x \in \mathbb{R}^d : f(x) \neq 0\} < (2\pi)^d.$$

Then

$$\operatorname{meas}_d\{ x \in K(\pi) : g(x, t) \neq 0 \} < (2\pi)^d$$

for all $t \in \mathbb{R}^d$. Here $K(L)$ is the cube $\{x \in \mathbb{R}^d : |x| < L\}$.[2] Therefore, for a.e. $t \in \mathbb{R}^d$ the function $g(t, x)$ vanishes at a.e. x's. The identity $\hat{f}(t) = 0$ is an immediate consequence of this fact.

We want to note two points in the argument just given. The first is that the supports of f and $\hat{f}$ are not mentioned in it. Only the sets $S(f) = \{x \in \mathbb{R}^d : f(x) \neq 0\}$ and $S(\hat{f}) = \{t \in \mathbb{R}^d : \hat{f}(t) \neq 0\}$ appear, and these

1991 *Mathematics Subject Classification.* Primary 42B10.

[1]Theorem A improves, in the particular case $G = \mathbb{R}^d$, a result of [2] on Fourier transforms on locally compact groups G.

[2]We use the norm $|x| = \max\{|x_1|, \dots, |x_d|\}$ in $\mathbb{R}^d$.

sets could a priori be, e.g., dense open sets. The second is that Theorem A differs from most uniqueness theorems for Fourier transforms by the use of the lacunary condition instead of the condition of fast decrease of $f(x)$ as $|x| \to \infty$.

In the present paper we maintain the scheme of the proof of Theorem A but investigate the more general case when $g(t, x)$ is the sum of lacunary (in some sense) Fourier series. This investigation is carried out in the second section, and in the first one we prove some uniqueness theorems for lacunary Fourier series of functions of several variables needed for this approach. Even for the well-known case $d = 1$ (see [3–5]; there is a detailed bibliography in [4]) some results of ours are new. In the third section there are applications of the results just obtained.

Acknowledgment. We wish to thank our colleague M. L. Sodin for his inspiring attention to this work.

§1. Uniqueness theorems for lacunary Fourier series

The following uniqueness theorem for lacunary in the sense of Hadamard trigonometric series of one variable is well known [6, p. 206].

THEOREM B. *Let the series*

$$\sum_k (a_k \cos n_k t + b_k \sin n_k t),$$

where $n_k \in \mathbb{N}$ *satisfies the inequality* $n_{k+1}/n_k \geqslant q > 1$ *for all* k, *converge to* 0 *on a set of positive measure. Then it vanishes identically.*

For multiple trigonometric series we need a slightly more general definition of lacunarity.

DEFINITION 1. A set $A \subset \mathbb{Z}^d$ is called *L-lacunary* (or *lacunary*) if for every $p \in \mathbb{Z}^d \setminus \{0\}$ the set $\{n \in A : n + p \in A\}$ has no more than L points.

EXAMPLE 1. Let $A \subset \mathbb{Z}^d$ have the following property: for some $\alpha > 0$ and all $n, m \in A$, $n \neq m$, the inequality holds

$$|n - m| \geqslant \alpha \max\{|n|, |m|\}.$$

Let us show that this set is L-lacunary for some finite L depending on α and d only. Put $p \in \mathbb{Z}^d \setminus \{0\}$ and let $n, m \in A$ satisfy

$$n - m = p. \tag{2}$$

Assume also that $|n| > |m|$. Then we have

$$|n| \geqslant |n - m|/2 = |p|/2, \qquad |n| \leqslant \alpha^{-1}|n - m| = \alpha^{-1}|p|.$$

Let $n', m' \in A$, $n' \neq n$, $|n'| > |m'|$, satisfy (2). In this case

$$|n - n'| \geqslant \alpha \max\{|n'|, |n|\} \geqslant \alpha|p|/2, \qquad |n'| \leqslant \alpha^{-1}|p|.$$

Therefore, the number of solutions of equation (2) is less than some constant which does not depend on p.

DEFINITION 2. A series

$$\sum_{n\in\mathbb{Z}^d} c_n \exp\{i\langle n, x\rangle\}$$

is called *lacunary* if the set $\{n \in \mathbb{Z}^d : c_n \neq 0\}$ is lacunary.

We formulate the multiple analog of Theorem B in a form convenient for our purposes.

THEOREM 1. *Let $f \in L^1(K(\pi))$ vanish on a set of positive (Lebesgue) measure. If the Fourier series of this function is lacunary, then $f(x) = 0$ a.e.*

The assertion of Theorem 1 for $d = 1$ was known before [6, p. 379]. In our case the general outline of the proof is the same as in [6] and uses several lemmas.

LEMMA 1. *Suppose we are given a set $E \subset K(\pi)$ of positive measure and $L \in \mathbb{N}$. Then there exists an integer $r = r(E, L)$ such that for any L-lacunary series $\sum_n c_n \exp\{i\langle n, x\rangle\}$ of a function belonging to $L^2(K(\pi))$ we have*

$$\sum_{|n|>r} |c_n|^2 \leqslant 2\{\operatorname{meas}_d E\}^{-1} \int_E \left| \sum_{|n|>r} c_n \exp\{i\langle n, x\rangle\}\right|^2 dx.$$

LEMMA 2. *If E is any measurable set in $K(\pi)$, then there exists a sequence of positive numbers $h_j \to 0$ such that, for almost all $x \in E$ and all $j \geqslant j(x)$, the vectors $x + h_j e_s$ are in E for every element e_s of the natural basis in $\mathbb{R}^d$.*

The proofs of these lemmas repeat the proofs of the corresponding one-dimensional assertions 6.5 and 6.14 of [6] almost literally.

LEMMA 3 (cf. Theorem 6.4 from [6]). *Let $\sum_{n\in\mathbb{Z}^d} c_n \exp\{i\langle n, x\rangle\}$ be a lacunary trigonometric series and $k_{n,j}$, $n \in \mathbb{Z}^d$, $j \in \mathbb{N}$, be a sequence of complex numbers such that $k_{n,j} \to 1$ as $j \to \infty$. If for any sequence $N \to \infty$ the limit*

$$\lim_{j\to\infty} \lim_{N\to\infty} \sum_{|n|<N} k_{n,j} c_n \exp\{i\langle n, x\rangle\}$$

is finite on a set $E \subset K(\pi)$ of positive measure, then

$$\sum_{n\in\mathbb{Z}^d} |c_n|^2 < \infty.$$

PROOF. We can find a set $E_0 \subset E$ such that $\operatorname{meas}_d E_0 > 0$ and for all $j \geqslant j_0$ and $x \in E_0$

$$\lim_{N\to\infty} \left| \sum_{|n|<N} k_{n,j} c_n \exp\{i\langle n, x\rangle\}\right| \leqslant C < \infty.$$

For every $j \geqslant j_0$ it is possible to choose a number $N(j) \in \mathbb{N}$ and a set $E_j \subset E_0$ such that $\operatorname{meas}_d E_j < 2^{-j} \operatorname{meas}_d E_0$, and for all $x \in E_0 \setminus E_j$

$$\left| \sum_{|n|<N(j)} k_{n,j} c_n \exp\{i\langle n, x\rangle\}\right| \leqslant C + 1.$$

According to Lemma 1, there exists a number $r \in \mathbb{N}$ which does not depend on j, such that

$$\begin{aligned}\sum_{r<|n|<N(j)} |k_{n,j}|^2 |c_n|^2 &\leqslant 2\{\operatorname{meas}_d \widetilde{E}\}^{-1} \int_{\widetilde{E}} \left| \sum_{r<|n|<N(j)} k_{n,j} c_n \exp\{i\langle n, x\rangle\}\right|^2 dx \\ &\leqslant 2(C+1)^2 + 4\max\left\{\left| \sum_{|n|<r} k_{n,j} c_n \exp\{i\langle n, x\rangle\}\right|^2 : x \in K(\pi)\right\}.\end{aligned}$$

Here $\widetilde{E}$ denotes the set $E_0 \setminus \bigcup_{j=j_0}^{\infty} E_j$. Therefore, we can find a constant C' such that for every s and all sufficiently large j the inequality

$$\sum_{r<|n|<s} |k_{n,j}|^2 |c_n|^2 \leqslant C'$$

holds.

Letting first j and then s tend to infinity, we complete the proof of the lemma.

LEMMA 4. *Let $f \in L^2(K(\pi))$ have the lacunary Fourier series*

$$\sum_{n\in\mathbb{Z}^d} c_n \exp\{i\langle n, x\rangle\}.$$

Further let f coincide on a set E of positive measure with a function g analytic in some $\mathbb{C}^d$-neighborhood of E. Then the series $\sum c_n \exp\{i\langle n, x\rangle\}$ converges absolutely on the tubular domain $\{z \in \mathbb{C}^d : |\operatorname{Im} z| < \delta\}$ for some $\delta > 0$.

PROOF. Take a sequence $N \to \infty$ such that for almost all $x \in K(\pi)$

$$f(x) = \lim_{N\to\infty} \sum_{|n|<N} c_n \exp\{i\langle n, x\rangle\}.$$

Let $E' \subset E$ be the set from Lemma 2. At all its points we have

$$\begin{aligned}\frac{\partial g}{\partial x_1} &= \lim_{j\to\infty} \frac{f(x+h_j e_1) - f(x)}{h_j} \\ &= \lim_{j\to\infty} \lim_{N\to\infty} \sum_{|n|<N} \frac{\exp\{in_1 h_1\} - 1}{in_1 h_1} ic_n \exp\{i\langle n, x\rangle\} n_1,\end{aligned}$$

where n_1 is the first coordinate of the multi-index n. Since $\operatorname{meas}_d E' > 0$, the series $\sum |n_1|^2 |c_n|^2$ converges by virtue of Lemma 3. Consequently, the series $\sum in_1 c_n \exp\{i\langle n, x\rangle\}$ is the Fourier series of the function $\partial f/\partial x_1 \in L^2(K(\pi))$ coinciding with $\partial g/\partial x_1$ on E'. Repeating this argument, we finally get that $f(x)$ is infinitely differentiable and coincides with $g(x)$ on E' together with all its partial derivatives[3]. By virtue of Lemma 1, for every $k \in (\mathbb{Z}_+)^d$, we have the inequality

$$\sum_{|n|>r} |n_1|^{2k_1} \cdots |n_d|^{2k_d} |c_n|^2 \leqslant 2\{\operatorname{meas}_d E'\}^{-1} \int_{E'} \left| \frac{\partial^{k_1+\cdots+k_d} g_1(x)}{\partial x_1^{k_1} \cdots \partial x_d^{k_d}} \right|^2 dx, \quad (3)$$

where

$$g_1(x) = g(x) - \sum_{|n|\leqslant r} c_n \exp\{i\langle n, x\rangle\}.$$

The function g_1 is analytic in some $\mathbb{C}^d$-neighborhood of E. Therefore,

$$\left| \frac{\partial^{k_1+\cdots+k_d} g_1(x)}{\partial x_1^{k_1} \cdots \partial x_d^{k_d}} \right| \leqslant M k_1! \cdots k_d! \, \varepsilon^{-(k_1+\cdots+k_d)}$$

for some $M < \infty$, $\varepsilon > 0$ and all $x \in E$, $(k_1, \dots, k_d) \in (\mathbb{Z}_+)^d$. For this reason and by virtue of (3), for arbitrary $n \in \mathbb{Z}^d$, $|n| > r$, we have the estimate

$$|n_1|^{k_1} \cdots |n_d|^{k_d} |c_n| \leqslant 2M(k_1/\varepsilon)^{k_1} \cdots (k_d/\varepsilon)^{k_d}.$$

Taking here $k_j = [e^{-1}\varepsilon|n_j|]$, $j = 1, \dots, d$, we get the inequality

$$|c_n| \leqslant 2M \exp\{-\varepsilon(|n_1| + \cdots + |n_d|)/2\}$$

valid for every $n \in \mathbb{Z}^d$, $|n| > r$. Lemma 4 is proved.

Now, to prove Theorem 1, it is sufficient to set $g(x) \equiv 0$ in Lemma 4.

THEOREM 2. *Let $f \in L^2(K(\pi))$ have the Fourier series*

$$\sum_{n\in\mathbb{Z}^d} c_n \exp\{i\langle n, x\rangle\}.$$

Further let Γ be a convex proper cone[4] *in $\mathbb{R}^d$, let the set $\{n \in \Gamma \cap \mathbb{Z}^d : c_n \neq 0\}$ be lacunary and the set $\{n \in \mathbb{Z}^d \setminus (\Gamma \cup -\Gamma) : c_n \neq 0\}$ be finite. If f vanishes on some open set $E \subset K(\pi)$, then $f(x) = 0$ a.e.*

PROOF. Denote

$$F_1(z) = \sum_{n\in\Gamma} c_n \exp\{i\langle n, z\rangle\}, \qquad F_2(z) = -\sum_{n\in\mathbb{Z}^d\setminus\Gamma} c_n \exp\{i\langle n, z\rangle\}.$$

---

[3]Every new differentiation requires a new application of Lemma 2. Hence this assertion is correct for a subset E'' of E' only. But the difference between E' and E'' is a set of zero measure.

[4]A cone $\Gamma \subset \mathbb{R}^d$ with vertex at 0 is called *proper* if there exist a $y^0 \in \mathbb{R}^d$, $\langle y^0, y^0\rangle = 1$, and $\alpha > 0$ such that $\Gamma \subset \{x \in \mathbb{R}^d : \langle x, y^0\rangle \geqslant \alpha\langle x, x\rangle^{1/2}\}$.

It is easy to check that $F_1(z)$ is holomorphic in the tubular domain $\mathbb{R}^d + i\Gamma'$, where $\Gamma' = \{y \in \mathbb{R}^d : \langle x, y\rangle > 0 \text{ for all } x \in \Gamma\}$ is the conjugate cone to Γ. The second function $F_2(z)$ is holomorphic on $\mathbb{R}^d - i\Gamma'$. Since by the conditions of our theorem these functions coincide on some domain $E \subset \mathbb{R}^d + i\{0\}$, the edge-of-the-wedge theorem implies that the function

$$F(z) = \begin{cases} F_1(z), & z \in \mathbb{R}^d + i\Gamma', \\ F_2(z), & z \in \mathbb{R}^d - i\Gamma', \end{cases}$$

has a holomorphic continuation onto some $\mathbb{C}^d$-neighborhood of E. At points of E we have $F_1(x) = F(x)$. Hence, according to Lemma 4, F_1 has a holomorphic continuation to the tubular domain $|\operatorname{Im} z| < \delta$, $\delta > 0$. Therefore, $F_1(z) - F_2(z) \equiv 0$ on this domain. For this reason $c_n = 0$ for all $n \in \mathbb{Z}^d$. The theorem is proved.

Another type of lacunarity of Fourier series is dealt with in the following theorem.

THEOREM 3. *Let $f \in L^1(K(\pi))$ have the Fourier series*

$$\sum_{n\in\mathbb{Z}^d} c_n \exp\{i\langle n, x\rangle\}.$$

If the following conditions hold:

(i) $c_n = O(|n|^{-d})$;

(ii) $c_n = 0$ *for* $k_j \leqslant |n| \leqslant l_j$ *and* $n \in \mathbb{Z}^d \setminus (\mathbb{Z}_+)^d$, *where the sequences* $\{k_j\}$ *and* $\{l_j\}$ *of integers satisfy* $k_1 < l_1 < k_2 < l_2 < \cdots$ *and the series*

$$\sum_{j=1}^{\infty} \left(\frac{1}{k_j} - \frac{1}{l_j}\right) \ln l_j$$

is divergent;

(iii) $\operatorname{meas}_d\{x \in K(\pi) : f(x) = 0\} > 0$,

then $f(x) = 0$ *a.e.*

PROOF. To begin with suppose that $d = 1$ and denote $\sum_{n<-r} |c_n|^2$ by S_r^2, $r \in \mathbb{N}$. The lacunarity of the series implies that for any $r \in [k_j, l_j]$ we have the equality $S_r^2 = S_{l_j}^2$. Therefore,

$$\sum_{r=1}^{\infty} \frac{\ln S_r^2}{2} \leqslant C + \sum_{j=1}^{\infty}\sum_{r=k_j}^{l_j} \frac{\ln S_r^2}{2} \leqslant C - \sum_{j=1}^{\infty} \left(\frac{1}{k_j} - \frac{1}{l_j}\right) \ln l_j = -\infty$$

and by the well-known quasianalytic theorem of Beurling [7], $f(x) = 0$ a.e.

Let $d > 1$. We shall show now that for almost all $x' = (x_1, \dots, x_{d-1}) \in K'(\pi)$ such that

$$\operatorname{meas}_1\{E \cap \{(x_1, \dots, x_d) : (x_1, \dots, x_{d-1}) = x', |x_d| < \pi\}\} > 0$$

the function $f(x', y)$ vanishes a.e. as a function of y. After this fact is proved, we can transfer, step by step, the variables x_{d-1}, x_{d-2} and so on, from the first argument x' to the second argument y. In d steps we obtain $f(x) = 0$ a.e. on $K(\pi)$. Thus, we must prove the above-mentioned fact.

Since $f \in L^2(K(\pi))$, the function $f(x', y)$ belongs to $L^2(-\pi, \pi; dy)$ for almost all $x' \in K'(\pi)$. Its Fourier coefficients are

$$b_m(x') = \sum_{n' \in \mathbb{Z}^{d-1}} c_{(n', m)} \exp\{i\langle n', x'\rangle\}.$$

For every $m \in \mathbb{Z}$ we have

$$|b_m(x')| \leqslant \sum_{n' \in \mathbb{Z}^{d-1}} |c_{(n', m)}| \leqslant C_1 \sum_{n' \in \mathbb{Z}^{d-1}} |(n', m)|^{-d} \leqslant C_2 |m|^{-1},$$

and for $m < 0$, provided that $k_j \leqslant -m \leqslant l_j$, a more precise estimate is valid, namely $|b_m(x')| \leqslant \sum_{n' \in \mathbb{Z}^{d-1}} |c_{(n', m)}| \leqslant C_3 l_j^{-1}$. Therefore, for every $r \in \mathbb{N}$, $S_r^2(x') = \sum_{m<-r} |b_m(x')|^2 = O(r^{-1})$. If, besides that, $r \in [k_j, l_j]$, we have the stronger estimate

$$S_r^2(x') \leqslant S_{l_j}^2(x') + \sum_{m=r+1}^{l_j} |b_m(x')|^2 \leqslant C_4 l_j^{-1} + C_3(l_j - r) l_j^{-2} = O(l_j^{-1}).$$

To complete the proof, we need only apply the Beurling theorem.

NOTE. If $d = 1$, we can replace the condition (i) by a weaker one: $c_n = o(|n|^{-\alpha})$ with $\alpha > 1/2$. In the general case this change is possible too, but only for $\alpha > (d+1)/2$.

The following result is a trivial consequence of the uniqueness theorem for the Hardy class H^1.

THEOREM 4. *Let $f \in L^1(K(\pi))$ and let the Fourier series*

$$\sum_{n \in \mathbb{Z}^d} c_n \exp\{i\langle n, x\rangle\}$$

of f be such that the set $\{n \in \mathbb{Z}^d \setminus (\mathbb{Z}_+)^d : c_n \neq 0\}$ is finite. If $\operatorname{meas}_d\{x \in K(\pi) : f(x) = 0\} > 0$, *then $f(x) = 0$ a.e.*

§2. Theorems on lacunarity of Fourier transforms

Let $f \in L^1(\mathbb{R}^d)$, let $\hat{f}(t)$ be its Fourier transform and let the function $g(t, x)$ be defined by (1). Then

$$\int_{K(\pi)} \sum_{k \in \mathbb{Z}^d} |f(x - 2\pi k)|\, dx = \int_{\mathbb{R}^d} |f(x)|\, dx < \infty,$$

and

$$\begin{aligned}(2\pi)^{-d}&\int_{K(\pi)} g(t,x)\exp\{-i\langle n,x\rangle\}\,dx\\&=\sum_{k\in\mathbb{Z}^d}(2\pi)^{-d}\int_{K(\pi)} f(x-2\pi k)\exp\{-i\langle t+n,x-2\pi k\rangle\}\,dx\\&=(2\pi)^{-d}\int_{\mathbb{R}^d} f(x)\exp\{-i\langle t+n,x\rangle\}\,dx.\end{aligned}$$

It follows now that $g(t,x)\in L^1(K(\pi);dx)$ and its Fourier series is

$$g(t,x)\sim\sum_{n\in\mathbb{Z}^d}\hat{f}(t+n)\exp\{i\langle n,x\rangle\}. \tag{4}$$

It is easy to verify that the condition $f\in L^2(\mathbb{R}^d)$ implies

$$g(t,x)\in L^2(K(\pi);dx)$$

for all fixed t. Besides, (1) shows that for almost all $x\in K(\pi)$

$$g(t,x)\exp\{i\langle t,x\rangle\}=\sum_{k\in\mathbb{Z}^d} f(x-2\pi k)\exp\{i\langle t,2\pi k\rangle\} \tag{5}$$

and the series on the right-hand side converges absolutely.

Using (4) and (5), it is possible to reduce the proof of uniqueness theorems for functions f with sparse sets $S(f)$ and $S(\hat{f})$ to that of uniqueness theorems for lacunary Fourier series just given.

DEFINITION 3. Let us say that a function $f:\mathbb{R}^d\to\mathbb{C}$ *vanishes on an a-periodic set* if $\mathbb{R}^d\setminus S(f)$ contains a subset $E+a\mathbb{Z}^d$ for some $E\subset K(a/2)$.

DEFINITION 4. A set $A\subset\mathbb{R}^d$ is called *finite over* $E\subset K(a/2)$ if every point of E is covered at most by a finite number of sets $A+an$, $n\in\mathbb{Z}^d$. In other words, the set $A\cap\{x+a\mathbb{Z}^d\}$ is finite for all $x\in E$. A set $A\subset\mathbb{R}^d$ is called *finite a.e. over* $K(a/2)$ if it is finite over some subset of this cube of complete measure.

In the above-mentioned article [1] Benedicks noted that every set of finite measure is finite a.e. over $K(a/2)$ for arbitrary $a>0$. Another example of a set finite a.e. over $K(a/2)$ is the union

$$\bigcup_{n\in\mathbb{Z}^d}\{x:|x-an|<r(n)\}$$

provided that $r(n)\to 0$ as $|n|\to\infty$.

We shall need some properties of periodic and finite a.e. sets.

LEMMA 5. *If $A\subset\mathbb{R}^d$ is finite a.e. over $K(a/2)$, then A is finite a.e. over $K(ra/2)$, where $r>0$ is an arbitrary rational number.*

PROOF. To begin with, suppose that $r \in \mathbb{N}$. For every $y \in K(ra/2)$ we can find $n \in \mathbb{Z}^d$ such that $|n| < r$ and $y + an \in K(a/2)$. Hence, $y + ra\mathbb{Z}^d \subset y + an + a\mathbb{Z}^d$, which proves Lemma 5 for $r \in \mathbb{N}$.

If $r^{-1} \in \mathbb{N}$, the lemma is the consequence of the relation

$$y + ra\mathbb{Z}^d = a\mathbb{Z}^d + \bigcup_{k\in\mathbb{Z}^d,\, |k|<r^{-1}} (y + rak),$$

which is true for all $y \in K(ra/2)$.

These two particular cases give the proof in the general case.

LEMMA 6. *If $A \subset \mathbb{R}^d$ is finite over $E \subset K(b/2)$ and $\operatorname{meas}_d E > 0$, then for some $N \in \mathbb{N}$ the set $\mathbb{R}^d \setminus A$ contains an Nb-periodic subset of a positive measure.*

PROOF. Let us choose an integer N such that the set

$$E_1 = \{x \in K(b/2) : x + bn \notin A \text{ for any } n \in \mathbb{Z}^d \text{ with } |n| \geqslant (N-1)/2\}$$

has positive measure. Then the set $E_2 = E_1 + (N-1)(b/2)\mathbb{1}$, $\mathbb{1} = (1, \ldots, 1) \in \mathbb{Z}^d$, is a subset of $K(Nb/2)$ of positive measure and $(E_2 + Nb\mathbb{Z}^d) \cap A = \varnothing$. Lemma 6 is proved.

THEOREM 5. *Let $f \in L^1(\mathbb{R}^d)$ and let a and b be positive numbers such that $ab/(2\pi)$ is rational. Each of the following conditions implies that $f(x) = 0$ a.e.:*

(i) *$S(f)$ is finite a.e. over $K(a/2)$ and $\hat{f}(t)$ vanishes on the set $\widehat{E} + b\mathbb{Z}^d$, where $\widehat{E} \subset K(b/2)$ is a uniqueness set for the class of trigonometric polynomials*[5];

(ii) *$S(f) \setminus (\mathbb{R}_+)^d$ is finite a.e. over $K(a/2)$ and $S(\hat{f})$ is finite over $\widehat{E} \subset K(b/2)$ with $\operatorname{meas}_d \widehat{E} > 0$;*

(iii) *$S(\hat{f}) \setminus (\mathbb{R}_+)^d$ is finite a.e. over $K(b/2)$ and $S(f)$ is finite over $E \subset K(a/2)$ with $\operatorname{meas}_d E > 0$;*

(iv) *$f(x) \geqslant 0$, the set $S(\hat{f}) \cap b(\mathbb{Z}^d \setminus (\mathbb{Z}_+)^d)$ is finite and $S(f)$ is finite over $E \subset K(a/2)$ with $\operatorname{meas}_d E > 0$.*

PROOF. (i) Changing variables $x \mapsto x/b$, we get $b = 1$. In the new variables $S(f)$ is finite over $K(a'/2)$, where a' is a rational multiple of 2π. According to Lemma 5, we can assume, without loss of generality, that $a' = 2\pi$. Therefore, for a.e. $x \in K(\pi)$ the series (5) is a trigonometric polynomial of t. By virtue of (4), the function $g(t, x)$ vanishes for every $t \in E$. Consequently, $g(t, x) = 0$ for almost all $x \in K(\pi)$, $t \in K(1/2)$. By this reason $f(x) = 0$ a.e. on $\mathbb{R}^d$.

(ii) Using Lemma 6, we can replace b by $b' = Nb$, $N \in \mathbb{N}$, in such a way that $\hat{f}(t)$ vanishes on the b'-periodic set $\widehat{F} + b'\mathbb{Z}^d$, where $\widehat{F} \subset K(b'/2)$

[5] In the case $d = 1$ this condition means that the set E is infinite.

is the set of positive measure. Using the same argument as in (i), we get $b' = 1$, $a' = 2\pi$. Then the representation (3) implies that $g(t, x) = 0$ for $t \in \widehat{F}$ and for almost all $x \in K(\pi)$. Besides, the series

$$\sum_{k\in\mathbb{Z}^d\setminus(\mathbb{Z}_+)^d} f(x-2\pi k)\exp\{i\langle t, 2\pi k\rangle\}$$

is, indeed, a finite sum for almost all $x \in K(\pi)$. The application of Theorem 4 shows that $g(t, x) = 0$ for all $t \in K(1/2)$ and almost all $x \in K(\pi)$. So $f(x) = 0$ a.e. on $\mathbb{R}^d$.

(iii) The only difference between this and the preceding cases is the interchange of restrictions on the behavior of $f(x)$ and $\hat{f}(t)$, in particular the interchange of parts of the representation (4) and (5).

(iv) In this case the series (4) satisfies the conditions of Theorem 4 for $t = 0$ only. Thus, $g(0, x) = 0$ a.e. Therefore, the sum of the series (5) is equal to zero only when $t = 0$. But for such t's every term of (5) is nonnegative and, hence, vanishes. For this reason $f(x) = 0$ a.e. The proof is over.

COROLLARY. *Let $f \in L^1(\mathbb{R}^d)$ and let Γ be a convex proper cone in $\mathbb{R}^d$. Every one of the following conditions implies that $f(x) = 0$ a.e.:*

(i) $\operatorname{meas}_d(S(f)\setminus\Gamma) < \infty$ *and* $\operatorname{meas}_d(S(\hat{f})) < \infty$;

(ii) $\operatorname{meas}_d(S(\hat{f})\setminus\Gamma) < \infty$ *and* $\operatorname{meas}_d(S(f)) < \infty$.

PROOF. A linear transformation of coordinates reduces the general case to $\Gamma \subset (\mathbb{R}_+)^d$. Since any set of finite measure is finite a.e. over an arbitrary cube, we can use Theorem 5.

DEFINITION 5. A set A is called *lacunary over* $E \subset K(a/2)$ if for almost all $x \in E$ there exists a number $N = N(x) \in \mathbb{N}$ such that for every $p \in \mathbb{Z}^d \setminus \{0\}$ the point x is covered by at most N sets belonging to the family $(A + a(n+p)) \cap (A+n)$, $n \in \mathbb{Z}^d$.

The lacunarity just defined means that for almost all $x \in E$ and all $p \in \mathbb{Z}^d \setminus \{0\}$ the number of representations of p in the form $m - n$, where $m, n \in \mathbb{Z}^d$, $x - an \in A$ and $x - am \in A$, does not exceed $N = N(x)$ which does not depend on p. In other words, the sets $\{n \in \mathbb{Z}^d : x + an \in A\}$ are lacunary (in the sense of Definition 1) for almost all $x \in E$.

EXAMPLE 2. Let $\{x^{(r)}\}_{r=1}^{\infty}$ be a sequence of vectors of $\mathbb{R}^d \setminus K(1)$ and let for some $\alpha > 0$ and all $r, s \in \mathbb{N}$, $r \neq s$,

$$|x^{(r)} - x^{(s)}| \geqslant \alpha \max\{|x^{(r)}|, |x^{(s)}|\}.$$

Define the set A generated by this sequence

$$A = \bigcup_{r=1}^{\infty}\{x : |x - x^{(r)}| < 1\}.$$

It turns out that A is lacunary over $K(1/2)$. For the proof it is sufficient, due to Example 1 of §1, to show that for some $\beta > 0$ the set

$$B = \{n \in \mathbb{Z}^d : x + n \in A\}$$

satisfies for a.e. $x \in K(1/2)$ the following condition

$$\forall m, n \in B: \quad |n - m| \geqslant \beta\{|m|, |n|\}. \tag{6}$$

If $x + n \in A$, $x + m \in A$ and $m \neq n$, then we have $|(x+n) - (x+m)| \geqslant 1$. Hence $x^{(r)}$ and $x^{(s)}$ such that $|x + n - x^{(r)}| < 1/2$ and $|x + m - x^{(s)}| < 1/2$ are different. It is clear that $|n - x^{(r)}| < 1$ and $|m - x^{(s)}| < 1$. Thus,

$$\begin{aligned} |n - m| \geqslant |x^{(r)} - x^{(s)}| - 2 &\geqslant \alpha \max\{|x^{(r)}|, |x^{(s)}|\} - 2 \\ &\geqslant \alpha \max\{|n|, |m|\} - \alpha - 2. \end{aligned}$$

Inequality (6) is valid for all pairs (m, n) with $\max\{|m|, |n|\}$ sufficiently large. For other pairs and sufficiently small β it is evident.

THEOREM 6. *Let $f \in L^1(\mathbb{R}^d)$. Each of the following conditions implies that $f(x) = 0$ a.e.:*

(i) *$S(f)$ is lacunary over $K(\pi)$ and $\hat{f}(t)$ vanishes on a 1-periodic set of positive measure;*
(ii) *$S(\hat{f})$ is lacunary over $K(1/2)$ and $f(x)$ vanishes on a 2π-periodic set of positive measure;*
(iii) *$S(\hat{f})$ is lacunary over the set $\widehat{F} \subset K(1/2)$ whose closure is a set of positive measure, $S(f)$ is lacunary over $K(\pi)$, and, besides, $f(x)$ vanishes on a 2π-periodic set;*
(iv) *$f(x) \geqslant 0$ and vanishes on a 2π-periodic set of positive measure, and $S(\hat{f}) \cap \mathbb{Z}^d$ is a lacunary subset of $\mathbb{Z}^d$.*

PROOF. (i) Let $E + \mathbb{Z}^d$ be the above-mentioned 1-periodic set. Equation (4) implies $g(t, x) = 0$ for $t \in \widehat{E}$. Since the series (5) is lacunary for almost all $x \in K(\pi)$, we have $g(t, x) = 0$ a.e. on $K(1/2) \times K(\pi)$ (due to Theorem 1). Therefore, $f(x) = 0$ a.e.

(ii) The proof of this assertion can be obtained by the method used in the preceding theorem.

(iii) Let $f(x)$ vanish on the 2π-periodic set $E + 2\pi\mathbb{Z}^d$. Equation (5) implies $g(t, x) = 0$ for all t and $x \in E$. Since the series (4) is lacunary for $t \in \widehat{F}$, the function $g(t, x)$ equals 0 for $t \in \widehat{F}$ and almost all $x \in K(\pi)$. Since for a.e. x the function $g(t, x)$ is continuous with respect to t, it is equal to zero for almost all $x \in K(\pi)$ and for all t from some set of positive measure. For this reason conditions (ii) are satisfied, which complete the proof in case (iii).

(iv) The series (4) is lacunary. Replacing the argument used in (ii), we get $g(0, x) = 0$ a.e. Since $f(x) \geqslant 0$, (5) implies that $f(x) = 0$ a.e.

The theorem is proved.

THEOREM 7. *Let $f \in L^1(\mathbb{R}^d)$ and let Γ be a convex proper cone in $\mathbb{R}^d$. Each of the following conditions implies that $f(x) = 0$ a.e.:*

(i) *$S(f) \cap \Gamma$ is a lacunary set over $K(\pi)$, $S(f) \setminus (\Gamma \cup -\Gamma)$ is finite over $K(\pi)$ and $\hat{f}(t)$ vanishes on a 1-periodic open set;*

(ii) *$f \in L^2(\mathbb{R}^d)$ and vanishes on a 2π-periodic open set, $S(\hat{f}) \cap (\Gamma \cup -\Gamma)$ is finite a.e. over $K(1/2)$.*

The proof of this theorem differs from the proof of the preceding one only by the application of Theorem 2 instead of Theorem 1. It is necessary, however, to note that the set $\{n \in \mathbb{Z}^d \setminus (\Gamma \cup -\Gamma) : c_n \neq 0\}$ must not be finite in this case. But this condition is evidently satisfied for an arbitrary convex proper cone $\Gamma' \Supset \Gamma$.

THEOREM 8. *Let $f \in L^1(\mathbb{R}^d)$. Each of the following conditions implies $f(x) = 0$ a.e.:*

(i) $\operatorname{ess\,sup}\{|f(x)|(1 + |x|^d) : x \in \mathbb{R}^d\} < \infty$,

$$f(x) = 0 \quad \text{for } x \in \bigcup_j \{x \in \mathbb{R}^d \setminus (\mathbb{R}_+)^d : k_j \leqslant |x| \leqslant l_j\},$$

where the monotonically increasing sequences $\{k_j\}$ and $\{l_j\}$ are such that $\sum_j (1/k_j - 1/l_j) \ln l_j = \infty$, and $f(t)$ vanishes on some periodic set of positive measure;

(ii) *$f^{(k)} \in L^1(\mathbb{R}^d)$ for all $k \in (\mathbb{Z}_+)^d$ such that $k_1 + \cdots + k_d \leqslant d$, $f(x) = 0$ on some set of positive measure, and $\hat{f}(t)$ vanishes on the same set as $f(x)$ does in* (i).

PROOF. (i) Without loss of generality, assume that $\hat{f}(t) = 0$ on some 1-periodic set. For almost all $x \in K(\pi)$, the conditions of Theorem 3 are satisfied by the series (5). Therefore, $g(t, x) = 0$ a.e. and, hence, $f(x) = 0$ a.e.

(ii) We have $\hat{f}(t) = o(|t|^{-d})$ as $|t| \to \infty$. Thus, using Theorem 3 for the series (4), we get the desired result.

§3. Some applications

Let A and B be sets in $\mathbb{R}^d$ and let $\operatorname{meas}_d A > 0$. If a system

$$\Lambda_B = \{\exp\{i\langle\lambda, x\rangle\} : \lambda \in B\}$$

is not complete in the space $L^1(A)$, then there exists $f \in L^1(A)$, $f \neq 0$, such that $S(f) \subset A$ and $S(\hat{f}) \subset \mathbb{R}^d \setminus B$. It turns out that every theorem of §2 implies a completeness theorem for some system of exponentials. For example, the following result is an immediate consequence of assertion (i) of Theorem 6.

THEOREM 9. *Let the set A be lacunary over $K(\pi)$ and the set B contain a 1-periodic subset of positive measure. Then the system Λ_B is complete in $L^1(A)$.*

Now we would like to give another application dealing with the Radon transform.

Let $f \in L^d(\mathbb{R}^d)$. For $(\omega, \tau) \in \mathbb{S}^{d-1} \times \mathbb{R}$ set

$$\tilde{f}(\omega, \tau) = \int_{\langle s, \omega\rangle = \tau} f(s)\, ds.$$

It is known that $\tilde{f}(\omega, \tau) \in L^1(\mathbb{R}; d\tau)$ for every $\omega \in \mathbb{S}^{d-1}$. The following connection between Fourier and Radon transforms is very important:

$$\hat{f}(\lambda\omega) = \int_{-\infty}^{\infty} \tilde{f}(\omega, \tau) \exp\{-i\lambda\tau\}\, d\tau.$$

It permits us to prove uniqueness theorems for the Radon transform as consequences of results of §2. For example, the particular case $d = 1$ of Theorem 6(i) implies

THEOREM 10. *Let $f \in L^1(\mathbb{R}^d)$ be such a function that for all ω from a dense set in $\mathbb{S}^{d-1}$ its Radon transform $\tilde{f}(\omega, \tau)$ is not zero for ω from a set finite over $[-a(\omega), a(\omega)]$, $a(\omega) < \infty$. Let $\hat{f}(\lambda\omega)$ vanish on a $b(\omega)$-periodic set $E + b(\omega)\mathbb{Z}$ for the same set of ω. If E is an infinite set and $a(\omega)b(\omega)/(2\pi)$ is a rational number, then $f(x) = 0$ a.e.*

REFERENCES

1. M. Benedicks, *On Fourier transforms of functions supported on a set of finite Lebesgue measure*, J. Math. Anal. Appl. **106** (1985), 180–183.

2. T. Matolsci and J. Szücs, *Intersection des mesures spectrales conjugées*, C. R. Acad. Sci. Paris Ser. A **277** (1973), 841–843.

3. P. P. Kargaev, *Nonlocal almost differential operators and interpolation by functions with sparse spectrum*, Mat. Sb. **128** (1985), 133–142; English transl. in Math. USSR-Sb. **56** (1987).

4. B. Ya. Levin, *Density of some systems of functions, quasianalyticity and subharmonic majorants*, Zap. Nauchn. Sem. Leningrad. Otdel. Mat. Inst. Steklov. (LOMI) **170** (1989), 102–156; English transl. in J. Soviet Math. (to appear).

5. P. Koosis, *The logarithmic integral*, 1, Cambridge University Press, Cambridge, 1988.

6. A. Zygmund, *Trigonometric series*, vol. 1, Cambridge University Press, Cambridge, 1959.

7. A. Beurling, *On quasianalyticity and general distributions*, Mimeographed Lecture Notes, Stanford University, 1961.

KHARKOV STATE UNIVERSITY, 4 INDEPENDENCE SQUARE, KHARKOV, 310077 UKRAINE

Translated by V. N. LOGVINENKO

ADVANCES IN SOVIET MATHEMATICS
Volume 11, 1992

Parametrization of Entire Functions of Sine-Type by Their Critical Values

A. E. EREMENKO AND M. L. SODIN

An entire function $f(z)$ is called a *sine-type* function if it satisfies the following conditions:

(a) $f(z)$ has exponential type σ $(0 < \sigma < \infty)$ in each of half-planes $\mathbb{C}_+ = \{\operatorname{Im} z > 0\}$ and $\mathbb{C}_- = \{\operatorname{Im} z < 0\}$;
(b) all zeros of the function $f(z)$ are located in some strip $\{|\operatorname{Im} z| \leqslant h < \infty\}$;
(c) the inequality $0 < K_1 \leqslant |f(x + iH)| \leqslant K_2 < \infty$ is valid for some $H > h$ and every $x \in \mathbb{R}$.

In other words, $K_1 e^{\sigma|y|} \leqslant |f(x+iy)| \leqslant K_2 e^{\sigma|y|}$, where the upper estimate holds everywhere, and the lower estimate holds outside some exceptional set lying in the strip $\{|\operatorname{Im} z| \leqslant h_1 < \infty\}$.

This class of functions was introduced by B. Ya. Levin, and numerous applications of this class were found by him and his coauthors (see, for example, [1–6]).

B. Ya. Levin and I. V. Ostrovskiĭ posed the problem of parametric description of zeros of sine-type functions [7]. They noticed that only real functions with real zeros need be considered, since conditions (a)–(c) remain valid when zeros of $f(z)$ move in some strip perpendicularly to the real axis, and only the constants h, K_1, K_2 may change under such motion.

In this paper we shall give a parametrization of sine-type functions using their critical values. This parametrization leads to a parametrization of zeros of sine-type functions.

We shall make use of some results of G. R. MacLane [8]. Their modern exposition was given by E. B. Vinberg in [9]. We agree with E. B. Vinberg, who wrote in [9] that these results deserve to be known better.

Let $\mathfrak{LP}$ be the Laguerre-Polya class of real entire functions approximable by real polynomials with real zeros uniformly on every compact set. It is well known that this class is invariant with respect to differentiation.

1991 *Mathematics Subject Classification.* Primary 30D15.

For any function $f \in \mathfrak{LP}$ we consider its critical points, i.e. zeros of the derivative $f'(z)$, and number them in increasing order, taking into account their multiplicities. Without loss of generality, we may suppose that the numbers of all points of maximum are even. Denote by c_n the value of $f(z)$ at the nth critical point. If the sequence $\{\dots, c_{-1}, c_0, c_1, \dots\}$ is finite from the left (from the right) and if c_m is its first (last) number, then we set

$$c_{m-1} = \lim_{x\to\infty} f(x) \qquad \left(c_{m+1} = \lim_{x\to\infty} f(x)\right).$$

One or both of these values may be infinite. The set

$$\operatorname{Cr} f = \{\dots, c_{-1}, c_0, c_1, \dots\},$$

is the sequence of critical values of the function $f(z)$. This sequence has alternating signs: $(-1)^n c_n \geqslant 0$.

Let $s = \{c_n\}$ be an arbitrary alternating sequence. An elegant way of constructing a function $f \in \mathfrak{LP}$ such that $\operatorname{Cr} f = s$ was proposed in [8, 9]. Define a simply connected comb-like domain

$$\Omega(s) = \mathbb{C} \setminus \bigcup_n \{\zeta = \xi + in\pi : -\infty < \xi \leqslant \log|c_n|\}. \tag{1}$$

If $c_n = 0$, then the nth cut is missing. If the first (the last) member of the sequence is infinite, then we consider the domain lying in the half-plane $\{\operatorname{Im}\zeta > (m-1)\pi\}$ (in $\{\operatorname{Im}\zeta > (m+1)\pi\}$). Let $\varphi(z)$ be the conformal map of the lower half-plane $\mathbb{C}_-$ onto the domain $\Omega(s)$ such that the real axis is mapped onto the consecutive cuts, i.e. $\operatorname{Re}\varphi(iy) \to +\infty$ as $y \to -\infty$. Then the function

$$f(z) = e^{\varphi(z)} \tag{2}$$

can be continued across the real axis by the Reflection Principle. Thus $f(z)$ is an entire function. It may be proved that $f \in \mathfrak{LP}$ and $\operatorname{Cr} f = s$. Conversely, an arbitrary function of the class $\mathfrak{LP}$ can be represented in the form (2), where $\varphi(z)$ is the conformal map of the upper half-plane onto the comb-like domain $\Omega(s)$, $s = \operatorname{Cr} f$.

THEOREM 1. *A real function $f(z)$ is a sine-type function if and only if it can be represented in the form* (2) *and the sequence $s = \operatorname{Cr} f$ satisfies the conditions*:

(α) $\sup\{\log|c_n| : n \in \mathbb{Z}\} \leqslant M_1 < \infty$;
(β) *there exist a natural number N and a real number M such that* $\max\{\log|c_n| : n \leqslant k \leqslant n+N\} \geqslant M_2 > -\infty$ *for every* $n \in \mathbb{N}$.

PROOF. Suppose that the sequence s satisfies conditions (α) and (β). First, we check that any function of the form (2) is of exponential type.

Let $L(r)$, $0 < r < \infty$, be a connected arc of the intersection $\Omega(s) \cap \{|\zeta| = r\}$ that contains the point $\zeta = r$, and let $\theta(r)$ be the radian measure of $L(r)$. It follows from conditions (α) and (β) that

$$|\theta(r) - \pi| \leqslant K_3/r. \tag{3}$$

By virtue of the well-known Distortion Theorem [10, Chapter VI] (see also [11]), the estimate (3) implies that

$$\max\{|z| : z \in J(r)\} \approx \min\{|z| : z \in J(r)\} \approx \sigma r, \qquad r \to \infty, \tag{4}$$

where $J(r)$ is the preimage of $L(r)$ under the mapping $\varphi(z)$ and $\sigma > 0$. It follows from (4) and (2) that $f(z)$ is an entire function of exponential type.

By condition (α), the function $f(z)$ is bounded on the real axis. Now we prove the estimate from below on a line parallel to the real axis.

Let $\chi(\zeta) = \varphi^{-1}(\zeta)$ be the inverse mapping, then $u(\zeta) = -\operatorname{Im}\chi(\zeta)$ is a positive harmonic function in the domain $\Omega(s)$. We shall prove that this function is bounded from above on the line $\zeta = M_1 + 1 + it$, $-\infty < t < \infty$. Without loss of generality, we assume that $M_1 > 1$. Let $L_t(r)$, $-\infty < t < \infty$, $0 < r < \infty$ be the connected arc of the intersection

$$\Omega(s) \cap \{|\zeta - M_1 - 1 - it| = r\}$$

that contains the point $\zeta_1 = M_1 + 1 + it + r$, and let $\theta_t(r)$ be the radian measure of this arc. By conditions (α) and (β), the value $\theta_t(r)$ satisfies the estimate (3) with the same constant K_3 that does not depend on t. Denote by $\Omega_{t,r}(s)$ the simply connected subdomain of $\Omega(s)$ bounded by the cuts and the arc $L_t(r)$ and such that points $\zeta = M_1 + 1 + it + r + \varepsilon$, $\varepsilon > 0$, do not belong to $\Omega_{t,r}(s)$. By $\omega_t(r)$ we denote the harmonic measure of the arc $L_t(r)$ at the point $\zeta = M_1 + 1 + it$ with respect to the domain $\Omega_{t,r}(s)$.

By the well-known estimate of harmonic measure (see for example [12]), we have

$$\omega_t(r) \leqslant C_1 \exp\left[-\pi \int_1^r \frac{d\tau}{\tau\,\theta_t(r)}\right] \leqslant \frac{C_2}{r}, \qquad r \geqslant 3,$$

with the same constant C_2 for every value t, $-\infty < t < \infty$. (We have used the estimate (2) in the second inequality.) Choosing r sufficiently large, we obtain

$$u(M_1 + 1 + it) \leqslant \omega_t(r)\max\{u(\zeta) : \zeta \in L_t(r)\} \leqslant (C_2/r)(\sigma + o(1))r \leqslant C_3.$$

Thus the image of $\Omega(s) \cap \{\zeta : \operatorname{Re}\zeta \leqslant M_1 + 1\}$ under the mapping $\chi(\zeta)$ is located in the strip $\{-C_3 \leqslant \operatorname{Im} z < 0\}$. Therefore,

$$\operatorname{Re}\varphi(x - (C_3 + 1)i) \geqslant M_1 + 1$$

for all x from the real axis. The lower estimate for $f(z)$ is proved.

Conversely, let $f(z)$ be a sine-type function. It is evident that the sequence $s = \operatorname{Cr} f$ satisfies condition (α). We shall show that condition (β) holds as well.

Suppose that (β) does not hold. Without loss of generality we assume that $M_1 = 0$. Then for every sufficiently large value $T < \infty$ there exists a value $t \in (-\infty, \infty)$ such that the domain $\Omega(s)$ does not intersect the square

$$Q_t(T) = \{\zeta = \xi + i\eta : -2T \leqslant \xi \leqslant 0,\ |\eta - t| \leqslant T\}.$$

It follows from Lindelöf variational principle (see, for example, [13]) that there exists a real quantity x^* such that

$$\operatorname{Re}\varphi(x^* - i) \leqslant -1/2. \tag{5}$$

But then (2) and (5) contradict condition (c) from the definition of sine-type functions. The theorem is proved.

NOTE 1. If in the definition of sine-type functions one requires that the exponential type of $f(z)$ equals π, then the conformal mapping $\varphi(z)$ has to be normalized: $\operatorname{Re}\varphi(z) \approx \pi|y|$ as $y \to -\infty$.

NOTE 2. Suppose that a sequence s satisfies condition (α). Consider the translations $\Omega(s)+it$, $-\infty < t < \infty$. Let us notice that Carathéodory kernels of this set of simply connected domains do not contain the whole complex plane if and only if condition (β) is fulfilled. In this case every kernel domain has the form (1) for some sequence s that satisfies the conditions (α) and (β) with the same constants M_1 and M_2.

This observation and a version of Carathéodory Convergence Theorem allow us to give another proof of Theorem 1 without using the Distortion Theorem and estimates of the harmonic measure.

Sometimes the following condition of separation of zeros of $f(z)$:

(d) $$|z_n - z_m| \geqslant \delta > 0 \quad \text{as } n \neq m$$

is included in the definition of sine-type functions (see, for example, [2]). In this case one can replace conditions (α) and (β) by the condition

(α^*) $$-M \leqslant \inf\{\log|c_n| : n \in \mathbb{Z}\} \leqslant \sup\{\log|c_n| : n \in \mathbb{Z}\} \leqslant M.$$

Indeed, as proved in [2], the estimate

$$\inf\{|f'(z_n)| : n \in \mathbb{Z}\} \geqslant m > 0 \tag{6}$$

holds for sine-type functions with separated zeros. The derivatives of sine-type functions are bounded on the real axis. So (6) yields $|f'(x)| \geqslant m/2$ for some $\varepsilon > 0$ and for any $n \in \mathbb{Z}$, when $z_n - \varepsilon \leqslant x \leqslant z_n + \varepsilon$. Hence condition (α^*) is fulfilled.

We shall give a different parametrization of real sine-type functions with real separated zeros. To this end, using the sequence $s = \{c_n\}$, we define a simply connected comb-like domain of another type:

$$G(s) = \mathbb{C}_+ \setminus \bigcup_n \{\zeta = \pi n + i\eta : 0 \leqslant \eta \leqslant \cosh^{-1}|c_n|\}.$$

Denote by $\theta(z)$ the conformal mapping of $\mathbb{C}_+$ onto $G(s)$ that maps infinity onto itself. Then the function

$$f(z) = \cos\theta(z) \tag{7}$$

can be continued to an entire function. It is not difficult to show that this function has finite or zero exponential type and all its $+1$-points are real. Conversely, every real entire function with real $+1$-points can be represented in the form (7) (see [14, §1]).

THEOREM 2. *A real function $f(z)$ is a real sine-type function with separated zeros if and only if the function $cf(z)$ can be represented in the form* (7), *for some sequence s bounded from above, where $c = \sup_{n\in\mathbb{Z}}\{|c_n|^{-1}\}$.*

PROOF. Let s be a sequence bounded from above. Then the function $f(z) = \cos\theta(z)$ is an entire function of exponential type, bounded on the real axis. The lower estimate and the fact of separation of zeros can be deduced from the Lindelöf variational principle (see [14, §2], where similar arguments were used).

Conversely, let $f(z)$ be a real sine-type function with real separated zeros. As was proved above, the value $\sup\{|c_n|^{-1} : n \in \mathbb{Z}\}$ is finite. Consider the function $cf(z)$. One can choose a subsequence of its 1-points interlacing with its zeros. Standard arguments (see [14], §1) show that this subsequence contains, in fact, all 1-points of $cf(z)$. The same conclusion is also true for all -1-points of the function $cf(z)$. So this function can be represented in the form (7). The theorem is proved.

NOTE 3. It seems interesting to describe some natural subclasses of the class $\mathfrak{LP}$ imposing some additional conditions on the sequence $\operatorname{Cr} f$. Some results in this direction were obtained in [15] and in [14, §2]. Here is one situation that we would like to touch briefly.

Suppose that the sequence $s = \operatorname{Cr} f$ is bounded. Then the function $f(z) = e^{\varphi(z)}$ is bounded on the real axis and has the order of growth $\rho_f \geqslant 1$. One can ask in this case whether the function $f(z)$ has an exponential type or not. It is not difficult to notice that $f(z)$ has exponential type if and only if the angular derivative of the mapping $\varphi(z)$ exists. This means that for any $\varepsilon > 0$ the limit of $\varphi(z)/z$ as $z \to \infty$ and $-\pi + \varepsilon \leqslant \arg z \leqslant \pi - \varepsilon$ exists. After this observation the criterion of B. Rodin and S. Warschawski can be applied.

The authors thank A. A. Goldberg and I. V. Ostrovskiĭ for useful critical remarks.

REFERENCES

1. B. Ya. Levin, *On bases of exponential functions in* L_2, Zap. Mekh. Mat. Fak. Kharkov. Gos. Univ. i Kharkov. Mat. Obshch. (4) **27** (1961), 39–48. (Russian)

2. ———, *Interpolation by entire functions of exponential type*, Mat. Fiz. i Funktsional. Anal. **1** (1969), 136–146. (Russian)

3. B. Ya. Levin and Din Than Hoa, *On interference operators over entire functions of exponential type*, Funktsional. Anal. i Prilozhen. **3** (1969), no. 1, 48–61; English transl. in Functional Anal. Appl. **3** (1969).

4. V. E. Katsnelson, *On bases of exponential functions in* L_2, Funktsional. Anal. i Prilozhen. **5** (1971), no. 1, 37–47; English transl. in Functional Anal. Appl. **5** (1971).

5. Yu. I. Lyubarskiĭ, *System of exponents on a curve*, Sibirsk. Mat. Zh. **30** (1989), no. 2, 108–121; English transl. in Siberian Math. J. **30** (1989).

6. B. Ya. Levin and I. V. Ostrovskiĭ, *On small perturbation of the set of functions of sine type*, Izv. Akad. Nauk SSSR Ser. Mat. **43** (1979), no. 1, 87-110; English transl. in Math. USSR-Izv. **14** (1980).

7. B. Ya. Levin and I. V. Ostrovskiĭ, *Zero sets of sine-type functions*, Linear Complex Analysis Problem Book, Lecture Notes in Math., vol. 1043, Springer-Verlag, Berlin and New York, 1984, pp. 605–608.

8. G. R. MacLane, *Concerning the uniformization of certain Riemann surfaces*, Trans. Amer. Math. Soc. **62** (1947), 99–111.

9. E. B. Vinberg, *Real entire functions with prescribed critical values*, Problems in Group Theory and in Homological Algebra, Yaroslav. Gos. Univ., Yaroslavl′, 1989, pp. 127–138. (Russian)

10. H. Wittich, *Neuere Untersuchungen über eindeutige analytische Funktionen*, Springer-Verlag, Berlin, 1955.

11. A. A. Goldberg, *On ramified values of entire functions*, Sibirsk. Mat. Zh. **14** (1973), no. 4, 862–866; English transl. in Siberian Math. J. **14** (1973).

12. M. Tsuji, *Potential theory in modern function theory*, Maruzen, Tokyo, 1959.

13. M. A. Lavrent′ev and B. V. Shabat, *Methods of the theory of functions of a complex variable*, "Nauka", Moscow, 1965; German transl., VEB Deutscher Verlag Wiss., Berlin, 1967.

14. V. A. Marchenko and I. V. Ostrovskiĭ, *A characterization of the spectrum of Hill's operator*, Mat. Sb. **97** (1975), 540–606; English transl. in Math. USSR-Sb. **26** (1975).

15. A. A. Goldberg, *Certain asymptotic properties of meromorphic functions*, Uchen. Zap. L′vov. Univ. **36** (1956), no. 5, 54–73. (Russian)

16. B. Rodin and S. E. Warschawski, *Extremal length and univalent functions*, Math. Z. **153** (1977), 1–17.

Institute for Low Temperature Physics & Engineering, 47 Lenin pr., Kharkov 310164 Ukraine

Translated by M. L. SODIN

ADVANCES IN SOVIET MATHEMATICS
Volume 11, 1992

On Stability Estimates in the Marcinkiewicz Theorem and Its Generalization

L. B. GOLINSKIĬ AND G. P. CHISTYAKOV

The well-known Marcinkiewicz theorem asserts that

$$\varphi(t) = \exp\{Q(t)\}, \qquad Q(t) = \sum_{k=1}^{n} q_k t^k,$$

is a characteristic function (c.f.) for the distribution function (d.f.) F only if F is a normal d.f. (possibly, singular). The result is applied in a number of problems of mathematical statistics. When these problems are studied from the point of view of stability, the question of the stability of the Marcinkiewicz theorem itself arises, i.e., one may ask what can be said of the coefficients of the polynomial $Q(t)$ if the function $\varphi(t)$ is near, in a certain sense, to the c.f. This problem was stated by N. A. Sapogov [1] who obtained the first estimates of the coefficients of $Q(t)$. Later, these estimates were significantly refined by L. B. Golinskiĭ [2].

I. V. Ostrovskiĭ, answering a question posed by Yu. V. Linnik [4, p. 255], obtained a natural generalization of the Marcinkiewicz theorem. He showed that the assertion of this theorem remains valid if we require that $Q(t)$ be an entire function of order not exceeding one and of minimal type. Note that the example of the c.f. for Poisson's d.f. shows that Ostrovskiĭ's result cannot be improved. As in Sapogov's approach, a similar question concerning the stability of the Ostrovskiĭ theorem arises. This note is aimed at obtaining estimates of the latter theorem's stability which, in the case when $Q(t)$ is a polynomial, should refine known estimates.

Further on, the letters $c, c_1, c_2, \dots$ will denote positive absolute constants, c without a subscript being used for different constants when their numeric value is not of interest.

1991 *Mathematics Subject Classification*. Primary 60E10.

Let

$$Q(z) = Q(z, \varepsilon, T) = \sum_{j\geqslant 1} q_j z^j \qquad (z \in \mathbb{C}) \tag{1}$$

be an entire function, $Q(0) = 0$, and for the c.f. $\varphi(t; F)$ of a d.f. F we have

$$|\exp\{Q(t)\} - \varphi(t; F)| \leqslant \varepsilon, \qquad t \in [-T, T]. \tag{2}$$

We assume that the function $Q(z)$ admits a majorant

$$\max_{|z|\leqslant r} |Q(z)| \leqslant M(r) = \exp\{\delta(r)\}, \qquad r \geqslant c,$$

where $M(r)$ is independent of ε and T, and $M(r) \geqslant c_1 r$, $r \geqslant c$. We shall also assume that the normalization conditions

$$\operatorname{Im} q_1 = 0, \qquad \operatorname{Re} q_2 = -1/2, \tag{3}$$

are fulfilled. The problem consists in obtaining estimates of the coefficients q_j, $j \geqslant 3$ for sufficiently small ε and $T \geqslant c$.

To state the main result of the present work, we shall introduce the following quantities:

$M^{(-1)}(r)$ is the function inverse to $M(r)$; $f(r) = \ln M^{(-1)}(r)$;

$$kf(k) = \ln(1/\varepsilon), \qquad T_1 = \min\left\{T, \sqrt{M^{(-1)}(k)}\right\}, \tag{4}$$

and we shall determine a parameter R from the equality

$$M(2R) \ln R = c_2 T_1 f(k). \tag{5}$$

THEOREM. *Let the entire function $Q(z)$ defined by* (1) *satisfy relation* (2) *and the normalization condition* (3). *Then*

$$|q_j| \leqslant \frac{c}{R_1^{j-2}}, \qquad j \geqslant 3, \; R_1 = \frac{\pi}{64} \frac{R}{\delta(R/4)}, \tag{6}$$

where the value R is determined from (5).

NOTE. Assume that $\delta(r) r^{-1} \to 0$, $r \to +\infty$ (i.e., consider the case when estimates (6) are meaningful). Then $q_j \to 0$ as $\varepsilon \to 0$. This theorem gives quantitative estimates of stability in the Ostrovskiĭ theorem.

EXAMPLE. Let $M(r) = cr^n$. This case corresponds to the stability in the Marcinkiewicz theorem studied in [2]. If we put $T = (\ln(1/\varepsilon))^\beta$, $\beta > 0$, in this theorem, we arrive at the estimates

$$|q_j| \leqslant c \left\{ \frac{(\ln\ln(1/\varepsilon))^{n+1}}{(\ln(1/\varepsilon))^\gamma} \right\}^{(j-2)/n}, \quad \gamma = \min\left(\beta, \frac{1}{2n}\right); \; j = 3, 4, \dots, n.$$

The theorem also includes the case of a fixed interval ($T = 1$, see [2, p. 21]). Calculations yield the estimates

$$|q_j| \leqslant c \left\{ \frac{(\ln\ln\ln(1/\varepsilon))^{n+1}}{\ln\ln(1/\varepsilon)} \right\}^{(j-2)/n}, \qquad j = 3, 4, \dots, n.$$

So, the theorem proved permits refining the results of [2].

PROOF OF THE THEOREM. **1**. Estimation of the tail of the d.f. F. Here we shall follow the argument of [2, pp. 15–17]. Denote $\varphi_1(t) = \exp\{Q(t)\}$. By virtue of Sapogov's inequality [5],

$$\begin{aligned} W_p(2\pi/s) &= 1 - F(2\pi/s) + F(-2\pi/s + 0) \\ &\leqslant \frac{2\pi}{4^k I_{2k} s} \int_0^s |\Delta_u^{(2k)}(\varphi_1; 0)|\, du + \frac{2\pi\varepsilon}{I_{2k}} = W_1 + W_2, \end{aligned}$$

where $s > 0$, k is a natural number and $ks \leqslant T$;

$$I_{2k} = \sqrt{\pi}\frac{\Gamma(k+1/2)}{\Gamma(k+1)} > \frac{c}{\sqrt{k}}, \qquad \Delta_u^{(2k)}(\varphi_1; 0) = \sum_{j=0}^{2k}(-1)^j \binom{2k}{j} \varphi_1((j-k)u).$$

Further on, we shall assume that $ks = T_1 \leqslant T$, T_1 as chosen in (4). As in [2, p. 16], for $k \leqslant s$ and $r > T_1$ we have

$$|\Delta_u^{(2k)}(\varphi_1; 0)| \leqslant (2k)!\, u^{2k} \frac{r\exp(M(r))}{(r - T_1)^{2k+1}},$$

so that

$$W_1 \leqslant 2\pi \frac{(2k)!\, s^{2k}}{4^k I_{2k}(2k+1)} \frac{r\exp(M(r))}{(r-T_1)^{2k+1}} \leqslant c\frac{(2ks)^{2k}}{(2e)^{2k}} \frac{r\exp(M(r))}{(r-T_1)^{2k+1}}.$$

If $r \geqslant 2T_1$, then

$$W_1 \leqslant c\exp\{-2k\ln(er/(2T_1)) + M(r)\}, \qquad W_2 \leqslant c\varepsilon\sqrt{k}. \tag{7}$$

In the latter inequalities k may be regarded as a positive number (integer or not), which will lead only to changing the constants c.

Let the quantities k and T_1 be found from equalities (4). Set $r = M^{(-1)}(k)$ in (7). For a sufficiently small ε, it is obvious that

$$r > e\sqrt{M^{(-1)}(k)} \geqslant eT_1,$$

hence

$$-2k\ln\frac{er}{2T_1} + M(r) = -2k\ln\frac{er}{2T_1} + k \leqslant -2k\ln\frac{r}{T_1} + k < -k\ln\frac{r}{T_1}.$$

Therefore

$$W_F(2\pi k/T_1) \leqslant c\{\exp(-k\ln(r/T_1)) + \varepsilon\sqrt{k}\} = \varepsilon_1. \tag{8}$$

2. Let F^* be the truncated d.f. on the level

$$N = \frac{2\pi k}{T_1} = \frac{2\pi M(r)}{T_1} \geqslant 2\pi c_1 \frac{M^{(-1)}(k)}{T_1} \geqslant 2\pi c_1\sqrt{M^{(-1)}(k)} > c, \tag{9}$$

$\varphi^*(t) = \varphi(t; F^*)$. Then $|\varphi^*(t) - \varphi(t; F)| \leqslant 2W_F(N)$, $|\varphi^*(t) - \varphi_1(t)| \leqslant 3\varepsilon_1$, $|t| \leqslant T$. Choose a natural number q from the condition

$$q = [c_3 k T_1^{-1} M(2R)] \tag{10}$$

and let us approximate the functions $\varphi^*(z)$ and $\varphi_1(z)$ by portions of length q of their Taylor series:

$$|\varphi^*(z) - Q^*(z)| \leqslant \frac{(N|z|)^q}{q!}\exp(N|z|), \qquad \deg Q^*(z) \leqslant q-1,$$

$$|\varphi_1(z) - Q_1(z)| \leqslant 2(|z|/r_1)^q \exp(M(r_1)), \qquad r_1 \geqslant 2|z|, \ \deg Q_1(z) \leqslant q-1.$$

Then on the segment $|t| \leqslant 1$, we have

$$|Q^*(t) - Q_1(t)| \leqslant 3\varepsilon_1 + (1/q!)N^q e^N + 2r_1^{-q}\exp(M(r_1)) = \varepsilon_2, \qquad r_1 \geqslant 2.$$

By the Bernstein theorem [6, Chapter 2, §1] in the disk $\{|z| \leqslant R_2 = R/4\}$, R chosen as in (5), we have $|Q^*(t) - Q_1(t)| \leqslant \varepsilon_2 R^{q-1}$. Therefore, for $|z| \leqslant R_2$,

$$|\varphi^*(z) - \varphi_1(z)| \leqslant \varepsilon_2 R^{q-1} + \frac{(NR_2)^q}{q!}\exp(NR_2) + 2(R_2/r_2)^q\exp(M(r_2))$$
$$= \varepsilon_3, \qquad r_2 \geqslant 2R_2.$$

Let us estimate ε_3.

1) Let the constant c_3 satisfy the condition $c_3\ln(c_3/(2\pi e)) > 2\pi/c_1$. By virtue of the choice of q (10), for a sufficiently small ε we have:

$$\begin{aligned}\frac{(NR_2)^q}{q!}\exp(NR_2) &\leqslant \frac{(NR_2)^q}{q!}\exp(NR)\\ &\leqslant c\exp\{q\ln(NR) + NR + q - q\ln q\}\\ &= c\exp\left\{-q\ln\frac{qT_1}{2\pi ekR} + 2\pi\frac{kR}{T_1}\right\}\\ &\leqslant c\exp\left\{-q\ln\frac{c_3M(2R)}{4\pi eR} + 2\pi\frac{kR}{T_1}\right\}\\ &= c\exp\left\{-q\ln\frac{c_3}{2\pi e} + 2\pi\frac{kR}{T_1} - q\ln\frac{M(2R)}{2R}\right\}\\ &< c\exp\left\{-q\ln\frac{M(2R)}{2R}\right\}.\end{aligned}$$

2) Setting $r_2 = 2R$, we obtain by using (10)

$$\begin{aligned}2(R_2/r_2)^q\exp(M(r_2)) &\leqslant 2(R/r_2)^q\exp(M(r_2)) = 2\exp\{M(2R) - q\ln 2\}\\ &\leqslant 2\exp\{M(2R) - (1/2)c_3(\ln 2)kT_1^{-1}M(2R)\}\\ &< 2\exp\{-c_4kT_1^{-1}M(2R)\}.\end{aligned}$$

3) We have

$$\varepsilon_2 R^{q-1} = 3\varepsilon_1 R^{q-1} + (1/q!)N^q e^N R^{q-1} + 2r_1^{-q}R^{q-1}\exp(M(r_1)).$$

The last two terms in the right-hand side of this equality (at $r_1 = r_2$) have been estimated in 1) and 2). Further, (8) implies

$$\begin{aligned}\varepsilon_1 R^{q-1} &= c\exp\{-k\ln(r/T_1) + (q-1)\ln R\}\\ &\quad + c\exp\{-\ln(1/\varepsilon) + (\ln k)/2 + (q-1)\ln R\}, \qquad r = M^{(-1)}(k).\end{aligned}$$

Since $rT_1^{-1} \geqslant \sqrt{M^{(-1)}(k)}$, we have $-k\ln(rT_1^{-1}) \leqslant -(1/2)kf(k)$, and by virtue of the choice of R and q (see (5), (10)),

$$c\exp\{-k\ln(r/T_1) + (q-1)\ln R\} \leqslant c\exp\{-(1/2 - c_2c_3)kf(k)\}.$$

Combining (4), (5) and (10), for a sufficiently small ε we have

$$\begin{aligned} &c\exp\{-\ln(1/\varepsilon) + (1/2)\ln k + (q-1)\ln R\} \\ &\quad \leqslant c\exp\{-kf(k) + (1/2)\ln k + c_2c_3kf(k)\} \\ &\quad \leqslant c\exp\{-(1-2c_2c_3)kf(k)\}. \end{aligned}$$

Let us choose the constant c_2 from the condition $c_2c_3 = 1/4$. Then

$$\varepsilon_1 R^{q-1} \leqslant c\exp\{-(1/4)kf(k)\} = c\varepsilon^{1/4}.$$

Comparing the estimates obtained in 1), 2), and 3), we arrive at the following result

$$\varepsilon_3 \leqslant c\exp\{-ckT_1^{-1}M(2R)\}. \tag{11}$$

4) For the function $\varphi^*(z)$ in the disk $\{|z| \leqslant R_2\}$ and for a sufficiently small ε, we have the estimate from below

$$\begin{aligned} |\varphi^*(z)| &\geqslant |\varphi_1(z)| - |\varphi^*(z) - \varphi_1(z)| \\ &\geqslant \exp(-M(R_2)) - \varepsilon_3 > (1/2)\exp(-M(R_2)). \end{aligned}$$

The latter inequality follows from (11) and the choice of the parameters k, T_1. Therefore $\varphi^*(z) \neq 0$, $|z| \leqslant R_2$, so that the function

$$g(z) = \ln\varphi^*(z) = iaz - \sigma^2z^2/2 + \sum_{j\geqslant 3} g_jz^j$$

is holomorphic. Further we have

$$\begin{gathered} |\exp\{Q(z)\}|\,|\exp\{g(z) - Q(z)\} - 1| \leqslant \varepsilon_3, \\ |\exp\{g(z) - Q(z)\} - 1| \leqslant \varepsilon_3\exp(M(R_2)) = \varepsilon_4, \qquad |z| \leqslant R_2, \end{gathered}$$

whence

$$\operatorname{Re}\{g(z) - Q(z)\} \leqslant c\varepsilon_4, \qquad |z| \leqslant R_2. \tag{12}$$

The Schwarz formula for the disk yields

$$|g_j - q_j| \leqslant c\varepsilon_4 R_2^{-j}, \qquad j \geqslant 1, \tag{13}$$

so it only remains to estimate the coefficients g_j.

5) For a sufficiently small ε (12) implies

$$\operatorname{Re} g(z) \leqslant M(R_2) + c\varepsilon_4 \leqslant 2M(R_2), \qquad |z| \leqslant R_2, \tag{14}$$

whence, applying the Schwarz formula, it is easy to get the following estimates for the derivatives:

$$|g^{(j)}(z)| \leqslant 4M(R_2)j!, \qquad |z| \leqslant R_2 - 1,\ j \geqslant 1. \tag{15}$$

Let $U(z) = \operatorname{Re} g(z)$; we shall consider the function

$$U_\xi(x, y) = U(x + iy + \xi) - U(-x + iy + \xi)$$

where the parameter ξ satisfies the condition

$$0 \leqslant \xi \leqslant R_1 = \frac{\pi}{16} \frac{R_2}{\delta(R_2)}.$$

(The idea of using the function $U_\xi(x, y)$ for studying entire ridge functions is due to A. M. Ulanovskiĭ.) The function $U_\xi(x, y)$ is harmonic in the rectangle $\{0 \leqslant x \leqslant 2\xi, |y| \leqslant R_2/2\}$, and $U_\xi(0, y) = 0$, $U_\xi(\xi, y) = U(2\xi + iy) - U(iy) \leqslant 0$, according to the ridge property. In the rectangle $\Gamma\{0 \leqslant x \leqslant 2\xi, |y| \leqslant R_2/2\}$ (and, in particular, on its horizontal sides) the estimate (14) is valid:

$$U_\xi(x, y) \leqslant 4M(R_2) = 4\exp\{\delta(R_2)\}, \qquad |y| = R_2/2.$$

Thus, $U_\xi(x, y) \leqslant 4\exp\{\delta(R_2)\}\omega(z, \Gamma)$, where $\omega(z, \Gamma)$ is the harmonic measure of the sides $\{0 \leqslant x \leqslant \xi, |y| = R_2/2\}$ of the rectangle Γ with respect to Γ. Known estimates of the harmonic measure (see [7, Chapter 8, §4]) enable us to write

$$U_\xi(x, y) \leqslant 4\exp\{\delta(R_2) + (\pi/\xi)|y| - \pi R_2/(2\xi)\}, \qquad (x, y) \in \Gamma,$$

so in the smaller rectangle $\{0 \leqslant x \leqslant \xi, |y| \leqslant R_2/4\}$ we have

$$U_\xi(x, y) \leqslant 4\exp\{\delta(R_2) - \pi R_2/(4R_1)\} = 4\exp\{-3\delta(R_2)\}.$$

We set $x = (1/2)\exp(-2\delta(R_2))$. By the mean value theorem

$$\begin{aligned} 4\exp\{-3\delta(R_2)\} \geqslant U_\xi(x, y) &= U(x + iy + \xi) - U(-x + iy + \xi) \\ &= \exp\{-2\delta(R_2)\} \cdot \frac{\partial U}{\partial x}(\eta_y + iy + \xi), \end{aligned}$$

and $|\eta_y| \leqslant (1/2)\exp\{-2\delta(R_2)\}$. Expand the harmonic function $\partial U/\partial x$ into a series in a neighborhood of the point $\eta_y + iy + \xi$:

$$\frac{\partial U}{\partial x}(iy + \xi) = \frac{\partial U}{\partial x}(\eta_y + iy + \xi) + \sum_{n \geqslant 1} \frac{\partial^{n+1} U}{\partial x^{n+1}}(\eta_y + iy + \xi) \frac{(-\eta_y)^n}{n!}.$$

By virtue of (15), we have

$$\left| \frac{\partial^{n+1} U}{\partial x^{n+1}}(\eta_y + iy + \xi) \right| \leqslant 4M(R_2)(n+1)!$$

and, therefore,

$$\begin{aligned} \frac{\partial U}{\partial x}(iy + \xi) &\leqslant 4\exp\{-\delta(R_2)\} + 4M(R_2) \sum_{n \geqslant 1} \frac{n+1}{2^n} \exp\{-2n\delta(R_2)\} \\ &\leqslant 4\exp\{-\delta(R_2)\} + 8M(R_2)\exp\{-2\delta(R_2)\} \\ &= 12\exp\{-\delta(R_2)\}. \end{aligned}$$

Thus

$$\frac{\partial U}{\partial x}(x+iy) \leqslant 12\exp\{-\delta(R_2)\}, \qquad 0 \leqslant x \leqslant R_1, \ |y| \leqslant \frac{1}{4}R_2. \tag{16}$$

We shall apply the Schwarz-Nevanlinna formula [8, pp. 16–17] to the function $g_1(z) = \exp\{g'(z)\}$ which is holomorphic and nowhere equal to zero. As $\operatorname{Re} g'(iy) = 0$, then $\ln|g_1(iy)| = 0$ and

$$\begin{aligned} \ln g_1(z) &= g'(z) \\ &= \frac{1}{2\pi}\int_{-\pi/2}^{\pi/2} \ln|g_1(R_1e^{it})| \left\{\frac{R_1e^{it}+z}{R_1e^{it}-z} - \frac{R_1e^{-it}-z}{R_1e^{-it}+z}\right\} dt + ic(R_1), \end{aligned} \tag{17}$$

where $c(R_1)$ is a real constant. We denote

$$h(t) = \ln|g_1(R_1e^{it})| = \operatorname{Re} g'(R_1e^{it}) = \frac{\partial U}{\partial x}(R_1e^{it}) = h^+(t) - h^-(t),$$

where, as usual, $h^+(t)$ ($h^-(t)$) denotes the positive (negative) part of the function $h(t)$. Let us differentiate (17) n times with respect to z:

$$\frac{g^{(n+1)}(z)}{(n+1)!} = \frac{R_1}{\pi(n+1)}\int_{-\pi/2}^{\pi/2} h(t)\left\{\frac{e^{it}}{(R_1e^{it}-z)^{n+1}} - \frac{(-1)^n e^{-it}}{(R_1e^{-it}+z)^{n+1}}\right\} dt.$$

At $z = 0$, we have

$$g_{2m+1} = \frac{R_1}{\pi(2m+1)}\int_{-\pi/2}^{\pi/2} h(t)\frac{2\sin 2mt}{iR_1^{2m+1}}\, dt,$$

$$g_{2m+2} = \frac{R_1}{\pi(m+1)}\int_{-\pi/2}^{\pi/2} h(t)\frac{\cos(2m+1)t}{R_1^{2m+2}}\, dt.$$

Let us single out the case of $n = 1$:

$$g_2 = -\frac{\sigma^2}{2} = \frac{1}{\pi R_1}\int_{-\pi/2}^{\pi/2} (h^+(t) - h^-(t))\cos t\, dt,$$

whence, by virtue of (16), we have

$$\begin{aligned} \int_{-\pi/2}^{\pi/2} h^-(t)\cos t\, dt &= (\pi\sigma^2/2)R_1 + \int_{-\pi/2}^{\pi/2} h^+(t)\cos t\, dt \\ &\leqslant (\pi\sigma^2/2)R_1 + c\exp\{-\delta(R_2)\}. \end{aligned} \tag{18}$$

Combining (18) with the elementary inequalities

$$\left|\frac{\cos(2m+1)t}{\cos t}\right| \leqslant 2m+1, \qquad \left|\frac{\sin 2mt}{\cos t}\right| \leqslant 2m,$$

we obtain the following estimates of the coefficients g_j:

$$\begin{aligned}|g_{2m+2}| &\leqslant \frac{1}{\pi(m+1)R_1^{2m+1}}\int_{-\pi/2}^{\pi/2}(h^+(t)+h^-(t))|\cos(2m+1)t|\,dt\\ &\leqslant \frac{1}{\pi(m+1)R_1^{2m+1}}\left\{c\exp\{-\delta(R_2)\}\right.\\ &\qquad\left.+\int_{-\pi/2}^{\pi/2}h^-(t)\cos t\left|\frac{\cos(2m+1)t}{\cos t}\right|\,dt\right\}\\ &\leqslant \frac{c(2m+1)}{(m+1)R_1^{2m+1}}\{\exp\{-\delta(R_2)\}+\sigma^2R_1\}.\end{aligned}$$

From the normalization condition (3) and relation (13) with $j=2$, it follows that $\sigma^2=1+o(1)$, $\varepsilon\to 0$, so that $|g_{2m+2}|\leqslant cR_2^{-2m}$. Similar estimates hold for g_{2m+1}. The total estimate of the coefficients g_j is

$$|g_j|\leqslant cR_1^{2-j},\qquad j\geqslant 3.$$

Comparing the last inequalities with relation (3), we arrive at (6). The theorem is completely proved.

References

1. N. A. Sapogov, *The stability problem for the Marcinkiewicz theorem*, Zap. Nauchn. Sem. Leningrad. Otdel. Mat. Inst. Steklov. (LOMI) **87** (1979), 104–124; English transl. in J. Soviet Math. **17** (1981), no. 1.

2. L. B. Golinskiĭ, *Stability estimates in Marcinkiewicz's theorem*, Problems of Stability of Stochastic Models: Trudy Seminara Vsesoyuzn. Inst. Sistemnykh Issled. (1988), 8–24, Moscow. (Russian)

3. I. V. Ostrovskiĭ, *On entire functions satisfying some special inequalities associated with the theory of characteristic functions of probabilistic laws*, Zap. Mekh. Mat. Fak. Kharkov. Gos. Univ. i Kharkov. Mat. Obshch. (4) **29** (1963), 145–168. (Russian)

4. Yu. V. Linnik, *Decompositions of probability distributions*, Izdat. Leningrad. Univ., Leningrad, 1960; English transl., Dover, New York , Oliver & Boyd, London, 1964.

5. N. A. Sapogov, *A problem on stability in the theorem of uniqueness of a characteristic function analytic in the vicinity of the origin* $t=0$, Problems of Stability of Stochastic Models: Trudy Seminara Vsesoyuzn. Inst. Sistemnykh Issled. (1980), 88–94, Moscow. (Russian)

6. S. N. Bernstein, *Extremal properties of polynomials*, GITTL, Leningrad and Moscow, 1937.

7. M. A. Evgrafov, *Analytic functions*, "Nauka", Moscow, 1968; English transl., Saunders, Philadelphia, 1966.

8. A. A. Goldberg and I. V. Ostrovskiĭ, *Distribution of values of meromorphic functions*, "Nauka", Moscow, 1970. (Russian)

Institute for Low Temperature Physics & Engineering, 47 Lenin pr., Kharkov 310164 Ukraine

Translated by V. A. TKACHENKO

ADVANCES IN SOVIET MATHEMATICS
Volume 11, 1992

Limit Sets and Multiplicators of Entire Functions

V. S. AZARIN AND V. B. GINER

Introduction

0.1. Let $\Phi(\lambda)$ be an entire function of finite order ρ and of normal type $(\Phi \in A(\rho))$. We recall that the indicator of Φ is determined by the relation (see, e.g., [1, p. 72])

$$h_\Phi(\varphi) = \overline{\lim_{r\to\infty}} \ln|\Phi(re^{i\varphi})|r^{-\rho}$$

and is a 2π-periodic ρ-trigonometrically convex function (ρ-t.c.f.).

Let $H(\varphi)$ be a ρ-t.c.f. A function $g \in A(\rho)$ is called an *H-multiplicator* of Φ if the indicator of the product gH satisfies $h_{g\Phi}(\varphi) \leqslant H(\varphi)$ for all φ. In some questions (for instance, when we investigate the completeness of functional systems in a complex domain or the solvability of convolution equations [2–5], [24]) we must be able to determine from the function Φ whether it has a multiplicator.

We shall study this problem in terms of the limit set[1] of Φ. First we recall some definitions ([6], see also [7]).

Let $D' := D'(\mathbb{C})$ be the space of L. Schwartz distributions, i.e., the space of distributions over the space $D(\mathbb{C})$ of infinitely differentiable functions with compact support, and let $u(\lambda) := \ln|\Phi(\lambda)|$; this is a subharmonic function in $\mathbb{C}$, while the transformation $(\cdot)_t$ is given by $u_t(\lambda) = u(\lambda t)t^{-\rho}$. The family $\{u_t : t > 0\}$ is precompact as $t \to \infty$ in the following sense: for each sequence $t_j \to \infty$ there exists a subsequence t'_j and a subharmonic function v such that $u_{t'_j} \to v$ in the D'-topology.

The limit set $\operatorname{Fr}\Phi$ is the set of all such limits, i.e.,

$$\operatorname{Fr}\Phi = \{v : (\exists t_j \to \infty)(v = D'\text{-}\lim u_{t_j})\}.$$

The properties of the limit set were studied and applied in [6–17]. We recall some of these properties.

1991 *Mathematics Subject Classification*. Primary 30D20.

[1]In [25] this set was called "the cluster set".

Let SH be the set of subharmonic functions in the plane. We denote

$$U[\rho, \sigma] = \{v \in SH : v(0) = 0,\ v(\lambda) \leqslant \sigma|\lambda|^\rho\ \forall\lambda \in \mathbb{C}\},$$
$$U[\rho] = \bigcup_{\sigma>0} U[\rho, \sigma].$$

The relation $U \Subset U[\rho]$ means that $\exists\sigma : U \subset U[\rho, \sigma]$.

The limit set $\operatorname{Fr}\Phi$ is closed in the D'-topology and is connected, i.e., cannot be decomposed into a sum of disjoint closed sets, $\operatorname{Fr}\Phi \Subset U[\rho]$, and is invariant with respect to the transformation $(\cdot)_t$.

0.2. Let $v \in U[\rho]$. Consider the function $m(\lambda, v, H) = H(\lambda) - v(\lambda)$, where $\lambda = re^{i\theta}$, $H(\theta) = H(\varphi)r^\rho$.

The maximal subharmonic minorant (m.s.m.) m belonging to $U[\rho]$ will be denoted by $\mathscr{S}_H v$, while the domain of definition of the operator $\mathscr{S}_H$ will be denoted by D_H.

Though $m(0, \cdot) = 0$, the m.s.m. m can differ from zero at zero (as was remarked by A. E. Eremenko and M. L. Sodin), but if the m.s.m. $m(\lambda, \cdot)$ equals zero at zero, then it belongs to $U[\rho]$.

THEOREM 1. *$\Phi \in A(\rho)$ has an H-multiplicator if and only if*

$$\operatorname{Fr}\Phi \subset D_H. \tag{1}$$

0.3. Suppose that some H-multiplicator $g = g(\lambda, \Phi, H)$ of the function Φ is found. We examine the function $\Psi = g\Phi$. The structure of its limit set is described by the following statement.

PROPOSITION 2. *Any $v_\psi \in \operatorname{Fr}(g\Phi)$ can be written as $v_\psi = v + w_1$, where $v \in \operatorname{Fr}\Phi$ and $w_1 \in U[\rho]$ with the condition*

$$w_1(\lambda) \leqslant \mathscr{S}_H v(\lambda), \qquad \forall\lambda \in \mathbb{C}, \tag{2}$$

and, conversely, for any $v \in \operatorname{Fr}\Phi$ there exists a v_g, $v_g \leqslant \mathscr{S}_H v$ such that

$$v + v_g \in \operatorname{Fr}(g\Phi).$$

An H-multiplicator G of the function Φ will be called *ideally completing*[2] if it fulfills the condition

$$\operatorname{Fr} G\Phi = \{v_\psi = v + \mathscr{S}_H v : v \in \operatorname{Fr}\Phi\}.$$

If a multiplicator is ideally completing, then the equality is achieved in (2) for all $v \in \operatorname{Fr}\Phi$. This makes the multiplicator optimal in another respect.

We recall that the *type* (or the *type value*) of an entire function is given by the equality (see, e.g., [1, p. 11] or [18, p. 61])

$$\sigma_f = \overline{\lim_{r\to\infty}} \ln M(r, f)r^{-\rho},$$

[2]"Perfectly complementary" in [25].

where $M(r, f)$ is the maximum of the modulus of the function f on the circle $\{|z| = r\}$. If $\sigma_f = 0$, then f is said to be a function of *minimal* (or *zero*) type. For such a function the limit set is trivial: it consists of zero functions only.

PROPOSITION 3. *Let $G = G(\cdot, \Phi, H)$ be an ideally completing H-multiplicator of a function Φ. Then each H-multiplicator of the function $\Psi = G\Phi$ is of minimal type.*

A function Φ is said to be *ideally completable*[3] if for each H condition (0.1) implies that Φ has an ideally completing H-multiplicator.

For instance, if Φ is a function of completely regular growth (see [1]), then it is ideally completable.

0.4. Let $v \in U[\rho]$ and the transformation $v_t(\lambda) = v(t\lambda)t^{-\rho}$ be a periodic function of period p with respect to $\ln t$. We shall consider the set $U[v] = \{v_t : 1 \leqslant t \leqslant e^p\}$. $U[v]$ is a limit set for some entire function and is called *periodic*.

Functions with periodic limit set are a generalization of functions of completely regular growth obtained under the condition $v(\lambda) = h(\varphi)r^\rho$. They were studied in [8, 10, 15]; see [14] for details. They form a sufficiently broad class, since the limit set of any function can be approximated by periodic limit sets in the following sense.

We shall say that for U_n, $U \Subset U[\rho]$, the relation

$$\lim_{n\to\infty} U_n = U \tag{3}$$

holds if the following conditions are met:

(a) if $v_j \in U_{n_j}$ and $D'\text{-}\lim v_j = v$, then $v \in U$;

(b) for all $v \in U$ there exists a sequence $v_n \in U_n$ such that $D'\text{-}\lim v_n = v$.

THEOREM G [10]. *The closed set $U \Subset U[\rho]$ is the limit set for a function $\Phi \in A(\rho)$ if and only if there exists a sequence of periodic limit sets U_n satisfying condition* (3).

THEOREM 4. *Any function Φ with a periodic limit set is ideally completable.*

0.5. Let $\{h(\varphi, c) : 0 \leqslant c \leqslant 1\}$ be a set of ρ-t.c.f. continuous with respect to the parameter c. We set

$$U_{\text{ind}} = \{v(re^{i\varphi}) = h(\varphi, c)r^\rho : 0 \leqslant c \leqslant 1\}. \tag{4}$$

Such a set is called an *indicator set*. It always is the limit set for some entire function. Functions with such a limit set can be regarded as a generalization of functions of completely regular growth (which are obtained when the family consists of one function only).

[3]"Perfectly complemented" in [25].

THEOREM 5. *Any function Φ whose limit set is an indicator set is ideally completable.*

0.6. There exists a function $\Phi \in A(\rho)$ which is not ideally completable. The authors are indebted to A. E. Erëmenko and M. L. Sodin for the idea on which the example is based.

There is a weaker requirement than ideal completability, though it is, in a sense, also extremal with respect to property (0.2). A function Φ is said to be *completable*[4] if condition (0.1) implies the existence of an H-multiplicator g such that for any $v \in \operatorname{Fr}\Phi$ we have $v + \mathscr{S}_H v \in \operatorname{Fr}(g\Phi)$. The multiplicator itself will be called a *completing multiplicator*[5]. The latter meets a condition similar to Proposition 3. We recall that the *lower type* of a function $f \in A(\rho)$ is given by the limit

$$\underline{\sigma}_f = \varliminf_{r\to\infty} \ln M(r, f) r^{-\rho}.$$

PROPOSITION 6. *Let g be a completing H-multiplicator of Φ. Then any H-multiplicator of the function $\Psi = g\Phi$ has zero lower type.*

THEOREM 7. *Any function $\Phi \in A(\rho)$ is completable.*

0.7. Let us return to the property of ideal completability. The existence of an ideally completing H-multiplicator depends, of course, both on $\Phi \in A(\rho)$ or its limit set and on H.

THEOREM 8. *Let Φ and H be such that condition* (1) *is satisfied. The function Φ has an ideally completing H-multiplicator if and only if the operator G_H is continuous on* $\operatorname{Fr}\Phi$.

We shall say that the maximum principle for $U[\rho]$ is valid in the domain G (which is, generally speaking, unbounded), if the conditions $w \in U[\rho]$, $w(\lambda) = 0$, $\lambda \notin G$ imply that $w(\lambda) = 0$.

Let us denote by $\mathscr{H}_w$ the region of harmonicity of $w \in U[\rho]$, i.e., a region where the conditions "w is harmonic in G" and "$G \supset \mathscr{H}_w$" imply $G = \mathscr{H}_w$.

We remark that $\mathscr{H}_w$ is a connected component of the open set on which w is harmonic. Generally, it is not unique.

The image of U will be denoted by $\mathscr{S}_H U$, while its closure in the D'-topology will be denoted by $\operatorname{clos}\mathscr{S}_H U$.

THEOREM 9. *For any $w \in \operatorname{clos}\mathscr{S}_H U$ and in any $\mathscr{H}_w$ the maximum principle for $U[\rho]$ holds. Then $\mathscr{S}_H$ is continuous on U.*

The authors are indebted to I. F. Krasichkov-Ternovskiĭ for stating the problem which is considered in this work, to A. E. Eremenko, M. L. Sodin

[4]"Complementary" in [25].

[5]"Complementary multiplier" in [25].

and V. A. Tkachenko for fruitful discussions. We are grateful to L. Hörmander and R. Sigurdsson for sending us the preprint of their paper [21] before it was published.

This paper is organized as follows. We shall prove Propositions 2, 3 and 6 in §1. Note that the necessity of condition (0.1) follows from Proposition 2. Theorem 9 will be proved in §2. Section 3 contains the proof of Theorem 7. Its presentation has been strongly influenced by [21]. Theorems 8, 4 and 5 are proved in §4. Theorem 4 is deduced from Theorem 8, while Theorem 9 is used for proving Theorem 5. Section 5 presents an example of a function which for some H has no ideally completing H-multiplicator.

§1. Proof of Propositions 2, 3 and 6

1.1. PROOF OF PROPOSITION 2. Let g be a multiplicator of Φ, i.e.,

$$h_{g\Phi}(\varphi) \leqslant H(\varphi) \tag{1.1}$$

and let $v \in \operatorname{Fr}\Phi$. We shall choose a sequence $t_j \to \infty$ which satisfies the following conditions:

$$\ln|g\Phi|(t_j\lambda)t_j^{-\rho} \to v\psi(\lambda), \quad \ln|\Phi|(t_j\lambda)t_j^{-\rho} \to v(\lambda), \quad \ln|g|(t_j\lambda)t_j^{-\rho} \to w_1(\lambda),$$

where $v\psi$ is a function from $\operatorname{Fr}(\Phi g)$, w_1 is a function from $\operatorname{Fr} g$.

Then $v_\psi = v + w_1$. Now from (1.1) we obtain

$$v_\psi(\lambda) \leqslant H(\lambda), \tag{1.2}$$

since (see [20]) any function $f \in A(\rho)$ satisfies the relation

$$r^\rho h_f(\varphi) = \sup\{v(re^{i\varphi}) : v \in \operatorname{Fr}\Phi\}. \tag{1.3}$$

From (1.2) we get $w_1(\lambda) \leqslant H(\lambda) - v(\lambda) = m(\lambda, v, H)$. Since $w_1 \in U[\rho]$, the set $\mathscr{S}_H v$ exists according to the limit set property, and $w_1(\lambda) \leqslant \mathscr{S}_H v(\lambda)$. Proposition 2 is proved.

1.2. PROOF OF PROPOSITION 3. Let G be an ideally completing multiplicator of Φ. Let g be any multiplicator of the function $\Psi = G\Phi$. We denote

$$\theta(\lambda) = (gG\Phi)(\lambda). \tag{1.4}$$

Let $v_g \in \operatorname{Fr} g$. Let us choose $t_j \to \infty$ fulfilling the relations:

$$(\ln|g|)_{t_j} \to v_g, \quad (\ln|\Psi|)_{t_j} \to v_\Psi \in \operatorname{Fr}\Psi, \quad (\ln|\theta|)_{t_j} \to v_\theta \in \operatorname{Fr}\theta.$$

It follows from (1.4) that $v_\theta = v_g + v_\Psi$. Since gH is a multiplicator of Ψ, we have, according to (1.3),

$$v_\theta = v_g + v_\Psi \leqslant H(\lambda) = H(\varphi)r^\rho. \tag{1.5}$$

Since G is an ideally completing multiplicator, $v_\Psi = v + \mathscr{S}_H v$. So for all λ's, (1.5) implies $v_g + \mathscr{S}_H v \leqslant H - v$. Since $\mathscr{S}_H$ is a maximal subharmonic minorant, $v_g \equiv 0$. According to (1.3), we have $h_g(\varphi) = 0$ and, therefore, $\sigma_g = 0$. $\square$

1.3. Before proving proposition 6, we shall prove

LEMMA 1.1. *If* $v \in \operatorname{Fr}\Phi$ *and* $(\ln|\Phi|)_{t_j} \to v$, *then*

$$\overline{\lim_{j\to\infty}} \ln M(r_j, \Phi) r_j^{-\rho} = \max_{\varphi} v(e^{i\varphi}). \tag{1.6}$$

PROOF (see [19, Theorem 3.1.2.2]). By $\alpha(u)$ we denote the function

$$\alpha(u) = \begin{cases} ce^{-1/(1-u^2)}, & u \in (-1, 1), \\ 0, & u \notin (-1, 1), \end{cases}$$

where c is selected to satisfy the condition

$$\int_0^1 u\alpha(u)\,du = \frac{1}{2\pi}.$$

Further, we set

$$\alpha_\varepsilon(\lambda) = \varepsilon^{-2}\alpha(|\lambda|/\varepsilon),$$
$$u_\varepsilon(\lambda) = \int u(\zeta)\alpha_\varepsilon(\zeta + \lambda) l(d\zeta),$$

where l is the Lebesgue measure on the plane.

If $u_n \to v$ in D', then $(u_n)_\varepsilon \rightrightarrows v_\varepsilon$ uniformly on each compact set (see, e.g., [19, Theorem 2.7.2.3]). Let $u = \ln|\Phi|$. Since $u_{t_j} \to v$ in D', we have $(u_{t_j})_\varepsilon \rightrightarrows v_\varepsilon$ uniformly on each compact set. Hence

$$\max_{\varphi}(u_{t_j})_\varepsilon(re^{i\varphi}) \to \max_{\varphi} v_\varepsilon(re^{i\varphi}).$$

By the property of the operation $(\cdot)_\varepsilon$ (see, e.g., [19, Theorem 2.6.2.4]), we have $u(\lambda) \leqslant u_\varepsilon(\lambda)$. Hence

$$\max_{\varphi} u_{t_j}(re^{i\varphi}) = \ln M(rt_j, \Phi)t_j^{-\rho} \leqslant \max_{\varphi} v_\varepsilon(re^{i\varphi}).$$

This implies

$$\overline{\lim_{t_j\to\infty}} \ln M(rt_j, \Phi)t_j^{-\rho} \leqslant \max_{\varphi} v_\varepsilon(re^{i\varphi}).$$

We know (see, e.g., [19, Theorem 2.1.2.7]) that $\max_\varphi v_\varepsilon(re^{i\varphi}) \downarrow \max_\varphi v(re^{i\varphi})$ when $\varepsilon \downarrow 0$. Hence

$$\overline{\lim_{t_j\to\infty}} \ln M(rt_j, \Phi)t_j^{-\rho} \leqslant \max_{\varphi} v(re^{i\varphi}). \tag{1.7}$$

Let us prove the inverse inequality. We denote

$$\Psi(r) = \overline{\lim_{t_j\to\infty}} \ln M(rt_j, \Phi)t_j^{-\rho} = \overline{\lim_{t_j\to\infty}} \ln M(r, u_{t_j}).$$

According to the H. Cartan theorem (see, e.g., [19, Theorem 2.7.3.3]) $\Psi(|z|)$ differs from its upper semicontinuous regularization $\Psi^*(|z|)$ on a set of no more than zero capacity. But if the inequality $\Psi^*(|z|) \neq \Psi(|z|)$ were valid at one point, it would be valid for the entire circle, i.e., on a set of positive capacity. Therefore, $\Psi(|z|)$ is subharmonic and $\Psi(r)$ is convex with respect to $\ln r$, hence Ψ is continuous.

Therefore we have

$$\begin{aligned}\max v(re^{i\varphi}) &\leqslant \max_{\varphi} v_{\varepsilon}(re^{i\varphi}) = \lim_{t_j\to\infty} (u_{t_j})_{\varepsilon}(re^{i\varphi}) \\ &\leqslant \overline{\lim_{t_j\to\infty}} \ln M((r+\varepsilon)t_j, \Phi)t_j^{-\rho} = \Psi(r+\varepsilon).\end{aligned}$$

Since Ψ is continuous, we may pass to the limit as $\varepsilon \downarrow 0$ and obtain

$$\max_{\varphi} v(re^{i\varphi}) \leqslant \Psi(r). \tag{1.8}$$

Inequalities (1.8) and (1.7) imply (1.6). □

COROLLARY 1.2. *We have*

$$\underline{\sigma}_{\Phi} = \min\{\max_{\varphi} v(e^{i\varphi}) : v \in \operatorname{Fr}\Phi\}. \tag{1.9}$$

PROOF. It follows from the relation

$$\underline{\lim_{n\to\infty}} = \inf\Big(\overline{\lim_{n_j\to\infty}} a_{n_j}\Big),$$

where inf is taken over all the subsequences $n_j \to \infty$, and Lemma 1.1 that

$$\underline{\sigma}_{\Phi} \leqslant \inf\Big\{\max_{\varphi} v(e^{i\varphi}) : v \in \operatorname{Fr}\Phi\Big\}.$$

Let $r_j \to \infty$ be the subsequence on which $\underline{\lim}$ is achieved in the definition of $\underline{\sigma}_{\Phi}$. Let us choose it so that it satisfies, for some $v_0 \in \operatorname{Fr}\Phi$, the condition

$$\ln|\Phi(r_j\lambda)|r_j^{-\rho} \xrightarrow{D'} v_0.$$

Now we obtain from Lemma 1.1 $\underline{\sigma}_{\Phi} = \max_{\varphi} v_0(e^{i\varphi})$. □

1.4. PROOF OF PROPOSITION 6. Let $q(\cdot, v, H)$ be any completing H-multiplicator for Φ. Then there exists a subsequence $t_j \to \infty$ such that $(\ln|\Phi g|)_{t_j} \to v + \mathscr{S}_H v$. Let f be an H-multiplicator of Φg. We set $\Psi = f\Phi g$. Now, let us choose a subsequence from t_j (which we still denote by t_j) which satisfies the condition

$$(\ln|f|)_{t_j} \to v_f \in \operatorname{Fr} f.$$

Then, passing to the limit over this subsequence in the equality

$$(\ln|\Psi|_{t_j}) = (\ln|f|)_{t_j} + (\ln|g\Phi|)_{t_j},$$

we get $v_{\Psi} = v_f + \mathscr{S}_H v$, hence $v_f + \mathscr{S}_H v \leqslant H - v$, that is $v_f \equiv 0$.

From Corollary 1.2 we now obtain $\underline{\sigma}_f = 0$. □

§2. Proof of Theorem 9

2.1. In order to prove Theorem 9, we shall need a number of auxiliary statements.

LEMMA 2.1. *Let the maximum principle be valid in* G *for* $U[\rho]$ *and for some continuous functions* W_1, $W \in U[\rho]$. *The conditions:* (a) W *is harmonic in* G, (b) $W = W_1$ *outside* G *hold. Then*

$$W_1(\lambda) \leqslant W(\lambda), \qquad \lambda \in G. \tag{2.1}$$

PROOF. We set

$$W_0(\lambda) = \begin{cases} (W_1 - W)^+(\lambda), & \lambda \in G, \\ 0, & \lambda \notin G. \end{cases}$$

This function is continuous in $\mathbb{C}$ and, evidently, subharmonic both in G and in $\mathbb{C}\backslash\overline{G}$. Since $W_0(\lambda) \geqslant 0$, the inequality for the mean values

$$0 = W_0(\lambda) \leqslant \frac{1}{2\pi}\int_0^{2\pi} W_0(\lambda + \varepsilon r^{i\varphi})\, d\varphi, \qquad \lambda \in \partial G,$$

implies the subharmonicity on ∂G.

Since G satisfies the maximum principle for $U[\rho]$, we have $W_0 \equiv 0$, which is equivalent to (2.1). □

2.2. We shall dwell on some properties of maximal subharmonic minorants and, in particular, of $W \in \mathscr{S}_H v$ (cf. [23]). Let

$$E_v = \{\lambda \in \mathbb{C} : \mathscr{S}_H v(\lambda) = m(\lambda, v, H)\}. \tag{2.2}$$

We remark that $m(\lambda, v, H)$ is a δ-subharmonic function in $\mathbb{C}$ whose charge will be denoted by $\nu(\cdot, v)$, its positive and negative parts will be denoted by $\nu^+(\cdot, v)$ and ν^-.

Let us denote the μ_H-measure by $H(\lambda)$. It is decomposed into the product of measures

$$\mu_H = \Delta_H \otimes \rho r^{\rho-1}\, dr, \tag{2.3}$$

where Δ_H is the measure on the circle and $\rho r^{\rho-1}\, dr$ is the measure on the ray. It is obvious that

$$\nu^+(\cdot, v) \leqslant \mu_H(\cdot). \tag{2.4}$$

We shall denote the mass distribution of $W \in U[\rho]$ by μ_W.

The modulus of continuity of W (if W is continuous) will be denoted by $\omega_W(\lambda, h)$, $\lambda \in \mathbb{C}$, $h > 0$.

The following lemma lists various properties of $W \in \mathscr{S}_H U$ which will be useful in the sequel.

LEMMA 2.2. *Let* $W \in \mathscr{S}_H U$. *Then*

(1) $W \in U[\rho, \sigma]$, *where*

$$\sigma = [\max\{H(e^{i\varphi}) : \varphi \in [0, 2\pi)\} + 2\sigma_1]2^\rho \zeta,$$
$$\sigma_1 = \max\{v(\lambda)|\lambda|^{-\rho} : \lambda \in \mathbb{C},\ v \in U\};$$

(2) *the charge restriction* $\nu(\cdot, v)|_{E_v}$ *to* E_v *is nonnegative, i.e.*

$$\nu(\cdot, v)|_{E_v} = \nu^+(\cdot, v)|_{E_v};$$

(3) *outside* E_v *the function* W *is harmonic, i.e.,*

$$\mu_W|_{\mathbb{C}\setminus E_v} \equiv 0;$$

(4) *the measure* μ_W *is bounded from above by* $\nu^+(\cdot, v)$, *i.e.,*

$$\mu_W(\cdot) \leqslant \nu^+(\cdot, v);$$

(5) $\mathscr{S}_H U$ *is equicontinuous on each compact set, i.e.,*

$$\omega_W(\lambda, h) \leqslant C(R, \sigma, \rho)\sqrt{h}\ln(1/h),$$

where $C(R, \sigma, \rho)$ *is independent of* $W \in \mathscr{S}_H U$.

PROOF. Let us prove property (1). We have

$$\begin{aligned} T(r, w) : &= \frac{1}{2\pi}\int_0^{2\pi} W^+(re^{i\varphi})\,d\varphi \\ &\leqslant \frac{1}{2\pi}\left[r^\rho \int_0^{2\pi} H^+(e^{i\varphi})\,d\varphi + \int_0^{2\pi} v^+(re^{i\varphi})\,d\varphi + \int_0^{2\pi} v^-(re^{i\varphi})\,d\varphi\right]. \end{aligned}$$

Since $v(0) = 0$, we have

$$\frac{1}{2\pi}\int_0^{2\pi} v^-(re^{i\varphi})\,d\varphi \leqslant \frac{1}{2\pi}\int_0^{2\pi} v^+(re^{i\varphi})\,d\varphi.$$

Therefore

$$T(r, W) \leqslant [\max\{H(e^{i\varphi}) : \varphi \in [0, 2\pi)\} + 2\sigma_1]r^\rho.$$

It is known that (see, e.g., [18, p. 54]) $T(r) \leqslant M(r) \leqslant 3T(2r)$. So we conclude that

$$W(\lambda) \leqslant 3\cdot 2^\rho(\max H(e^{i\varphi}) + 2\sigma_1)|\lambda|^\rho = \sigma|\lambda|^\rho. \quad \square$$

Let us prove property (2). To this end we shall use

THEOREM A.F.G. ([22, part 1]). *Let* g *be a nonnegative* δ*-subharmonic function and* ν_g *be its charge. Then the restriction* $\nu_g|_E$ *to the set* $E = \{\lambda : \gamma(\lambda) = 0\}$ *is a measure.*

Applying this theorem to $g = m(\lambda, v, H) - \mathscr{S}_H v(\lambda)$, we get

$$\nu|_{E_v} \geqslant \mu_W|_{E_v}, \tag{2.5}$$

hence we obtain property (2).

Let us prove property (3). Since W and v are upper semicontinuous and H is continuous, the set $\{\lambda : (W + v)(\lambda) - H(\lambda) < 0\}$ is open.

Let us take a neighborhood of any point from this set and replace the function W within it by the Poisson integral constructed using this function, i.e., let us sweep out the mass from this neighborhood. The subharmonic function obtained would be strictly greater than the initial one, if the latter were not harmonic. This means that the initial W was not the maximal minorant. We have arrived at a contradiction, which proves property (3).

Property (4) immediately follows from property (3) and (2.5).

In order to prove property (5), we shall need auxiliary statements which will be stated as lemmas. Let

$$P(\lambda, \varphi, R) = \frac{1}{2\pi} \frac{R^2 - |\lambda|^2}{|\lambda - Re^{i\varphi}|}$$

be the Poisson kernel in the circle $K_R = \{\lambda : |\lambda| < R\}$.

LEMMA 2.3. *In the circle* $K_{R/2}$, *we have*

$$|\operatorname{grad}_\lambda P(\lambda, \varphi, R)| \leqslant C_1(R).$$

The proof is omitted. We shall introduce the notation for Green's function of the Laplace operator in the circle K_R:

$$G(\lambda, \zeta, R) = \ln\left|\frac{R^2 - \zeta\bar{\lambda}}{R(\lambda - \zeta)}\right|.$$

The circle $\{\zeta : |\lambda - z| < t\}$ will be denoted by $K_{\lambda, t}$.

LEMMA 2.4. *Let* $\lambda \in K_{R/2} \backslash K_{\zeta, \sqrt{h}}$. *Then for a small* h

$$|\operatorname{grad}_\lambda G(\lambda, \zeta, R)| \leqslant C_2(R)/\sqrt{h}.$$

The proof is omitted.

Let us denote $\mu(\lambda, t) = \mu(K_{\lambda, t})$.

LEMMA 2.5. *For* $\lambda \in K_{R/2}$, $0 < t < R/10$, *we have*

$$\mu_H(\lambda, t) \leqslant C_3(\sigma, R)\, t.$$

PROOF. Using the representation (2.3) of μ_H, we have

$$\mu_H(\lambda, t) \leqslant \Delta_H(\mathbb{T}^1)\rho \int_{|\lambda|-t}^{|\lambda|+t} r^{\rho-1}\, dr \leqslant C \max H(e^{i\varphi}) R^{\rho-1} t \leqslant C\sigma R^{\rho-1} t,$$

where $\Delta_H(\mathbb{T}^1)$ is the measure of the circle which, as is known, can be estimated via the indicator maximum (see [1, p. 79]). The last inequality proves the lemma.

LEMMA 2.6. *Let* $h < 1$ *and suppose that the monotonic function* $\mu(t)$ *satisfies the condition*

$$\mu(t) < ct. \tag{2.6}$$

Then

$$\int_0^{\sqrt{h}} \ln(1/t)\, d\mu(t) \leqslant (3/2) c\sqrt{h} \ln h.$$

The lemma is proved by integrating by parts with the use of (2.6).

LEMMA 2.7. *Let* $\lambda \in K_{R/2}$ *and* $\zeta \in K_R$. *Then*

$$|\ln|(R^2 - \bar{\lambda}\zeta)/R|| \leqslant C_4(R).$$

The proof is omitted.

Now we pass to the proof of assertion (5) from Lemma 2.2. According to the F. Riesz theorem, we represent W in the circle as

$$W(\lambda) = H(\lambda, W) - \int_{K_R} G(\lambda, \zeta, R)\, d\mu_W(\zeta), \tag{2.7}$$

where

$$H(\lambda, W) = \frac{1}{2\pi} \int_0^{2\pi} P(\lambda, \varphi, R) W(Re^{i\varphi})\, d\varphi.$$

It follows from Lemma 2.3 that

$$|\operatorname{grad}_\lambda H| \leqslant C_1(R) \frac{1}{2\pi} \int_0^{2\pi} |W|(Re^{i\varphi})\, d\varphi \leqslant C_1(R) 2\sigma \frac{R^\rho}{\rho}. \tag{2.8}$$

We split the integral in (2.7) into three terms

$$\Psi_1(\lambda, h) = \int_{K_R \setminus K_{\lambda_0, \sqrt{h}}} G(\lambda, \zeta, R)\, d\mu_W(\zeta),$$

$$\Psi_2(\lambda, h) = \int_{K_{\lambda_0, \sqrt{h}}} \ln|(R^2 - \bar{\lambda}\zeta)/R|\, d\mu_W(\zeta),$$

$$\Psi_3(\lambda, h) = \int_{K_{\lambda_0, \sqrt{h}}} \ln|\zeta - \lambda|\, d\mu_W(\zeta),$$

where λ_0 is an arbitrary fixed point in $K_{R/2}$.

Combining property (4) and inequality (2.4), we have

$$\mu_W(E) \leqslant \mu_H(E) \qquad \forall E. \tag{2.9}$$

For all $\lambda \in K_{\lambda_0, \sqrt{h}/2}$, Lemma 2.4 yields

$$|\operatorname{grad} \Psi_1(\lambda, h)| \leqslant C_2(R) \sigma R^\rho / \sqrt{h}. \tag{2.10}$$

Combining Lemmas 2.5 and 2.7 with inequality (2.9), we get

$$|\Psi_2(\lambda, h)| \leqslant C_4(R) C_3(\sigma, R) \sqrt{h}. \tag{2.11}$$

Further, from Lemmas 2.5 and 2.6, taking into account the fact that $\ln|\zeta - \lambda| < 0$, we obtain

$$|\Psi_3(\lambda, h)| < (3/2) C_3(\sigma, R) \sqrt{h} \ln h. \tag{2.12}$$

Now consider the difference $\Delta W = W(\lambda_0 + \Delta\lambda) - W(\lambda_0)$, $|\Delta\lambda| < h < \sqrt{h}/2$. It can be represented as

$$\Delta W = \Delta\Psi_1 + \Delta\Psi_2 + \Delta\Psi_3 + \Delta H. \tag{2.13}$$

Choosing h small enough, one may assume that $\lambda_0 + \Delta\lambda \in K_{\sqrt{h}/2, \lambda_0}$. Thus, according to (2.11),

$$|\Delta\Psi_2(\lambda_0, h)| \leqslant |\Psi_2(\lambda_0, h)| + |\Psi_2(\lambda_0 + h, h)| \leqslant C_6(\sigma, R)\sqrt{h}. \tag{2.14}$$

Likewise, (2.12) yields

$$|\Delta\Psi_3(\lambda_0, h)| \leqslant |\Psi_3(\lambda_0, h)| + |\Psi_3(\lambda_0 + h, h)| \leqslant C_7(\sigma, R)\sqrt{h}\ln h. \tag{2.15}$$

Finally, from (2.10) and (2.8), respectively, we obtain

$$|\Psi_1| \leqslant C_3(\sigma, R)\sqrt{h}, \qquad |\Delta H| \leqslant C_8(\sigma, R)h. \tag{2.16}$$

Substituting (2.14)–(2.16) into (2.13), we obtain relation (5) of Lemma 2.2. The lemma is proved.

2.3. Before proving Theorem 9, we shall prove

LEMMA 2.8. *Let* $W_n = \mathscr{S}_H v_n$, $v_n \xrightarrow{D'} v$ *and* $W_n \xrightarrow{D'} W_\infty$. *We set*

$$E_\infty = \{\lambda : W_\infty(\lambda) = H(\lambda) - v(\lambda)\}.$$

Then W_∞ *is harmonic in* $\mathbb{C}\backslash E_\infty$.

PROOF. Let $\lambda_0 \notin E_\infty$. Then there exists a $\delta > 0$ such that

$$W_\infty(\lambda_0) + v(\lambda_0) \leqslant H(\lambda_0) - 2\delta.$$

Since the function $b(\lambda) = W_\infty(\lambda) + v(\lambda) - H(\lambda)$ is semicontinuous, there exists an $\varepsilon = \varepsilon(\delta)$ such that $b(\lambda) < -\delta$ for all $\lambda \in \{\lambda : |\lambda - \lambda_0| < 2\varepsilon\}$. Let $(\mathfrak{M}_\varepsilon b)(\zeta)$ be the averaging operation defined in Lemma 1.1; it was denoted there as b_ε (see, e.g., [19, Theorem 2.6.2.1]). Then $\mathfrak{M}_\varepsilon b(\lambda) < -b$, $|\lambda - \lambda_0| < \varepsilon$ or $\mathfrak{M}_\varepsilon W_\infty(\lambda) + \mathfrak{M}_\varepsilon v(\lambda) < \mathfrak{M}_\varepsilon H(\lambda) - \delta$. The function $H(\lambda)$ is continuous, so it is uniformly continuous on the circle $|\lambda - \lambda_0| < 2\varepsilon$; hence $|\mathfrak{M}_{\varepsilon_1} H(\lambda) - H(\lambda)| < \delta/2$ for some $\varepsilon_1 < \varepsilon$.

Since $\mathfrak{M}_\varepsilon$ is monotonic in ε on subharmonic functions,

$$\mathfrak{M}_{\varepsilon_1} W_\infty(\lambda) + \mathfrak{M}_{\varepsilon_1} v(\lambda) < H(\lambda) - \delta/2, \qquad |\lambda - \lambda_0| < \varepsilon.$$

It is known (see, e.g., [19, Theorem 2.7.2.4]) that

$$\mathfrak{M}_{\varepsilon_1} W_n \rightrightarrows \mathfrak{M}_{\varepsilon_1} W$$

uniformly on each compact set, if $W_n \xrightarrow{D'} W$.

Therefore $\mathfrak{M}_{\varepsilon_1} W_n(\lambda) + \mathfrak{M}_{\varepsilon_1} v_n(\lambda) < H(\lambda) - \delta/4$, whence $W_n(\lambda) + v_n(\lambda) < H(\lambda) - \delta/4$, $|\lambda - \lambda_0| < \varepsilon$. Thus $\{|\lambda - \lambda_0| < \varepsilon\} \subset \mathbb{C}\backslash E_{v_n}$ (the definition of E_{v_n} appears in (2.2)). This means that W_n are harmonic on the circle $|\lambda - \lambda_0| < \varepsilon$.

The condition $W_n \xrightarrow{D'} W_\infty$ implies that W_∞ is harmonic in the circle $|\lambda - \lambda_0| < \varepsilon$. □

2.4. COMPLETION OF THE PROOF OF THEOREM 9. Let $v_n \xrightarrow{D'} v$. Then the set $\{W_n = \mathscr{S}_H v_n\}$ is equicontinuous by Lemma 2.2 and we can choose from it a subsequence uniformly converging to the continuous function W_∞. Let $W = \mathscr{S}_H v$, $E = E_W$, E_∞ being defined in Lemma 2.8.

Since

$$(W_\infty + v)(\lambda) \leqslant H(\lambda), \qquad (W + v)(\lambda) \leqslant H(\lambda) \qquad \forall \lambda,$$

and v is upper semicontinuous, while W and H are continuous, the sets E and E_∞ are closed.

Since $W_\infty(\lambda) \leqslant H(\lambda) - v(\lambda)$ we have

$$W_\infty(\lambda) \leqslant W(\lambda), \qquad \forall \lambda, \tag{2.17}$$

and, therefore,

$$E_\infty \subset E. \tag{2.18}$$

The function W is subharmonic in $\mathbb{C}\backslash E_\infty$, and W_∞ is harmonic in $\mathbb{C}\backslash E_\infty$ by Lemma 2.8. They take the same values on E_∞. As the maximum principle holds in $\mathscr{H}_{W_\infty}$ by assumption, we have, according to Lemma 2.1, the inequality $W(\lambda) \leqslant W_\infty(\lambda)$, for all λ. Inequalities (2.17) and (2.19) imply that $W(\lambda) \equiv W_\infty(\lambda)$, i.e., $\mathscr{S}_H$ is continuous. This completes the proof of Theorem 9.

3. Proof of Theorem 7

3.1. In order to prove this theorem, we shall use the general statement on the structure of limit sets proved by one of the authors (see [11, 12]). A similar proposition for plurisubharmonic functions was given in [21].

Let $U \Subset U[\rho]$ be a closed set. One can introduce here a metric generating the D'-topology, for instance, in the following way:

$$d(v_1, v_2) = \sum_{j=1}^{\infty} \frac{|\langle v_1 - v_2, \Psi_j \rangle|}{2^j(1 + |\langle v_1 - v_2, \Psi_j \rangle|)}, \qquad \text{where } \langle g, \Psi \rangle = \int g \Psi l(d\lambda),$$

$\{\Psi_j\}$ is a set dense in the space of test functions $D(\mathbb{C})$, $l(d\lambda)$ is the Lebesgue measure, the integral being taken over the entire plane $\mathbb{C}$.

THEOREM V.B. [12, 11]. *Let $U \Subset U[\rho]$. A function $f \in A(\rho)$ such that* $\operatorname{Fr} f = U$ *exists if and only if there exists a piecewise continuous mapping*

$$v(\cdot \mid t) \colon (0, \infty) \to U$$

which satisfies the conditions:

1) $d(v(\cdot \mid \tau t), v_\tau(\cdot \mid t)) \to 0$, $t \to \infty$, *uniformly in* $\tau \in [a, b] \Subset (0, \infty)$ (*for any segment* $[a, b]$);

2) $\operatorname{clos}\{v(\cdot \mid t) : t \in [a, \infty)\} = U$, $\forall a > 0$.

Here

$$v(\cdot \mid t) - \ln|f(\cdot t)| t^{-\rho} \xrightarrow{D'} 0 \qquad \text{as } t \to \infty. \tag{3.1}$$

If the set U is the homeomorphic image $v(\cdot, m)\colon M \to U$, $m \in M$, of a connected finite-dimensional compact set M, then the transformation $(\cdot)_t$ induces on M the dynamical system f^τ according to the formula

$$v_t(\cdot, m) = v(\cdot, f^\tau_m), \qquad \tau = \ln t,$$

while $v(\cdot \mid t)$ induces the pseudotrajectory $m(\tau)\colon (-\infty, \infty) \to M$ according to the formula $v(\cdot \mid e^\tau) = v(\cdot, m(\tau))$. This pseudotrajectory is asymptotically dynamical with the asymptotics f^τ_m and is dense in M, i.e., it satisfies the conditions

$$m(t+\tau) - f^\tau m(t) \to 0, \qquad t \to \infty,$$
$$\overline{\{m(t) : t \in [a, \infty)\}} = M.$$

Further we shall often denote the transformation $(\cdot)_t$ by T_t.

3.2. Since $\mathrm{Fr}\,\Phi \Subset U[\rho]$ and $\mathscr{S}_H(\mathrm{Fr}\,\Phi) \Subset U[\rho]$, there exist σ_Φ and σ_H such that $\mathrm{Fr}\,\Phi \subset U[\rho, \sigma_\Phi]$, $\mathscr{S}_H(\mathrm{Fr}\,\Phi) \subset U[\rho, \sigma_H]$. We shall denote the closure of the graph of $\mathscr{S}_H$ by

$$\mathrm{gr}(\mathscr{S}_H) = \mathrm{clos}\{(v, \mathscr{S}_H v) : v \in \mathrm{Fr}\,\Phi\}.$$

Obviously, $\mathrm{gr}(\mathscr{S}_H) \subset U[\rho, \sigma_\Phi] \otimes U[\rho, \sigma_H]$, where $\otimes$ denotes the topological product with the evident metrics.

LEMMA 3.0. *There exists a countable set* $X = \{(v_j, \mathscr{S}_H v_j) : v_j \in \mathrm{Fr}\,\Phi\}$ *such that* $\mathrm{clos}\, X = \mathrm{gr}(\mathscr{S}_H)$.

PROOF. $\mathrm{gr}(\mathscr{S}_H)$ is a compact set and the set $Y = \{(v, \mathscr{S}_H v) : v \in \mathrm{Fr}\,\Phi\}$ is dense in it by definition. Hence there exists a countable set dense in $\mathrm{gr}(\mathscr{S}_H)$ which is contained in Y. □

3.3. Let $\beta \in C^\infty(\mathbb{R})$, $0 \leqslant \beta(x) \leqslant 1$, $\beta(x) = 0$ for $x \leqslant 0$ and $\beta(x) = 1$ for $x \geqslant 1$. Let the subsequences r_k and σ_k $(k = 1, 2, \dots)$ satisfy

$$r_0 = 1, \qquad r_k < r_k\sigma_k < r_{k+1}/\sigma_{k+1} < r_{k+1}, \quad k = 0, 1, \dots, \qquad \lim_{k\to\infty} \sigma_k = \infty, \tag{3.2}$$

$$\lim_{k\to\infty} \frac{r_{k+1}}{\sigma_{k+1} r_k \sigma_k} = \infty. \tag{3.3}$$

We set

$$\psi_k(x) = \beta\left(\frac{\ln x - \ln(r_k/\sigma_k)}{\ln(\sigma_k r_k) - \ln(r_k/\sigma_k)}\right) - \beta\left(\frac{\ln x - \ln(r_{k+1}/\sigma_{k+1})}{\ln(\sigma_{k+1} r_{k+1}) - \ln(r_{k+1}/\sigma_{k+1})}\right),$$

$$\psi_0(x) = 1 - \beta\left(\frac{\ln x - \ln(r_1/\sigma_1)}{\ln(\sigma_1 r_1) - \ln(r_1/\sigma_1)}\right).$$

It is easy to check that the partition of unity $\{\psi_k\}$ has the following properties (here $\nu = 1, 2, \dots$):

$$\operatorname{supp}\psi_\nu \subset \{x \in \mathbb{R} : x \equiv (\tau_\nu/\sigma_\nu, r_{\nu+1}, \sigma_{\nu+1})\}, \tag{3.4}$$

$$\psi_\nu(x) = 1, \qquad x \in (r_\nu\sigma_\nu, r_{\nu+1}/\sigma_{\nu+1}), \tag{3.5}$$

$$\operatorname{supp}\psi_\nu \cap \operatorname{supp}\psi_\mu = \varnothing \quad \text{if } |\nu - \mu| > 1, \tag{3.6}$$

$$\lim_{\nu\to\infty}\sup_r \psi'_\nu(r)r = \lim_{\nu\to\infty}\sup_r \psi''_\nu(r)r^2 = 0. \tag{3.7}$$

We shall denote the "centers" of the segments $[r_\nu\sigma_\nu, r_{\nu+1}/\sigma_{\nu+1}]$ by t_ν:

$$\lim_{\nu\to\infty} r_{\nu+1}/(\sigma_{\nu+1}t_\nu) = \lim_{\nu\to\infty} t_\nu/(\rho_\nu\sigma_\nu) = \infty. \tag{3.8}$$

PROPOSITION 3.1. *For any sequence* $\{p_\nu\} \subset U[\rho, \sigma]$, *the curve*

$$W(\cdot, t) = \sum \psi_\nu(t)T_{t/t_\nu}p_\nu(\cdot)$$

is an asymptotically dynamical trajectory passing through the points p_ν.

PROOF. Let $\tau \in [a, b] \Subset (0, \infty)$. We have

$$W(\cdot, \tau t) - W_\tau(\cdot, t) = \sum[\psi_\nu(\tau t) - \psi_\nu(t)]T_{t/t_\nu}(p_\nu)_\tau.$$

Since the sum has no more than two terms for each t, we obtain, taking (3.7) into consideration, that $W(\cdot, \tau t) - W_\tau(\cdot, t) \to 0$ uniformly in $\tau \in [a, b]$.

We also remark that $W(\cdot, t_\nu) = p_\nu$.

The following lemma will be stated without proof.

LEMMA 3.2. *Let* $v^n, w^n \in U[\rho, \sigma]$, $D'\text{-}\lim_{n\to\infty}(v^n - w^n) = 0$. *Then* $v^n_\tau - w^n_\tau \to 0$ *in* D' *uniformly in* τ *on any segment* $[a, b]$ $(0 < a < b)$.

3.4. We shall state and prove two lemmas which will be needed for proving Theorem 7.

LEMMA 3.3. *Let* $v(\cdot, t)$ *be an arbitrary asymptotically dynamical trajectory and* $\{v^k\}$ *be an arbitrary sequence from the set*

$$\bigcap_{a>1}\operatorname{clos}\{v(\cdot, t) : t > a\} = U. \tag{3.9}$$

There exists a sequence $\{p^k\} \supset \{v^k\}$, $\{p^k\} \subset U$ *and alternating sequences* $r_k < t_k < r_{k+1}$,

$$\lim_{k\to\infty} t_k/r_k = \lim_{k\to\infty} r_{k+1}/t_k = \infty \qquad (k = 0, 1, \dots, r_0 = 1), \tag{3.10}$$

such that the trajectory given by the equality

$$w(\cdot, t) = T_{t/t_k}p^k, \qquad t \in [r_k, r_{k+1}), \tag{3.11}$$

satisfies the condition

$$w(\cdot, t) - v(\cdot, t) \xrightarrow{D'} 0, \qquad t \to \infty. \tag{3.12}$$

Proof. Let $\varepsilon_j \downarrow 0$, $j \to \infty$ and $b_j \uparrow \infty$, $j \to \infty$. For the sake of brevity, we shall denote the expression $d(v(\cdot, \tau t), v_\tau(\cdot, t))$ by $d(\tau, t)$. Since $v(\cdot, t)$ is an asymptotically dynamical trajectory, $d(\tau, t) \to 0$ uniformly in $\tau \in [a, b] \Subset (0, \infty)$ as $t \to \infty$.

Choose $\tau_{1,1} > 1 \cdot b_1$ so that for $t \geqslant \tau_{1,1}$ the inequality $d(t, \tau) < \varepsilon_1/2$ holds for $\tau \in [b_1^{-2}, b_1^2]$, which is possible because of (3.9) and Lemma 3.2.

Suppose the number $\tau_{j,2}$ has been chosen so that $\tau_{j,2} > \tau_{j-1,2} b_{j-1}^2$, and for $t \geqslant t_{j,2}$ and $\tau \in [b_j^{-2}, b_j^2]$ we have the inequalities

$$d(t, \tau) < \varepsilon_j/2, \tag{3.13}$$

$$d(v_\tau(\cdot, \tau_{j,2}), v_\tau^j) < \varepsilon_j/2. \tag{3.14}$$

Now choose a number $\tau_{j+1,1}$ so that $\tau_{j+1,1} \geqslant \tau_{j,2} b_{j+1}^2$ and for $t \geqslant r_{j+1}$, $\tau \in [b_{j+1}^{-2}, b_{j+1}^2]$ the inequality $d(t, \tau) < \varepsilon_{j+1}/2$ holds. Choose $\tau_{j+1,2} \geqslant \tau_{j+1,1}$ so that for $\tau \in [b_{j+1}^{-2}, b_{j+1}^2]$ the inequality

$$d(v_\tau(\cdot, \tau_{j+1,2}), v_\tau^{j+1}) < \varepsilon_{j+1}/2$$

also holds. Continuing in this manner, we shall construct a subsequence $\{\tau_{j,2}\}_{j=1}^\infty$ such that

$$\tau_{j+1,2} \geqslant \tau_{j,2} b_j^2 \tag{3.15}$$

and for any integer $j > 0$ for $t \geqslant \tau_{j,2}$, $\tau \in [b_j^{-2}, b_j^2]$ inequalities (3.13), (3.14) hold. We now set

$$n_j = \left[\frac{\ln \tau_{j+1,2} - \ln \tau_{j,2}}{\ln b_j}\right] \tag{3.16}$$

where $[\alpha]$ is the integer part of α. We set

$$\ln \gamma_j = (\ln \tau_{j+1,2} - \ln \tau_{j,2})/n_j. \tag{3.17}$$

It follows from (3.15) that $n_j \geqslant 1$. Besides, by definition of n_j (see (3.16)), we obtain

$$0 \leqslant (\ln \tau_{j+1,2} - \ln \tau_{j,2}) - n_j \ln b_j < \ln b_j,$$

whence

$$\ln b_j \leqslant \ln \gamma_j < \ln b_j^2, \quad \text{i.e., } \gamma_j \in [b_j, b_j^2]. \tag{3.18}$$

We set $c_j = \sqrt{\gamma_j}$. By definition, each segment $[\tau_{j,2}, \tau_{j+1,2}]$ is split into an even number $(2n_j)$ of nonintersecting intervals of length c_j. Let us change the notation for the ends of these intervals. To this end we set $t_1 = \tau_{1,2}$, $r_2 = t_1 c_1$, $r_{k+1} = t_k c_1$, $t_{k+1} = r_{k+1} c_1$, $k = 1, \dots, n_1 - 1$, where $t_{n_1} = \tau_{2,2} = r_{n_1} c_1$, and

$$\tau_{j,2} = t_{N_j}, \qquad N_j = \sum_{m=1}^{j-1} n_m, \tag{3.19}$$
$$r_{k+1} = t_k c_j, \quad t_{k+1} = r_{k+1} c_j, \qquad k = N_j, \dots, N_{j+1} - 1.$$

Now, let us construct the sequence $\{p_k\}$ in the following way:

$$\begin{aligned} &p_1 v^1, \quad p_m = v(\cdot\,, t_m), \qquad m = 2, \dots, n_1 - 1, \\ &p_{n_1} v^2, \quad p_m = v(\cdot\,, t_m), \qquad m = n_1 + 1, \dots, n_2 - 1, \\ &p_{n_j} v^j, \quad p_m = v(\cdot\,, t_m), \qquad m = n_{j-1} + 1, \dots, n_j - 1, \end{aligned} \tag{3.20}$$

and so forth for all $j = 1, 2, \dots$.

Let us define the curve $W(\cdot\,, t)$ by formula (3.11), taking it to be equal to v^1 for $0 < t < t_1$. Notice that conditions (3.13)–(3.14), (3.18)–(3.20) imply the validity of (3.12).

We remark, by the way, that the property of being asymptotically dynamical now follows for the curve $W(\cdot\,, t)$ as well. Lemma 3.3 is proved.

Now let $v(\cdot\,, t)$, $W(\cdot\,, t)$, $\{v_k\}$, $\{p_k\}$, $\{t_k\}$, and $\{r_k\}$ be sequences defined in the previous lemma. From the constructions in that lemma it follows that

$$d(T_{r_{k+1}/t_k} p_k\,, T_{t_{k+1}/r_{k+1}} p_{k+1}) \to 0, \qquad k \to \infty. \tag{3.21}$$

LEMMA 3.4. *There exists a sequence $\sigma_n \uparrow \infty$ such that*

$$\lim_{k\to\infty} t_k/(r_k\sigma_k) = \lim_{k\to\infty} r_{k+1}/(\sigma_{k+1}t_k) = \infty$$

with

$$d(T_{\tau r_{k+1}/t_k} p_k\,, T_{\tau t_{k+1}/r_{k+1}} p_{k+1}) \to 0, \qquad k \to \infty, \tag{3.22}$$

tending to zero uniformly with respect to $\tau \in [\sigma_k^{-1}, \sigma_k]$.

The assertion of this lemma follows from Lemma 3.2. We leave it without proof.

Now, using the sequences $\{t_k\}$, $\{r_k\}$ and $\{\sigma_k\}$, we shall construct the partition of unity $\{\psi_k\}$. It follows from Proposition 1 that the curve

$$v^\psi(\cdot\,, t) = \sup \psi_\nu(t) T_{t/t_\nu} p_\nu(\cdot) \tag{3.23}$$

is asymptotically dynamical. We shall show that

$$D'\text{-}\lim_{t\to\infty} [W(\cdot\,, t) - v^\psi(\cdot\,, t)] = 0. \tag{3.24}$$

Let $t \in [t_k\,, t_{k+1})$. Then

$$v^\psi(\cdot\,, t) = \psi_k(t) T_{t/t_k} p_k + \psi_{k+1}(t) T_{t/t_{k+1}} p_{k+1};$$

here, by definition, if $t \in [t_k\,, r_{k+1}/\sigma_{k+1}]$ or $t \in [r_{k+1}\sigma_{k+1}\,, t_{k+1}]$, then $v^\psi(t) = W(\cdot\,, t)$. Let us assume that $t \in (r_{k+1}/\sigma_{k+1}\,, r_{k+1})$, then $W(\cdot\,, t) = T_{t/t_k} p_k$, and so

$$\begin{aligned} v^\psi(\cdot\,, t) - W(\cdot\,, t) &= \psi_{k+1}(t) T_{t/t_{k+1}} p_k + (\psi_k(t) - 1) T_{t/t_k} p_k \\ &= \psi_{k+1}(t) T_{t/t_{k+1}} p_{k+1} - \psi_{k+1}(t) T_{t/t_k} p_k. \end{aligned}$$

Denote $\tau = t/r_{k+1}$, then in the case considered $\tau \in (1/\sigma_{k+1}, 1)$ and t/t_{k+1}, t/t_k will take the form

$$t/t_k = (t/r_{k+1})(r_{k+1}/t_k) = \tau(r_{k+1}/t_k), \qquad t/t_{k+1} = \tau(r_{k+1}/t_{k+1}).$$

Now, it follows from (3.22) that in this case

$$v^{\Psi}(\cdot\,, t) - W(\cdot\,, t) \to 0, \qquad t \in (r_{k+1}/\sigma_{k+1}, r_{k+1}),\ k \to \infty.$$

The case $t \in (r_{k+1}, r_{k+1}\sigma_{k+1})$ is treated in the same fashion.

This property, being combined with (3.12), yields

$$D'\text{-}\lim_{t\to\infty} [v(\cdot\,, t) - v^{\Psi}(\cdot\,, t)] = 0. \tag{3.25}$$

PROOF OF THEOREM 7. We shall use the set of first projections of elements of the set X (see Lemma 3.0) to form the sequence $\{v_j\}$ so that each function q_ν appears in it an infinite number of times. According to Theorem V.B. and Lemma 3.3, we can construct a sequence $\{p^k\} \supset \{v_j\}$ and the asymptotic trajectory $W(\cdot\,, t)$ of the form (3.11) which satisfies the relation

$$w(\lambda, t) - \ln|\Phi(\lambda t)|t^{-\rho} \xrightarrow{D'} 0, \qquad t \to \infty.$$

Let us construct $v_\Phi(\cdot\,, t)$ in the form (3.23) so that (3.24) holds.

Replace p_ν in relation (3.23) by $\mathscr{S}_H p_\nu$. We obtain the curve $v_g(\cdot\,, t)$ which is an asymptotically dynamic trajectory. Let us show that the entire function $g \in A(\rho)$ constructed from it in accordance with Theorem V.B. is a completing multiplicator.

To confirm this property it will suffice to check that, first,

$$\sup\{v_{g\Phi}(\lambda) : v_{g\Phi} \in \mathrm{Fr}(g\Phi)\} \leqslant H(\lambda), \tag{3.26}$$

and, secondly, that for each $v \in \mathrm{Fr}\,\Phi$ we have

$$v + \mathscr{S}_H v \in \mathrm{Fr}(g\Phi). \tag{3.27}$$

By construction we have

$$(\ln|g\Phi|)_t = v_\Phi(\cdot\,, t) + v_g(\cdot\,, t) + o(1), \qquad t \to \infty;$$

transforming the right-hand side, we obtain

$$v_\Phi(\lambda, t) + v_g(\lambda, t) = \sum \psi_\nu(t) T_{t/t_\nu}(p_\nu(\lambda) + \mathscr{S}_H p_\nu(\lambda)) \leqslant H(\lambda)$$

whence (3.26) follows.

Choose an arbitrary pair $(v, \mathscr{S}_H v)$, $v \in \mathrm{Fr}\,\Phi$, and let the sequence $(q_\nu, \mathscr{S}_H q_\nu)$ from the set X converge to it. If j_ν are the subscripts such that $p_{j_\nu} = q_\nu$, then, choosing the sequence $\tau_\nu = t_{j_\nu}$, we obtain

$$v_\Phi(\cdot\,, \tau_\nu) + v_g(\cdot\,, \tau_\nu) = q_\nu + \mathscr{S}_H q_\nu \to v + \mathscr{S}_H v, \qquad \nu \to \infty,$$

i.e., (3.27) holds.

Theorem 7 is proved.

§4. Proof of Theorems 8, 4 and 5

4.1. At first we shall prove

LEMMA 4.1. *We have the relation*

$$\mathcal{S}_H(v_\tau) = (\mathcal{S}_H v)_\tau. \tag{4.1}$$

PROOF. It clearly follows from the inequality $(\mathcal{S}_H v)(\lambda) \leqslant m(\lambda, v, H)$ that $(\mathcal{S}_H v)_\tau(\lambda) \leqslant m(\lambda, v_\tau, H)$, whence

$$(\mathcal{S}_H v)_\tau(\lambda) \leqslant \mathcal{S}_H(v_\tau)(\lambda). \tag{4.2}$$

Applying $(\cdot)_{\tau^{-1}}$ to both sides of (4.2), we obtain

$$\mathcal{S}_H v(\lambda) \leqslant (\mathcal{S}_H(v_\tau))_{\tau^{-1}} \leqslant m(\lambda, v, H).$$

Since $\mathcal{S}_H$ is a maximal subharmonic minorant, we have

$$\mathcal{S}_H v \equiv (\mathcal{S}_H v_\tau)_{\tau^{-1}},$$

whence (4.1) follows.

COROLLARY 4.2. *If v_τ is periodic in $\ln\tau$ with period p, then $(\mathcal{S}_H v)_\tau$ is periodic with the same period.*

4.2. PROOF OF THEOREM 8. Sufficiency. Let $v_\Phi(\cdot \mid t)$ be a mapping corresponding to the function Φ by virtue of Theorem V.B. Consider the mapping $v_g(\cdot \mid t) = \mathcal{S}_H v_\Phi(\cdot \mid t)$; it exists since $\operatorname{Fr}\Phi \subset D_H$. We shall show that it satisfies the conditions of Theorem V.B.

Indeed, using Lemma 4.1, we have

$$v_g(\cdot \mid \tau t) - (v_g(\cdot \mid t))_\tau = \mathcal{S}_H[v_\Phi(\cdot \mid \tau t) - (v_\Phi(\cdot \mid \tau t))_\tau]$$

and, because $\mathcal{S}_H$ is continuous, condition (1) of Theorem V.B. is fulfilled.

Since $\mathcal{S}_H$ is continuous, it also follows that

$$\operatorname{clos}\{v_g(\cdot \mid t) : t \in [a, \infty)\} = \mathcal{S}_H(\operatorname{Fr}\Phi).$$

It is easy to see that $\operatorname{Fr}(g\Phi) = \{v + \mathcal{S}_H v : v \in \operatorname{Fr}\Phi\}$, i.e., g is an ideally completing H-multiplicator. □

Necessity. Let there exist an ideally completing H-multiplicator of G. Let us show that $\mathcal{S}_H$ is continuous on $\operatorname{Fr}\Phi$. Assume that it is not so, i.e., there exists a sequence $v_j \to v$ such that $\mathcal{S}_H v_j \to W$; here $W \neq \mathcal{S}_H v$.

Since the limit set $\operatorname{Fr} G\Phi$ is closed, we have $v_j + \mathcal{S}_H v_j \to v + \mathcal{S}_H v$, $v_j \in \operatorname{Fr}\Phi$. On the other hand, $v_j + \mathcal{S}_H v_j \to v + W$. Thus, $W = \mathcal{S}_H v$. □

4.3. PROOF OF THEOREM 4. Let $\operatorname{Fr}\Phi$ be a periodic limit set, that is

$$\operatorname{Fr}\Phi = U[v] = \{v_t : 1 \leqslant t \leqslant e^p\},$$

where $v \in U[\rho]$. We shall show that $\mathcal{S}_H$ is continuous on $U[v]$. For each $W = v_t$ the relation $\mathcal{S}_H(v_t) = (\mathcal{S}_H v)_t$ holds. But the operation $(\cdot)_t$ is continuous for any finite t. Therefore, $\mathcal{S}_H$ is continuous on $U[v]$.

By virtue of Theorem 8 the function Φ is ideally completable. □

4.4. PROOF OF THEOREM 5. At first we shall prove

LEMMA 4.3. *Let* $M_n(\varphi)$, $\varphi \in [0, \pi]$ *be a sequence of functions which satisfies the conditions*:

(a) $M_n(\varphi) \geqslant 0$;
(b) $M_n(\varphi) \rightrightarrows M_\infty(\varphi) \geqslant A \sin \rho\varphi$, $A > 0$;
(c) $M_n'(\varphi - 0)$, $M_n'(\varphi + 0)$ *exist at any point and they coincide almost everywhere*;
(d) *for each* $\varepsilon > 0$ *and* n_0 *there exist* $n > n_0$ *and* φ_n *such that*

$$M_n(\varphi_n) < \varepsilon\varphi_n.$$

Then there exists a sequence (ζ_n, η_n) *of disjoint intervals and a subsequence* M_{k_n} *such that*

$$M_{k_n}'(\eta_n) - M_{k_n}'(\zeta_n) \geqslant A/2. \tag{4.3}$$

PROOF. Set $\varepsilon_0 = 1/2$, $\zeta_0 = \pi/4$ and choose the required sequence recurrently. Let ε_n, η_n, ζ_n be already chosen. Set $\varphi_{n+1} < \zeta_n$, $\varepsilon_{n+1} = \varepsilon_n/2$ and choose $k_0 = k_0(n)$ so that for $k > k_0$

$$M_k(\varphi_{n+1}) - A \sin \varphi_{n+1} > -\varepsilon_{n+1} A \sin \varphi_{n+1}.$$

This is possible because of (b).

Now, choose $\psi_{n+1} < \varphi_{n+1}$ and $k_{n+1} > k_0$ so that

$$M_{k_{n+1}}(\psi_{n+1}) < \varepsilon_{n+1}\psi_{n+1}.$$

This is possible because of (d). Then

$$\frac{M_{k_{n+1}}(\varphi_{n+1}) - M_{k_{n+1}}(\psi_{n+1})}{\varphi_{n+1} - \psi_{n+1}} - \frac{M_{k_{n+1}}(\varphi_{n+1}) - M_{k_{n+1}}(0)}{\psi_{n+1}} \geqslant 1 - \varepsilon_{n+1}.$$

On the interval $(0, \psi_{n+1})$ there is a point at which the derivative exists and the inequality

$$M_{k_{n+1}}'(\zeta_{n+1}) \leqslant \frac{M_{k_{n+1}}(\psi_{n+1}) - M_{k_{n+1}}(0)}{\psi_{n+1}}$$

is valid. On the interval $(\varphi_{n+1}, \psi_{n+1})$ there is a point where the derivative exists and the inequality

$$M_{k_{n+1}}'(\eta_{n+1}) \geqslant \frac{M_{k_{n+1}}(\varphi_{n+1}) - M_{k_{n+1}}(\psi_{n+1})}{\varphi_{n+1} - \psi_{n+1}}$$

is valid. Thus, on the interval $(\zeta_{n+1}, \eta_{n+1})$, we have

$$M_{k_{n+1}}'(\eta_{n+1}) - M_{k_{n+1}}'(\zeta_{n+1}) \geqslant A/2,$$

which proves the lemma.

4.5. Let W be a function representable as a difference of ρ-trigonometrically convex functions. We mention some of its properties, which are easily deduced from the properties of ρ-t.c.f. (see, e.g., [1, p. 77]):

(a) $W'(\varphi - 0)$ and $W'(\varphi + 0)$ exist at each point and are bounded on $[0, 2\pi]$;

(b) $W'(\varphi - 0) = W'(\varphi + 0)$ for all φ, except for perhaps a countable set;

(c) the charge Δ_W generated by the function

$$\Delta_W = W'(\varphi) + \rho^2 \int_0^\varphi W(\theta)\, d\theta$$

has the bounded variation $|\Delta_W|$; the variation $|\Delta_W|(\alpha, \beta)$ of the charge on the interval (α, β) and the variation of the derivative $\operatorname{Var} W'(\alpha, \beta)$ on the same interval satisfy the relation

$$|\Delta_W|(\alpha, \beta) \geqslant \operatorname{Var} W'(\alpha, \beta) + \rho^2(\beta - \alpha).$$

The class of ρ-t.c.f. $h(\varphi)$ for which $\max_\varphi h(\varphi) \leqslant \sigma$ will be denoted by $TC[\rho, \sigma]$, while the class of functions $W = h_1 - h_2$ where $h_1, h_2 \in TC[\rho, \sigma]$ will be denoted by $\delta TC[\rho, \sigma]$. We set

$$TC[\rho] = \bigcup_{\sigma > 0} TC[\rho, \sigma], \qquad \delta TC[\rho] = \bigcup_{\sigma > 0} \delta TC[\rho, \sigma].$$

We shall also remark that for $W \in \delta TC[\rho, \sigma]$ the condition

(d) $\max[|W'|(\varphi - 0), |W'|(\varphi + 0)] \leqslant C(\rho, \sigma)$ $\forall \varphi \in [0, 2\pi)$, where $C(\rho, \sigma)$ is a constant independent of W;

is fulfilled, implying the condition:

(e) if $r^\rho W_n \xrightarrow{D'} r^\rho W$ and $W_n, W \in \delta TC[\rho, \sigma]$, then $W_n \to W$ uniformly on $[0, 2\pi]$.

4.6. COMPLETION OF THE PROOF OF THEOREM 5. To complete the proof, one must show that the operator $\mathcal{S}_H$ is continuous on the set U_{ind}. It easily follows from Lemma 4.1 that

$$\mathcal{S}_H(r^\rho h) = (\widehat{\mathcal{S}}_H h)(\varphi) r^\rho,$$

where $\widehat{\mathcal{S}}_H h$ is the maximal ρ-t.c.f. minorant of $H - h$. So, taking into consideration property (e) from 4.5, to complete the proof, one must prove the following statement.

LEMMA 4.4. *The operator $\widehat{\mathcal{S}}_H$ is continuous on the set*

$$T = \{h(\varphi, C) : 0 \leqslant C \leqslant 1\}$$

in the uniform topology.

PROOF. Let $h_n \to h$, $h_n, h \in T$. Denote $W_n = \widehat{\mathcal{S}}_H h_n$, $W = \widehat{\mathcal{S}}_H h$, $W_\infty = \lim W_n$. We set

$$\widehat{M}_n = H - h_n - W_n, \qquad \widehat{M}_\infty = H - h - W_\infty, \qquad \widehat{M} = H - h - W.$$

Let (α_n, β_n) be a maximum interval where $\widehat{M}_n(\varphi) > 0$. We shall show that $\beta_n - \alpha_n \leqslant \pi/\rho$. Indeed, for a fixed n let us consider the function

$$\widehat{W}_n = W_n + \varepsilon_n L(\varphi - (\alpha_n - \beta_n)/2)$$

where L was defined in 4.5, and $\varepsilon_n > 0$ is small enough. If $\beta_n - \alpha_n > \pi/\rho$, then $\widehat{W}_n$ is also a ρ-t.c.f. minorant of $H - h_n$, i.e., W_n is not maximal.

If $\beta_n - \alpha_n = \pi/\rho$, then, to ensure that W_n is a maximal minorant, at least one the conditions

$$\varliminf_{\varphi \to \alpha_n + 0} \frac{\widehat{M}_n(\varphi)}{\varphi - \alpha_n} = 0, \qquad \varliminf_{\varphi \to \beta_n - 0} \frac{\widehat{M}_n(\varphi)}{\beta_n - \varphi} = 0, \tag{4.4}$$

must be satisfied.

Let us choose (and preserve the previous notation) a subsequence $\widehat{M}_n(\varphi)$ for which $\alpha_n \to \alpha$ and $\beta_n \to \beta$.

If $\beta - \alpha < \pi/\rho$, then the maximum principle for ρ-t.c.f. is valid on the interval (α, β). Repeating the argument of 2.4 for the sequence $r^\rho W_n$, we obtain $W_\infty = W$ for all φ, which proves the lemma for the case considered.

Consider the case when $\beta - \alpha = \pi/\rho$.

Set $q(\varphi) = (W - W_\infty)(\varphi)$. The function q is a t.c.f. on the interval (α, β) since W_∞ is a ρ-t.c.f. by Lemma 2.8. Besides, we have $q(\varphi) \geqslant 0$, $q(\alpha) = q(\beta) = 0$. From the properties of ρ-t.c.f. it follows (see [1, p. 77]) that

$$q(\varphi) = A \sin \rho(\varphi - \alpha), \qquad A > 0. \tag{4.5}$$

Since $\widehat{M}(\varphi) > 0$, we have $\widehat{M}_\infty = \widehat{M} + W - W_\infty \geqslant W - W_\infty$, whence

$$\widehat{M}_\infty \geqslant A \sin \rho(\varphi - \alpha). \tag{4.6}$$

Since $\beta_n - \alpha_n \leqslant \pi/\rho$, the interval $[\alpha, \beta]$ contains the infinite sequence α_n or β_n. Let us single out a subsequence, let it be, for example, $\alpha_n \to 0$, $\alpha_n \in [\alpha, \beta]$.

Examine the sequence $M_n(\varphi) = \widehat{M}_n(\varphi - \alpha)$. From the definition of M_n and from relation (4.6) it follows that conditions (a) and (b) of Lemma 4.3 are fulfilled. Condition (c) is fulfilled because of property (a) of ρ-t.c. functions from 4.5. Further, if $\alpha_n \not\equiv \alpha$, then condition (d) of Lemma 4.3 is trivially true, since $M_n(\alpha_n - \alpha) = 0$; otherwise, if $\alpha_n \equiv \alpha$, condition (d) follows from (4.4).

Applying Lemma 4.3, we obtain the union of intervals (ζ_n, η_n) satisfying (4.3).

The equality $H(\varphi) = \widehat{M}_n(\varphi) - h_n(\varphi) - W_n(\varphi)$ yields the following inequality for the measure Δ_H: $\Delta_H((\zeta_n, \eta_n)) \geqslant A/2$. Summing these inequalities and taking into account the fact that the intervals do not intersect, we obtain $\Delta_H(U(\zeta_n, \eta_n)) = \infty$ which is impossible.

Lemma 4.4 is proved, which completes the proof of Theorem 5.

4.7. An example of a function without ideally completing H-multiplicator. According to Theorem 8, to construct such an example it is sufficient to construct a limit set on which $\mathcal{S}_H$ is not continuous.

We put

$$L(y) = \begin{cases} \cos y, & |y| < \pi/(2\rho), \\ 0, & |y| > \pi/(2\rho). \end{cases}$$

Let us define $X \in C^\infty$ so that $X(x) = 1$ for $x < 0$ and $X(x) = 0$ for $x > \alpha$.

We set

$$\varkappa_1 = (1/8)\max[2\rho X' + X''], \qquad H_0(y) = h(y) + \varkappa_1. \tag{4.7}$$

We also set

$$v(z, c) = [H_0(y) - X(x-c)h(y)]e^{\rho x}$$

where $H_0(y)$ and $L(y)$ have been periodically extended from the interval $[-\pi, \pi]$ to $(-\infty, \infty)$.

LEMMA 4.5. *We have*

$$v[\ln\lambda, c) \in U[\rho, \sigma], \qquad \sigma = 1 + \varkappa_1, \tag{4.8}$$

$$\mathcal{S}_{H_0} v(\cdot, c) \equiv 0, \tag{4.9}$$

$$\lim_{c\to\infty} v(\ln\lambda, c) = \varkappa_1 r^\rho, \tag{4.10}$$

uniformly with respect to $\lambda \in K \Subset \mathbb{C}$, and

$$\mathcal{S}_{H_0}(\varkappa r^\rho) = L(\varphi) r^\rho. \tag{4.11}$$

PROOF. For Laplace operators in λ and in z it is true that $\Delta_\lambda = \Delta_z/|\lambda|^2$. Let us check that $v(\cdot, c)$ is subharmonic in z. We have

$$\begin{aligned}\Delta_z v &= ([H_0''(y) + \rho^2 H_0(y) - X(x-c)(L''(y) + \rho^2 L(y))] \\ &\qquad - [2\rho X'(x-c) + X''(x-c)L(y)])e^{\rho x} \\ &= ([1 - X(x-c)](L'' + \rho^2 L) \\ &\qquad + [\rho^2 X_1 - (2\rho X'(x-c) + X''(x-c))L(y)])e^{\rho x}\end{aligned}$$

which proves that $v(\ln\lambda, c)$ is subharmonic for all c. Since, obviously, $v(z, c) = (1 + \lambda_1)e^{\rho x}$, we have $v(\ln\lambda, c) \leqslant (1 + x_1)|\lambda|^\rho$. Finally, since $v(\cdot, c) \geqslant 0$, we have $v(0) = 0$. Thus (4.8) is proved.

Let us prove (4.9). We have

$$H_0(\varphi)r^\rho - v(\ln\lambda, c) = X(\ln r - c)L(\varphi)r^\rho.$$

Since $X = 0$ for $r > e^{c+\alpha}$, the maximal subharmonic minorant is zero by the maximum principle.

Relation (4.10) is obvious, since $X(x-c) \rightrightarrows 1$ on each left semiaxis $(-\infty, a]$.

Relation (4.11) follows from the equality $H_0(\varphi)r^\rho - \varkappa r^\rho = L(\varphi)r^\rho$, since $L(\varphi)r^\rho \in U[\rho]$.

Now we pass to the construction of the example. Examine the set

$$U_1 = \operatorname{clos}\{v(\ln\lambda, c) : c \in [0, \infty)\}.$$

It contains the function $D'\text{-}\lim_{c\to\infty} v(\ln\lambda, c) = \varkappa r^\rho$. Let us consider the minimal convex T_t-invariant set U containing U_1. The set is contained in $U[\rho, 1+\varkappa_1]$, by (4.8). The equality

$$T_t v(\ln\lambda, 0) = v(\ln\lambda, \lambda t)$$

implies that U is the limit set for a certain entire function Φ. Let us show that $\mathscr{S}_{H_0}$ is not continuous on $\operatorname{Fr}\Phi$. We take any sequence $c_j \to \infty$ and set $v_j(\lambda) = v(\ln\lambda, c_j) \in U$. Now $D'\text{-}\lim_{j\to\infty} v_j(\lambda) = \varkappa_1 r^\rho$ by (4.10) and $\mathscr{S}_{H_0} v_j(\lambda) = 0$, so $D'\text{-}\lim_{j\to\infty} \mathscr{S}_H v_j = 0$ but

$$\mathscr{S}_{H_0}(\lim v_j(\lambda)) = \mathscr{S}_{H_0}(\varkappa r^\rho) = L(\varphi)r^\rho \not\equiv 0$$

which shows the absence of continuity.

By virtue of Theorem 8 Φ is not ideally completable.

References

1. B. Ya. Levin, *Distribution of zeros of entire functions*, GITTL, Moscow, 1956; English transl., Amer. Math. Soc., Providence, R.I., 1964.

2. V. S. Azarin and V. B. Giner, *On the completeness of exponent systems in convex domains*, Dokl. Akad. Nauk SSSR **305** (1989), no. 1, 11–13; English transl. in Soviet Math. Dokl. **39** (1989).

3. V. V. Napalkov, *On a class of equations of convolution type*, Uspekhi Mat. Nauk **29** (1974), 217–218. (Russian)

4. O. V. Epifanov, *On the convolution epimorphism in a convex set*, Dokl. Akad. Nauk SSSR **217** (1974), no. 3, 18–19; English transl. in Soviet Math. Dokl. **15** (1974).

5. V. A. Tkachenko, *Equations of convolution type in analytic functional spaces*, Izv. Akad. Nauk SSSR Ser. Mat. **41** (1977), 378–392; English transl. in Math. USSR-Izv. **11** (1977).

6. V. S. Azarin, *On the asymptotic behavior of subharmonic functions of finite order*, Mat. Sb. **108** (1979), no. 2, 147–167; English transl. in Math. USSR-Sb. **36** (1980), 135–154.

7. R. Sigurdsson, *Growth properties of analytic and plurisubharmonic functions of finite order*, Math. Scand. **59** (1986), 235–304.

8. V. S. Azarin and V. B. Giner, *On the structure of limit sets of entire and subharmonic functions*, Teor. Funktsiĭ, Funktsional. Anal. i Prilozhen. **38** (1982), 3–11. (Russian)

9. V. B. Giner, L. R. Podoshev, and M. L. Sodin, *On summing lower indicators of entire functions*, Teor. Funktsiĭ, Funktsional. Anal. i Prilozhen. **42** (1984), 27–36. (Russian)

10. V. B. Giner, *On approximating limit sets of subharmonic and entire functions by periodic limit sets*, Manuscript No. 1033-Uk 87, deposited at the Ukrain. NIINTI, Kharkov, 1987. (Russian)

11. ———, *Construction of an entire function with a given limit set*, Manuscript No. 2283-Uk 87, deposited at the Ukrain. NIINTI, Kharkov, 1987. (Russian)

12. V. S. Azarin and V. B. Giner, *A criterion for the existence of an entire function with a given boundary set*, Dopovidi Akad. Nauk Ukrain. RSR Ser. A (1988), no. 5, 3–5. (Ukrainian)

13. M. L. Sodin, *A remark on limit sets of subharmonic functions of natural order*, Teor. Funktsiĭ, Funktsional. Anal. i Prilozhen. **39** (1983), 125–129. (Russian)

14. A. F. Grishin and M. L. Sodin, *Growth along a ray, distribution of roots over the arguments of an entire function of finite order and a uniqueness theorem*, Teor. Funktsiĭ, Funktsional. Anal. i Prilozhen. **50** (1988), 47–61; English transl. in J. Soviet Math. **49** (1990), no. 6.

15. A. E. Eremenko, M. L. Sodin, and D. F. Shea, *On the minimum of the modulus of an entire function on the Polya peak sequence*, Teor. Funktsiĭ, Funktsional. Anal. i Prilozhen. **45** (1986), 26–40; English transl. in J. Soviet Math. **48** (1990), no. 4.

16. E. D. Faĭnberg, *On the integral over the nonadditive measure and on estimates of the entire function indicator*, Sibirsk. Mat. Zh. **24** (1983), 175–186; English transl. in Siberian Math. J. **24** (1983).

17. L. R. Podoshev, *A complete description of the pair indicator-lower indicator of an entire function*, Dokl. Akad. Nauk SSSR **300** (1988), no. 1, 38–40; English transl. in Soviet Math. Dokl. **37** (1988).

18. A. A. Goldberg and I. V. Ostrovskiĭ, *Distribution of values of meromorphic functions*, "Nauka", Moscow, 1970. (Russian)

19. V. S. Azarin, *Theory of growth of subharmonic functions*, I, II, Kharkov University, Kharkov, 1978. (Russian)

20. V. S. Azarin and L. R. Podoshev, *Limit sets and indicators of an entire function*, Sibirsk. Mat. Zh. **25** (1984), no. 6, 4–16; English transl. in Siberian Math. J. **25** (1984).

21. L. Hörmander and R. Sigurdsson, *Limit sets of plurisubharmonic functions*, Math. Scand. **65** (1989), 308–320.

22. A. F. Grishin, *On sets of regular growth of entire functions*, I, Teor. Funktsiĭ, Funktsional. Anal. i Prilozhen. **40** (1983), 36–47; II **41** (1984), 39–55; III **42** (1984), 32–42. (Russian)

23. P. Koosis, *La plus petite majorante surharmonic et son rapport avec l'éxistence des fonctions entières de type exponentiel jouant le rôle de multiplicateurs*, Ann. Inst. Fourier **33** (1983), 67–107.

24. B. N. Khabibullin, *On the growth of entire functions of exponential type along the imaginary axis*, Mat. Sb. **180** (1989), no. 5, 706–719; English transl. in Math. USSR-Sb. **67** (1990).

25. V. S. Azarin and V. B. Giner, *On multiples of entire functions of finite order*, Dokl. Akad. Nauk SSSR **314** (1990), no. 5, 1033–1035; English transl. in Soviet Math. Dokl. **42** (1991).

KHARKOV INSTITUTE OF RAILWAY ENGINEERING, 7 FEIERBACH SQUARE, KHARKOV 310050 UKRAINE

Translated by V. A. TKACHENKO